Robust Design for Quality Engineering and Six Sigma

Robust Design for Quality Engineering and Six Sigma

Sung H. Park
Seoul National University, Korea

Jiju Antony
University of Strathclyde, UK

World Scientific

NEW JERSEY • LONDON • SINGAPORE • BEIJING • SHANGHAI • HONG KONG • TAIPEI • CHENNAI

Published by

World Scientific Publishing Co. Pte. Ltd.
5 Toh Tuck Link, Singapore 596224
USA office: 27 Warren Street, Suite 401-402, Hackensack, NJ 07601
UK office: 57 Shelton Street, Covent Garden, London WC2H 9HE

British Library Cataloguing-in-Publication Data
A catalogue record for this book is available from the British Library.

ROBUST DESIGN FOR QUALITY ENGINEERING AND SIX SIGMA

ISBN-13 978-981-277-867-3
ISBN-10 981-277-867-5

Printed by Fuisland Offset Printing (S) Pte Ltd. Singapore

Preface

Countless books have been published in the fields of both robust design and Six Sigma. However, only a few books have tried to combine robust design and Six Sigma to create synergistic effects for quality management. The primary scope of this book is to explain Taguchi's robust design principles and concepts for quality engineering in the earlier chapters, and to show the importance of robust design for implementation of Six Sigma — especially in the context of Design for Six Sigma (DFSS) — in the later chapters. The book also shows why robust design methodologies are working well in the industry, as demonstrated by several case studies taken from various manufacturing disciplines. We hope that this book will bridge the gap between quality engineering (in particular, robust design) and Six Sigma through value creation effort by the robust design methodology.

This book has been written primarily for engineers and researchers who wish to use statistical robust design methodology for quality engineering and Six Sigma, and for statisticians who wish to know about the wide range of applications of experimental design in the industry. This book will also be of interest to students, managers, quality improvement specialists and other Six Sigma professionals such as Black Belts and Green Belts with an interest in robust design methods (e.g. parameter design and tolerance design) as well as implementation of Six Sigma.

We have been actively involved for many years as professional teachers or consultants for quality management in the industry, and have personally experienced that both robust design methodology and Six Sigma strategy are extremely effective in the quest for process and quality improvements. We firmly believe that if both robust design and Six Sigma are combined, they would be much more powerful for quality improvement effort — hence, the motivation for us to write this book.

This book can be used as a textbook for courses in quality engineering in any engineering discipline, or for the experimental design module of a course in statistics for advanced undergraduate and graduate students. It can also be used for Six Sigma training courses, especially at Green Belt and Black Belt levels. It focuses on robust design and analysis of engineering problems from the standpoint of DFSS rather than on statistical design theory. Only statistical ideas relevant to the solution of the broad class of product and process design problems are discussed.

We would like to thank the editors Ms Gow Huey Ling and Mr Tan Yi Xin of World Scientific for making the publication possible, and for their assistance and guidance. We also would like to thank the graduate students under Professor Park for undertaking the heinous task of word-processing the lengthy manuscript.

We would like to dedicate this book to God for giving us the energy, health and inspiration to finish the manuscript. Finally, we are most grateful to our families for their kind understanding of our frequent absence from family activities, and their continuous encouragement in writing the final draft.

Sung H. Park
Jiju Antony

February 2008

Contents

1

Introduction to Quality Engineering

1.1 QUALITY

1.1.1 DEFINITION OF QUALITY

How is **quality** defined? It is interesting to observe how its definition varies according to the particular emphasis of quality activities. Juran (1964) defines quality as fitness for use. Crosby (1979) describes quality in terms of conformance to requirements. Deming (1986) says that quality is concerned with the present and future needs of the customer. For Feigenbaum (1983) quality is to do with the combined product characteristics of engineering and manufacture that determine the degree to which the product will meet the expectations of the customer. Taguchi (1986) defines quality as the loss a product causes society once it has been shipped, apart from any losses caused by its intrinsic functions. According to ISO 8402 (International Organization for Standardization, 1986), quality is the totality of features and characteristics of a product or service that have a bearing on its ability to satisfy stated or implied needs.

Recently, the most widely used definition is that of ISO 9001 (2000). It says that a quality is a characteristic that a product or service must have. For example, products much be reliable, useable, and repairable. These are some of the characteristics that a good quality product must have. Similarly, service should be courteous, efficient, and effective. These are some of the characteristics that a good quality service must have. In short, a quality is a desirable characteristic. However, not all qualities are equal. Some are more important than others. The most important qualities are the ones that customers want. These are the qualities that products and services must have. So providing quality products and services is all about meeting customer requirements. It is all about meeting the needs and expectations of customers. So a quality product or service is one that meets the needs and expectations of customers.

1.1.2 ACTIVITIES OF QUALITY

In the manufacturing industry, activities concerned with quality can be divided into six stages:

1. **Product planning**: planning for the function, price, life cycle, etc. of the product concerned.
2. **Product design**: designing the product to have the functions decided in product planning.
3. **Process design**: designing the manufacturing process to have the functions decided in the product design.
4. **Production**: the process of actually making the product so that it is of the designed quality.
5. **Sales**: activities to sell the manufactured product.
6. **After-sales service**: customer service activities such as maintenance and product services.

It is important to note that company-wide activities are necessary to improve quality and productivity at each of the six stages mentioned above. A company needs to build an overall quality system in which all activities interact to produce products of designed quality with minimum costs.

Note that there are three different characteristics of quality in an overall quality system in the manufacturing industry:

1. **Quality of design**: quality of product planning, product design and process design.

2. **Quality of conformance**: quality of production.
3. **Quality of service**: quality of sales and after-sales services.

Nowadays, these three aspects of quality are equally important in the manufacturing company. If any one of them is not up to the mark, then the overall quality system is unbalanced, and the company will face serious problems.

Although these definitions are somewhat different, some common ideas run through them. Quality involves developing specifications to meet customer needs (quality of design), manufacturing products which satisfy those specifications (quality of conformance), and then providing after-sales services. However, Taguchi's definition of product quality is unusual. The loss he refers to may be caused by variability of function, or by harmful side-effects. Hence, if a product costs society no loss, the product is of the best quality, and the poorer the product's quality is, the greater the cost of the product to the society.

An example of loss caused by variability of function would be an automobile tire that does not last long. The driver would suffer a loss if he replaced the flat tire in the middle of a highway at night because the tire has an unexpectedly short life. An example of loss caused by a harmful side-effect would be a cold medicine which causes drowsiness in the person who takes it. Then the person would suffer a loss if this drowsiness caused him to be unable to work.

In the next section we will explore further Taguchi's concept of quality engineering from the standpoint of how quality can be designed, manufactured and measured.

1.2 TAGUCHI'S APPROACH TO QUALITY ENGINEERING

A product's cost can be divided into two main parts: before sale and after sale to the customer. The costs incurred before sale are the manufacturing costs, and the costs incurred after sale are those due to quality loss. A defective product which is scrapped or reworked prior to shipment is viewed by Taguchi as a manufacturing cost to the company, but not a quality loss. **Quality engineering** is an interdisciplinary science which is concerned with not only producing satisfactory products for customers but also reducing the total loss (manufacturing cost plus quality loss). Hence, quality engineering involved engineering design, process operations, after-sales services, economics and statistics.

Taguchi's impact on the concept of quality control in the manufacturing industry has been far-reaching. His quality engineering system has been used successfully by many companies in Japan, the USA and elsewhere. Recently it is reported that several companies in Korea have used his methods with great success. Several case studies of his methods can be found in Chapters 6-9. He emphasizes the importance of designing quality control into the manufacturing processes. Also, he stresses that quality variation is the main enemy of quality engineering and that every effort should be made to reduce the variation in quality characteristics. Taguchi extensively uses experimental design primarily as a tool to design products more robust (which means less sensitive) to noise factors.

Robust design is an engineering methodology for optimizing the product and process conditions which are minimally sensitive to the various causes of variation, and which produce high-quality products with low development and manufacturing costs. Taguchi's parameter design is an important tool for robust design. His tolerance design can be also classified as a robust design. In a narrow sense robust design is identical to parameter design, but in a wider sense parameter design is a subset of robust design. Two major tools used in robust design are:

- signal-to-noise ratio, which measures quality with emphasis on variation;
- orthogonal arrays, which accommodate many design factors (parameters) simultaneously.

In the rest of this section, the major concepts of Taguchi's quality engineering are introduced.

1.2.1 IMPORTANCE OF DESIGN STAGE

Deming observed that 85% of poor quality is attributed to the manufacturing process for which managers are responsible, and only 15% to the workers. Similarly, Taguchi asserts that product and process design have a much greater impact on product quality than manufacturing and inspection. Quality should be designed into the product and not inspected into it.

1.2.2 TREATMENT OF QUALITY VARIATION

Taguchi emphasizes, as we have already mentioned, that quality variation is the main enemy of quality engineering. Quality is best achieved by minimizing the deviation from a target. He introduces the **loss function**

$$L(y) = k(y - m)^2 ,$$

which is to be minimized, where y is the quality characteristic, $\boldsymbol{m}$ is the target value and k is an appropriate constant. He also uses signal-to-noise (SN) ratios which are based in principle on the loss function.

1.2.3 OFF-LINE AND ON-LINE QUALITY CONTROL (QC)

As mentioned earlier, the typical quality system of a manufacturing company consists of three aspects: quality of design; quality of conformance; and quality of service. Taguchi divides the quality system into two parts as follows:

- **Off-line QC**: Activities for quality of design through market research and product/process development, which are QC efforts away from production lines.
- **On-line QC**: Activities for quality of conformance and quality of service through manufacturing care, inspection and customer service. These are QC efforts mainly on production lines.

Since the design stage carries greater emphasis in Taguchi's quality engineering, it is said that the off-line QC has the greater impact on quality. Taguchi's viewpoint on the importance of off-line QC is true in the Korean industry, as shown in Figure 1.1. It is believed that such phenomena hold in most other countries.

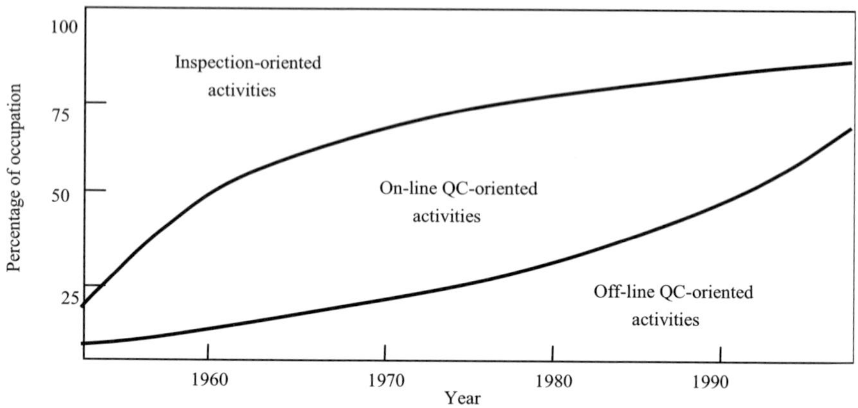

Figure 1.1 The change in importance of QC activities.

1.2.4 THREE STEPS IN PRODUCT DESIGN

During the product design and process design phases, there are three steps in each design phase. They are system design, parameter design and tolerance design. The three steps for product design are described below.

1. **System design** (primary design, functional design or concept design). This step involves the development of a prototype design which meets customer requirements, and the determination of materials, parts, components, assembly system, manufacturing technology, etc. The key emphasis here is on using the best available technology at the lowest cost to meet customer requirements obtained through quality function deployment (QFD). System design can play an important role in reducing sensitivity to noise factors as well as in reducing manufacturing costs.
2. **Parameter design** (secondary design). A parameter is the design variable or control factor which affects a product's

functional characteristics. In this step, we determine the levels (values) of design variables (control factors) that minimize the effect of noise factors on the product's quality, minimize the manufacturing cost, and get the mean quality of the product on target. In order to find the optimum levels, fractional factorial designs using tables of orthogonal arrays are often used because there are too many experimental combinations to be tested.

3. **Tolerance design** (tertiary design). During parameter design, we assume that low-grade components and materials which allow some tolerances for noise factors will be used, while minimizing the sensitivity to noise and the quality variation. Tolerance design applies if the reduction in quality variation achieved by parameter design is insufficient. In tolerance design, a trade-off is made between reduction in the quality variation and increase in manufacturing cost. That is, we selectively specify higher-grade parts, materials or components to reduce tolerances in the order of their cost-effectiveness. Since many choices are possible, experimental designs using tables of orthogonal arrays can also be effectively used in tolerance design.

1.2.5 THREE STEPS IN PROCESS DESIGN

The results of the three steps in product design by the design department are passed to the production department in the form of specifications. The production department then designs a manufacturing process that will produce adequate products. **Process design** also consists of three steps.

1. System design. This step selects the manufacturing process from knowledge of related manufacturing technology and cost such as automatic control, production cost and productivity.
2. Parameter design. This step decides the optimum operating conditions for each of the component processes and the optimum materials to purchase. The major goal of parameter design is to reduce quality variation (to improve process

capability) by minimizing the influence of noise. Experimental design is often used in this step.

3. Tolerance design. In this design, the tolerances of the process conditions and sources of variability are investigated. If the tolerances obtained by the parameter design are not sufficient, economic measures to suppress quality variation will be sought by directly removing its sources of variability. Experimental design is also a powerful tool in this step.

1.2.6 USE OF NOISE

The undesirable and uncontrollable sources that can cause deviation from target values in product's functional characteristics are called **noise**, and are divided into three types.

1. **External noise**: operating environment variables such as temperature and humidity, and conditions of use that disturb the functions of a product.
2. **Internal noise**: changes that occur when a product deteriorates during storage, by friction or by wearing out of parts during use.
3. **Unit-to-unit noise**: differences between individual products because of manufacturing-process imperfections such as variations in machine setting.

The overall quality system should be designed to produce a product that is robust with respect to all noise factors. In order to achieve robustness, QC efforts must begin in the product and process design (off-line QC) and must be continued through production operation (on-line QC).

Table 1.1 summarizes the ways to combat variability in functional quality. In this table, the product design stage is called "R&D" and the process design stage is called "production engineering".

Table 1.1 Quality control activities for dealing with noise factors.

Department countermeasure			Type of noise		
			External	Internal	Unit-to-unit
Off-line quality control	R&D	1. System design	◎	◎	◎
		2. Parameter design	◎	◎	◎
		3. Tolerance design	○	◎	◎
	Production engineering	1. Process diagnosis	×	×	◎
		2. Parameter design	×	×	◎
		3. Tolerance design	×	×	◎
On-line quality control	Production	1. Process diagnosis and adjustment	×	×	◎
		2. Prediction and correction	×	×	◎
		3. Measurement and action	×	×	◎
	Customer relations	After-sales service	×	×	×

Source: Taguchi (1986).
◎ Possible
○ Possible, but should be a last resort
× Impossible

In Table 1.1, Taguchi identifies three forms of on-line QC:

1. Process diagnosis and adjustment: the process is monitored at regular intervals, and necessary adjustment is made if needed.
2. Prediction and correction: past data is used to predict trends in the process. If the process is off target, it is corrected by engineering activities such as feedback control and feed-forward control.
3. Measurement and action: manufactured units are inspected and measured, and action taken (reworking or scrapping of defective units) as appropriate.

Table 1.1 shows that external and internal noise can be reduced most effectively at the R&D step. However, unit-to-unit noise can be handled in the overall off-line and on-line QC stages.

1.2.7 KEY POINTS IN QUALITY IMPROVEMENT PLANNING

Quality improvement efforts should be continuously carried out based on the followed key points:

1. The variation of product quality characteristics from their target values should be reduced. Taguchi views quality improvement as an ongoing effort to reduce variation from the target value. The primary aim of quality improvement is to achieve a population distribution as close to the target as possible. To accomplish this, the SN ratio is adopted.
2. The product and the process should be designed so that they are minimally sensitive to noise factors. Taguchi uses experimental designs as a tool to make products robust to noise factors, and to reduce the effects of variation on product and process quality characteristics. He especially uses constructed tables known as "tables of orthogonal arrays" in which he allocates the noise factors to the "outer array", and the design factors to the "inner array" in the parameter design. Classical applications of experimental design focused primarily on optimizing average product performance characteristics rather than considering effects on variation.
3. In designing the product and the process, the optimal levels of design factors should be determined to minimize the cost under the condition that the given quality tolerance from the targets are met for each quality characteristic concerned. To achieve this objective, tolerance design is often used.

Earlier, we defined quality engineering as an interdisciplinary science which is concerned with producing satisfactory products for customers while reducing the total cost. If we use the major concepts of Taguchi's quality engineering, we may define **quality engineering** as "an effective system of engineering management to minimize the total loss of products to society through off-line and on-line quality control activities". Figure 1.2 shows the major activities, methods and tools of quality engineering.

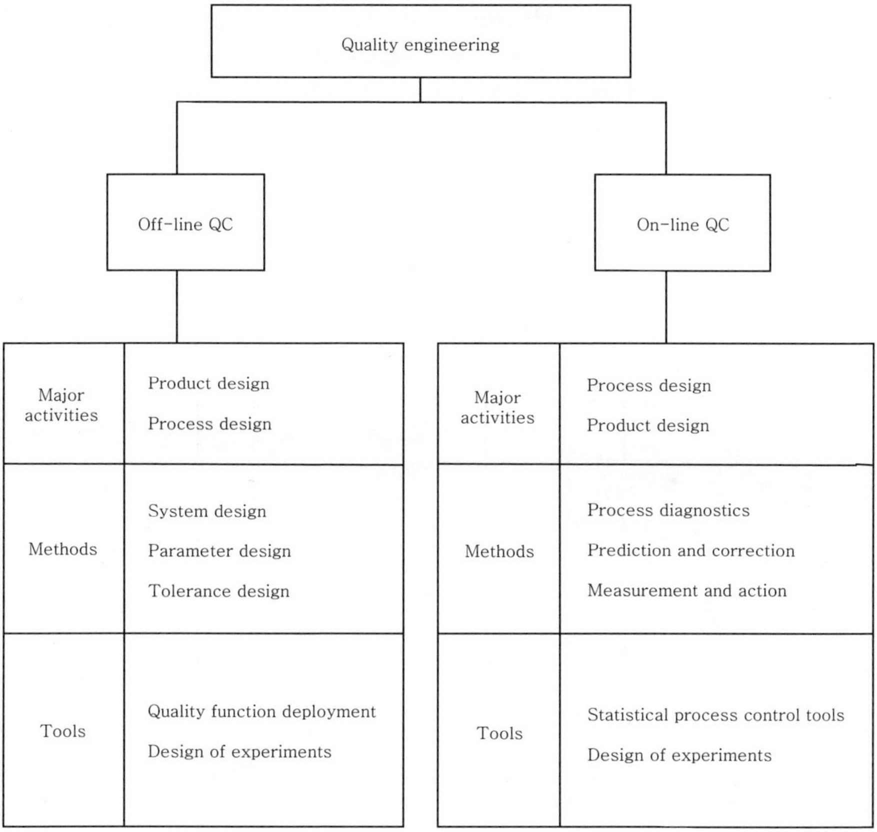

Figure 1.2 Overview of quality engineering.

1.3 STAGES OF NEW PRODUCT DEVELOPMENT

The stages of new product development are dependent on the type of product being developed. However, a general flow diagram for new product development is given in Figure 1.3.

The stages of new product development may be divided into two parts from the viewpoint of a quality system cycle: off-line and on-line quality systems. The off-line quality system starts with the survey of market information and ends with the set-up of managing methods for the manufacturing process. The on-line quality system begins with the production of sample products and ends with after-sales service activities. Note that three steps in design (system, parameter and tolerance) play a critical role in the off-line quality system. On the other hand, statistical process control (SPC) tools help keep manufacturing under control.

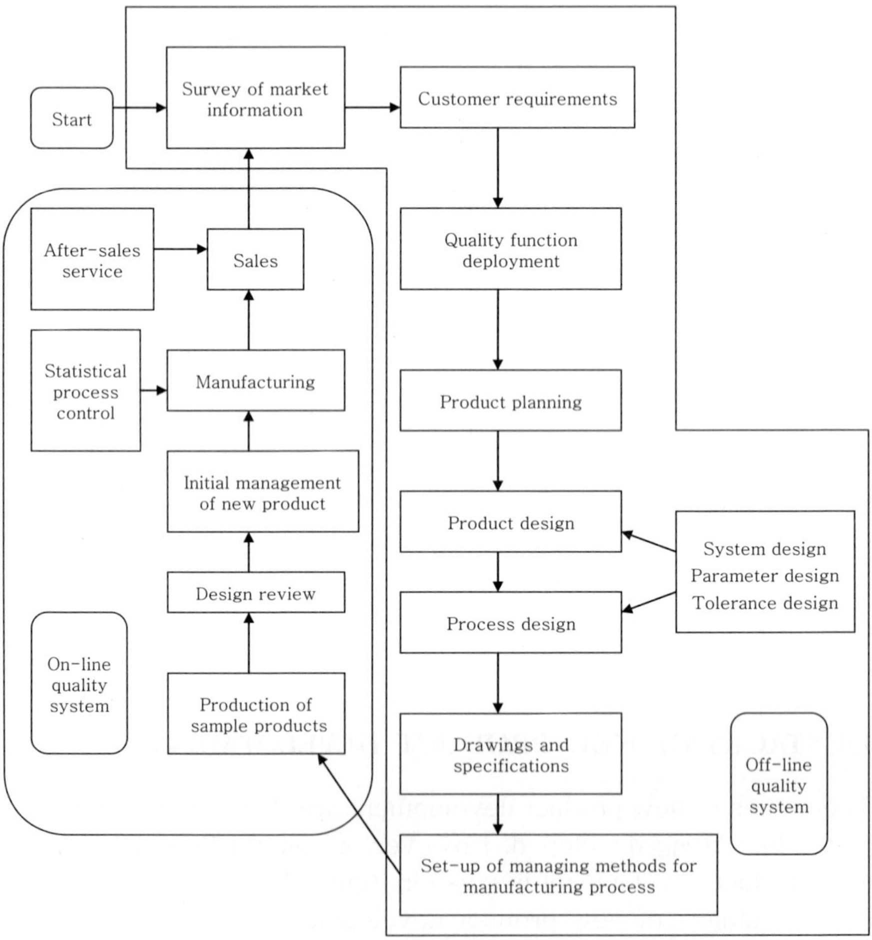

Figure 1.3 Stages of new product development and the quality system cycle.

1.4 QUALITY MANAGEMENT AND SIX SIGMA

We will explain the evolution of QC, quality engineering and quality management in the context of Korean industries. However, much of what we will say holds true in most other national contexts.

QC activity began in Korea in 1961 when the Industrial Standardization Law was announced. Based on this law, Korean Industrial Standards were laid down, and inspection activities dominated QC. When it was realized in the 1970s and early 1980s that inspection alone was not enough for QC, QC activities were stressed in manufacturing processes (on-line QC as shown in Figure 1.1) through quality circles, suggestion schemes, applications of statistic methods, etc.

From the late 1980s, as customer satisfaction (CS) became the critical focus in company management, R&D became based on customer opinion, product planning and product/process design (off-line QC) began to play a major role in quality management. **Quality management** (QM) is the overall management function that implements the quality policy (QP), quality control (QC), quality improvement (QI) and quality assurance (QA) for customer satisfaction under the leadership of senior management and with the full participation of all members of the organization. Here the abbreviations QP, QC, QI and QA have the following meaning:

QP (quality policy): the overall quality direction and operational philosophy of senior management as regards quality.

QC (quality control): the operational techniques and activities that are used to fulfill requirements for quality. This is the narrow concept of QC. When we speak of "company-wide QC" or "QC activities in a particular country", this is a broader concept of QC. Care should be taken to avoid confusion.

QI (quality improvement): the activities to increase the effectiveness of quality in the stages of R&D and manufacturing.

QA (quality assurance): the planned and systematic actions to provide adequate confidence that a product or service will meet customer satisfaction for quality.

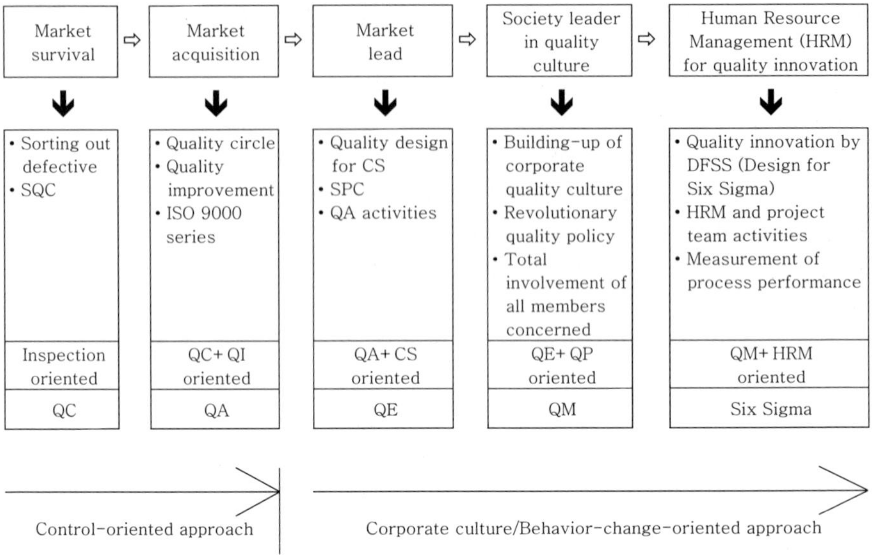

Figure 1.4 The development process of QM.

Since quality engineering (QE) is focused on product/process design and production for customer requirement, we may write QE and QM in the following equations:

$$QE = QC + QI + QA\ ,$$

$$\begin{aligned} QM &= QP + QC + QI + QA \\ &= QP + QE\ . \end{aligned}$$

Figure 1.4 shows the development process of QM. Total quality control (TQC) is very similar to QM. However, the emphasis is different. In QM, senior management involvement and customer satisfaction carry more weight, but in TQC, QC and QI by line workers and middle managers are more important. Total quality management (TQM) is actually the same as QM. However, when we add T to QM, we mean to emphasize company-wide efforts for QM activities.

Six Sigma as a management strategy was launched by Motorola in 1987. Six Sigma has evolved significantly and continues to expand to improve the performance of processes. It seems that recently Six Sigma begins to replace QM or TQM for management innovation. The authors believe that **Six Sigma** is a new strategic paradigm of management innovation for company survival in this 21st century, which implies three things: statistical measurement, management strategy and human resource management. It tells how good our products, services and processes are through statistical measurement of process performance. It is a new management strategy to create quality innovation and total customer satisfaction. For this purpose DFSS (Design for Six Sigma) plays an important role for product and service development. DFSS will be explained in detail in Chapter 13. It is also a wise human resource management, in which full scale training scheme by belt system and project team activities are effectively utilized. Park (2003) and Antony et al. (2006) give more details on Six Sigma and DFSS.

2

Analysis of Quality Information and Quality Improvement Effort

2.1 ASSESSMENT OF PROCESS CAPABILITY

Before getting into design problems, we need to know how to analyze the quality information we have obtained. First of all, let us consider process capability, which is an important basic tool for statistical quality control. **Process capability** measures the total variation in the process output against the predetermined quality specifications. It describes whether or not the process output is capable of meeting the specifications. There are three kinds of measure to assess process capability. In this section we introduce these measures and then compare them.

2.1.1 FRACTION DEFECTIVE

The **fraction defective**, denoted p, is the ratio of the number of defective items which are out of specification to the total number of items processed (or inspected). Fraction defective has been used in industry for a long time.

2.1.2 PROCESS CAPABILITY INDEX

There are two process capability indices, C_p and C_{pk}. The choice between C_p and C_{pk} depends on whether the process is centered (i.e. the mean is at the midpoint of the specification) and on whether the specification is bilateral (two-sided) or unilateral (one-sided, with only a maximum or minimum).

C_p should be used only if the process is centered and the specification is bilateral, as shown in Figure 2.1. C_p is defined as:

$$C_p = \frac{USL - LSL}{6\sigma} = \frac{Specification \quad tolerance}{6\sigma},$$

where USL denotes the upper specification limit, LSL denotes the lower specification limit, and σ is the process standard deviation. If we have a sample of size n and the quality characteristics are $y_1, y_2, \ldots, y_n$, then σ is estimated by the sample standard deviation:

$$\hat{\sigma} = s = \sqrt{\frac{\sum_{i=1}^{n}(y_i - \bar{y})^2}{n-1}}.$$

However, if we use an $\bar{x}-R$ control chart, in which there are k subgroups of size n, σ can be estimated by

$$s = \frac{\bar{R}}{d_2},$$

where $\bar{R} = \sum R_i / n$, and R_i is the range for each subgroup and d_2 is a constant value that depends on the sample size n. The values of d_2 can be found in Appendix H.

A value of 1 for C_p indicates that the distribution of 6σ of the process takes up 100% of the tolerance, and 99.730% (in the case of the normal distribution) of the process output just meets the specification. There is no room for the process mean to shift without making more products out of specification. For this reason, the minimum C_p required to process class A is 1.33 (see Figures 2.1 and 2.2). Note that if C_p=1.33, the distribution of 6σ of the process takes up 75% of the tolerance. Therefore, even if the process changes slightly, there is a high probability that it will not produce materials out of specification. In Motorola's Six Sigma effort, it is required that C_p should be equal to 2.0.

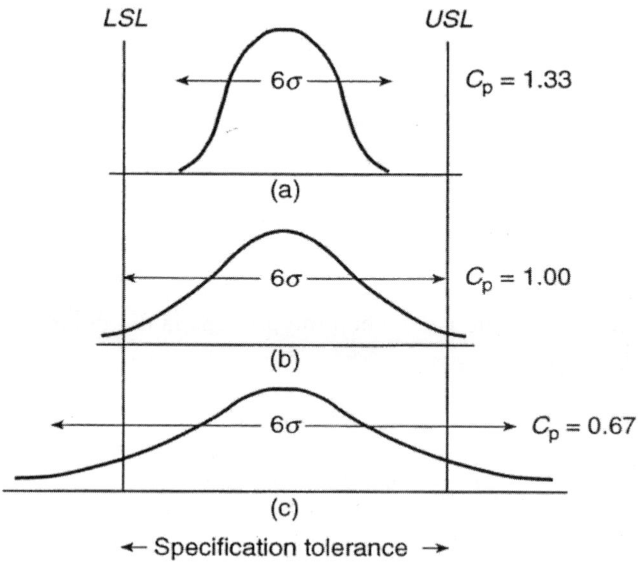

Figure 2.1 Specification and centered quality distribution.

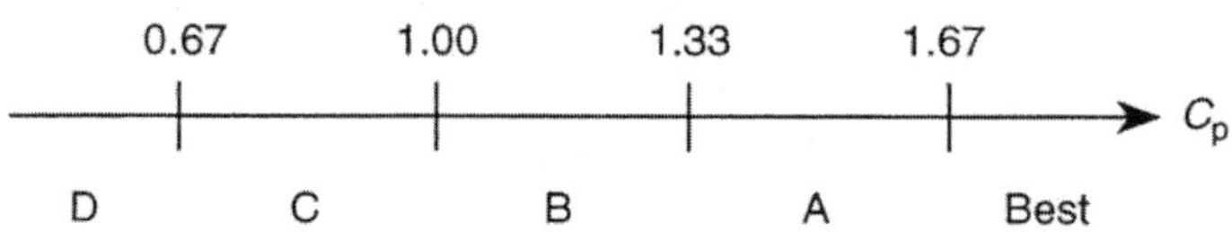

Figure 2.2 Evaluation of process capability.

If the process is not centered, as shown in Figure 2.3, C_{pk} should be used; this is defined by

$$C_{pk} = \min(C_{pu}, C_{pl}), \tag{2.1}$$

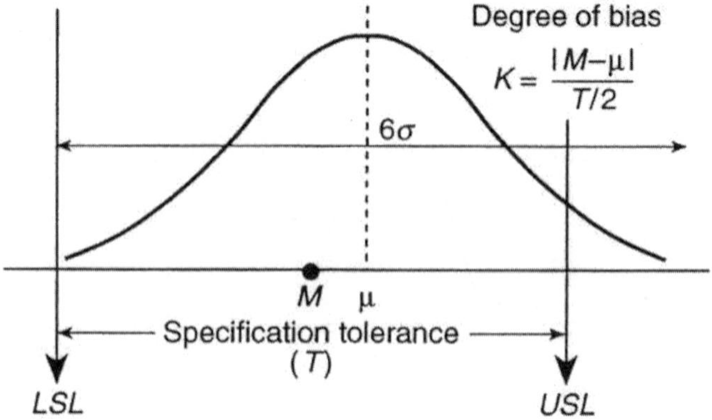

Figure 2.3 The case when the process is not centered.

where

$$C_{pu} = (USL - u)/3\sigma, \tag{2.2}$$

$$C_{pl} = (u - LSL)/3\sigma. \tag{2.3}$$

If the process is centered, then $C_{pu} = C_{pl}$ and $C_p = C_{pk}$. Otherwise $C_p > C_{pk}$. C_{pk} may be defined as

$$C_{pk} = (1 - K)C_p, \tag{2.4}$$

where K is the degree of bias,

$$K = \frac{|M - u|}{(USL - LSL)/2},$$

in which u is the process mean and M is the midpoint of the specification. The equation for C_{pk} in (2.1) is in fact equal to (2.4), and it is not difficult to prove their identity.

C_{pk} can also be used with one-sided specifications. If the specification is a maximum, then use $C_{pk} = C_{pu}$ in (2.2). If the specification is a minimum, use $C_{pk} = C_{pl}$ in (2.3).

2.1.3 LOSS FUNCTION

As stated in Chapter 1, Taguchi defined quality as the loss a product causes society after being shipped. If quality is seen in this way, it is natural to define a loss function to measure the quality of a product. Traditionally, product quality has been measured by comparing critical product characteristics to engineering specifications for the product. However, since it is the production processes which determine the quality of the finished product, it is important to control process capability. The loss function is a new approach to assessing process capability.

No matter how the quality of the product or the process is measured, the quality characteristics are divided into the following three characteristics:

- nominal-is-best,
- smaller-the-better,
- larger-the-better.

When the process under study has a finite target value, as for a battery of 9.0 volts, the measure of quality possesses **nominal-is-best** characteristics. Some measures of quality, for example, radiation leakage from a microwave oven, become worse as the value of the measure increases. Measures with this property are said to possess **smaller-the-better** characteristics. For other measures, as their values become larger, for example, the bond strength of adhesives, the quality is judged to be increasing. Such measures are described as possessing **larger-the-better** characteristics. The loss function is defined differently according to the way quality is measured.

(1) Nominal-is-best measures

Quality loss can be approximately evaluated by Taguchi's **loss**

function which unites the financial loss with the function specification through a quadratic relationship obtained as shown below. If y is the quality characteristics concerned, and m is the finite target value, then the loss $L(y)$ can be expanded as a Taylor series about m:

$$L(y) = L(m) + \frac{L'(m)}{1!}(y-m) + \frac{L''(m)}{2!}(y-m)^2 + \dots \quad (2.5)$$

Because $L(y)$ is a minimum at $y = m$, $L'(y) = 0$. $L(m)$ is always constant and is ignored since its effect is uniform to raise the value of $L(y)$ at all values of y. So the loss can be approximated by taking just the third term in

$$\begin{aligned} L(y) &= \frac{L''(m)}{2!}(y-m)^2 \\ &= k(y-m)^2 . \end{aligned} \quad (2.6)$$

Figure 2.4(a) shows the step-function approach to quality loss. Here, any value in the range $m \pm \Delta_0$ is equally good for the customer, but as soon as the range is breached, the product is equally bad. This is an inadequate way to quantify quality loss. Note that the use of the fraction defective is based on this concept of quality loss. Figure 2.4(b) shows the quadratic loss function, in which the loss increases slowly when the quality y is near m, but as it goes further from m, the loss increases more rapidly. This is a desirable property for a quality loss function.

The **quality loss coefficient** k can be determined on the basis of information about the losses in monetary terms caused by falling outside the customer tolerance limits as shown in Figure 2.4. Note here that we use the term **customer tolerance** instead of "specification tolerance" which was used in the process capability index, since we want to discuss the loss function from the customer's point of view. Normally the customer tolerance is wider than the manufacturer's specification tolerance. Let $m \pm \Delta_0$ represent the deviation at which functional failure of the product occurs, and let the loss at $m \pm \Delta_0$ be A_0^*. Then by

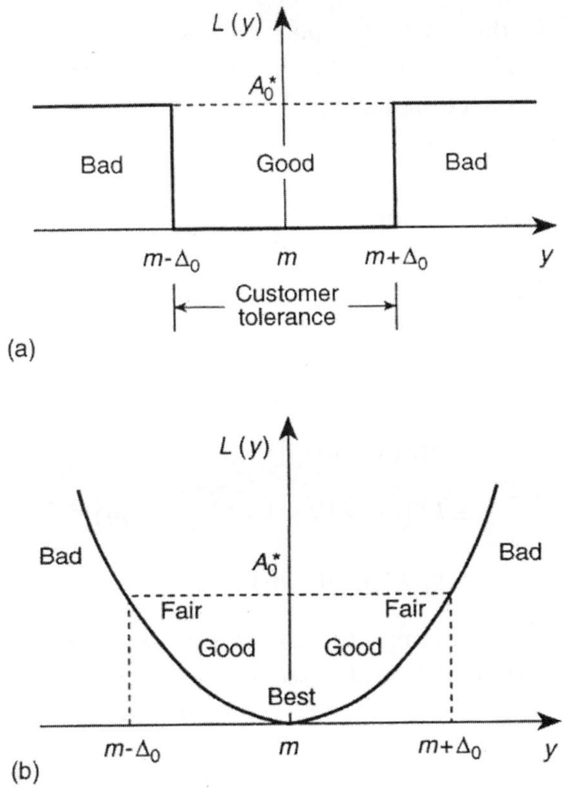

Figure 2.4 Types of quality loss function: (a) step function; (b) quadratic function.

substitution in equation (2.6), we obtain

$$k = \frac{A_0^*}{\Delta_0^2}, \tag{2.7}$$

and the final form of loss function may be written as

$$L(y) = k(y-m)^2 = \frac{A_0^*}{\Delta_0^2}(y-m)^2. \tag{2.8}$$

Note that A_0^* is the cost of repair or replacement of the product plus the cost of transporting the product from the customer to the repair center and back again, etc.

Equation (2.8) does not imply that every customer who receives a product of quality y will incur the same quality $L(y)$. Rather, we should understand that the average quality loss incurred by those customer is $L(y)$.

If the random variable y has mean μ and variance σ^2, then the expected loss of $L(y)$ is:

$$\begin{aligned} L = E(L(y)) &= E[k(y-m)^2] \\ &= kE[(y-m)^2] \\ &= kE[(y-E(y))+(E(y)-m)]^2 \\ &= k[\sigma^2+(\mu-m)^2]. \end{aligned} \tag{2.9}$$

Hence, in order to minimize the expected loss, we have to reduce the quality variation σ^2, and also reduce the difference between the quality mean μ and the quality target m.

(2) Smaller-the-better measures

The quality loss in this type of measure can be approximated by the following function, which is obtained from equation (2.6) by substituting $m = 0$:

$$L(y) = ky^2 = \frac{A_0^*}{\Delta_0^2} y^2. \tag{2.10}$$

Note this is a one-sided loss function because y cannot take negative values. This version of the quadratic loss function is plotted in Figure 2.5(a). As described in (2.7) and (2.8), the coefficient k can be determined in a similar way. The expected value of $L(y)$ is:

$$L = E[ky^2] = k(\sigma^2 + \mu^2). \tag{2.11}$$

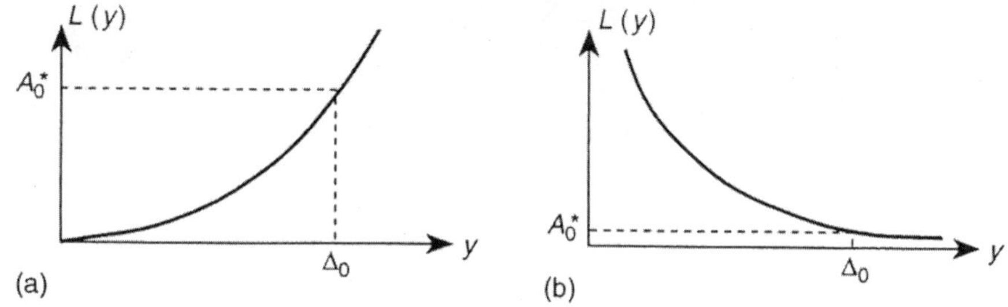

Figure 2.5 Quality loss function of one-sided characteristics: (a) smaller-the-better, (b) larger-the-better.

(3) Larger-the-better measures

The reciprocal of a larger-the-better measure has the same qualitative behavior as a smaller-the-better measure. Thus, we approximate the loss function for this type by substituting $1/y$ for y in equation (2.10):

$$L(y) = k(1/y^2). \tag{2.12}$$

To determine the constant k, we find the functional limit, Δ_0, as shown in Figure 2.5(b), and the corresponding loss A_0^*. Substituting Δ_0 and A_0^* in equation (2.12), and solving for k, we obtain

$$k = A_0^* \Delta_0^2. \tag{2.13}$$

The expected value of $L(y)$ is given by:

$$L = E[k(1/y^2)] = k(1/\mu^2)(1 + 3\sigma^2/\mu^2). \tag{2.14}$$

We obtained three loss functions for each type of quality measure. If we have a set of n observations $y_1, y_2, \ldots, y_n$ for y, we can estimate μ and σ^2 by $\bar{y}$ and V:

$$\bar{y} = \sum y_i / n ,$$

$$V = \sum (y_i - \bar{y})^2 / (n-1) . \qquad (2.15)$$

We can see Table 2.1 to compare loss functions in terms of theoretical and estimated expected loss.

Table 2.1 Comparison of loss functions.

Types of characteristic	Loss function	Expected loss	Estimate of expected loss
Nominal-is-best	$k_1(y-m)^2$	$k_1[\sigma^2 + (\mu - m)^2]$	$k_1[V + (\bar{y} - m)^2]$
Smaller-the-better	$k_1 y^2$	$k_1(\sigma^2 + \mu^2)$	$k_1[V + (\bar{y})^2]$
Larger-the-better	$\frac{k_2}{y^2}$	$\frac{k_2}{\mu^2}(1 + \frac{3\sigma^2}{\mu^2})$	$\frac{k_2}{\bar{y}^2}(1 + \frac{3V}{\bar{y}^2})$

Note: $k_1 = A_0^* / \Delta^2$, $k_2 = A_0^* \Delta_0^2$.

2.1.4 COMPARISON OF p, C_{pk} AND $L(y)$

We have considered fraction defective p, process capability index C_{pk} (if the distribution is centered, $C_{pk} = C_p$) and loss function $L(y)$ as measures of process capability or dispersion of quality characteristics. In order to compare the three measures, consider the case when there are two-sided specification limits and four typical distributions, as shown in Figure 2.6.

For the four types of distribution, we can obtain the fractions defective, process capability indices and loss functions as in Table 2.2, assuming that $k = A_0^* / \Delta_0^2 = 100$ monetary units, say, dollars. From this table distribution, type 1 has the worst fraction defective, the next best being 3, then 2, then finally 4. However, the process capability indices of distributions 2 and 3 are the same. For the expected loss, distribution type 1 gives the lowest figure, followed by 2, 3, and 4. Hence, p, C_{pk} and $L(y)$ give slightly different quality information, and the analysis of such information should be carefully used depending upon the purpose we have. Note that, if the customer's benefit is the first priority, the loss function approach may be the most suitable one.

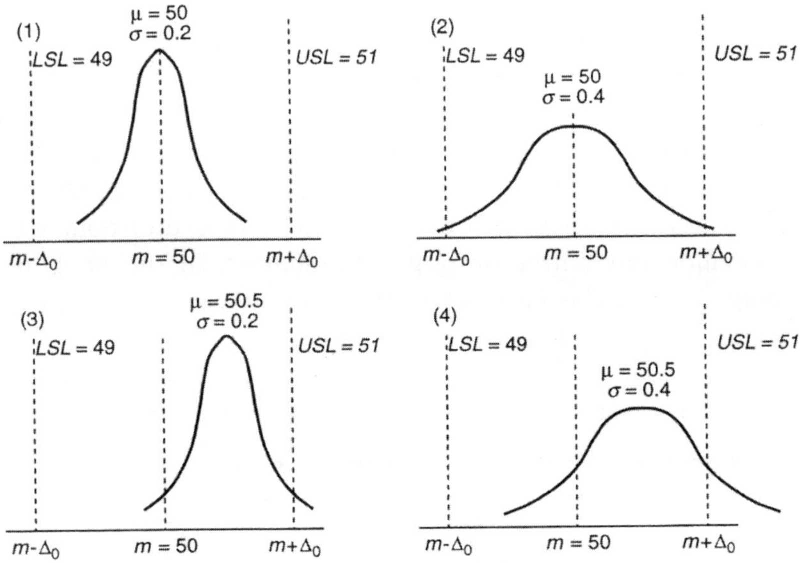

Figure 2.6 Four types of distribution for quality characteristics.

Table 2.2 Comparison of p, C_{pk} and $L(y)$.

Type	Fraction defective (p)	C_{pk}	Expected loss ($) $k[\sigma^2 + (\mu - m)^2]$
1	0.00%	1.67	4
2	1.24%	0.83	16
3	0.62%	0.83	29
4	10.56%	0.42	41

2.2 SIGNAL-TO-NOISE RATIO

In the field of communication engineering, a quantity called the **signal-to-noise** (SN) **ratio** has been used as the quality characteristic of choice. Taguchi, whose background is communication and electronic engineering, introduced this same concept into the design of experiments. Two of the applications in which the concept of SN ratio is useful are the improvement of quality via variability reduction and the improvement of measurement.

The control factors that may contribute to reduced variation and improved quality can be identified by the amount of variation present and by the shift of mean response when there are repetitive data. The SN ratio transforms several repetitions into one value which reflects the amount of variation present and the mean response. There are several SN ratios available depending on the type of characteristic: continuous or discrete; nominal-is-best, smaller-the-better or larger-the-better. In the section we will only discuss the case when the characteristic is continuous. The discrete case will be explained later.

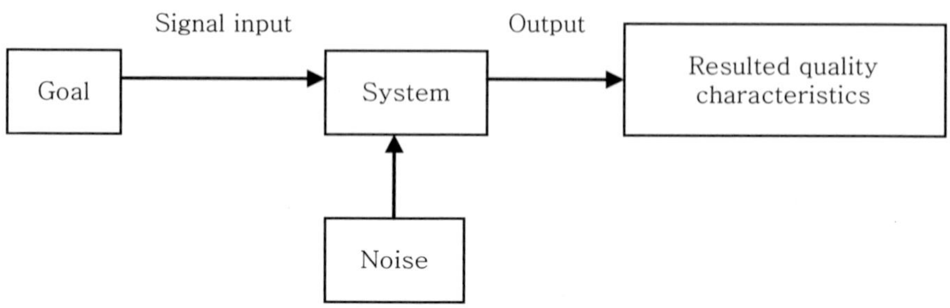

Figure 2.7 Functional relation in communication equipment.

2.2.1 NOMINAL-IS-BEST CHARACTERISTICS

To achieve the goal intended in communication equipment, signals enter the system and as output results accordingly, as shown in Figure 2.7. However, noise interferes with the system and the intended goal is not usually achieved. The ratio between the power of the signal and the power of the noise is called the signal-to-noise ratio:

$$\text{SN ratio} = \frac{power\ \ of\ \ signal}{power\ \ of\ \ noise} = \frac{\mu^2}{\sigma^2}. \tag{2.16}$$

The larger the numerical value of (2.16), the more desirable the system is.

To use this concept in a statistical experiment, the power of the signal and the power of the noise are replaced by the estimates $\hat{\mu}^2$ and $\hat{\sigma}^2$, respectively. If there is a random sample of size n which follows a probability distribution whose mean is μ and variance σ^2, we can find

$$\begin{aligned}
\hat{\sigma}^2 &= V = s^2 = \sum_{i=1}^{n}(y_i - \bar{y})^2 \Big/ (n-1), \\
\hat{\mu}^2 &= \frac{1}{n}(S_m - V), \\
S_m &= \frac{(\sum y_i)^2}{n},
\end{aligned} \tag{2.17}$$

since $E(S_m) = \sigma^2 + n\mu^2$ and $S_m = \hat{\sigma}^2 + n\hat{\mu}^2$. Substituting (2.17) into (2.16), we obtain

$$\text{SN ratio} = \frac{\frac{1}{n}(S_m - V)}{V}.$$

Taguchi recommends using the common logarithm of this SN ratio multiplied by 10, which expresses the ratio in decibels (dB); this has been used in communication for many years. Hence, for the case of continuous and nominal-is-best characteristics, the SN ratio is defined as

$$\text{SN} = 10\log\left(\frac{\frac{1}{n}(S_m - V)}{V}\right). \tag{2.18}$$

Note that as the power of the noise (V) decreases and the power of the signal $[(S_m - V)/n]$ increases, the value of SN increases.

We can observe that, since $S_m = \sum y_i^2 / n = n(\bar{y})^2$, (2.18) becomes

$$\mathrm{SN} = 10\log\left(\frac{\left(\bar{y}^2 - \frac{V}{n}\right)}{V}\right). \tag{2.19}$$

Also, if n is large enough, V/n becomes negligible, and the SN ratio may be expressed as

$$\begin{aligned}\mathrm{SN} &= 10\log\left(\frac{\bar{y}^2}{V}\right)\\ &= 10\log v\left(\frac{\bar{y}^2}{s^2}\right) = 20\log\left(\frac{\bar{y}}{s}\right).\end{aligned} \tag{2.20}$$

Note that $\bar{y}/s$ is the reciprocal of the coefficient of variation, which is a measure of quality dispersion.

2.2.2 SMALLER-THE-BETTER CHARACTERISTICS

Since the target value m is zero for this case, we cannot use the ratio (2.16). Note that the estimate u can be negative, and we cannot use the logarithmic transformation. Therefore, for this case Taguchi recommends using the expected loss in equation (2.11),

$$L = kE(y^2).$$

Suppose we have a set of characteristics $y_1, y_2, \ldots, y_n$. Then the natural estimate of $E(y^2)$ is

$$MSD = \frac{1}{n}\sum_{i=1}^{n}(y_i - 0)^2$$

$$= \frac{1}{n}\sum_{i=1}^{n} y_i^2, \qquad (2.21)$$

where MSD denotes the mean squared deviation from the target value of the quality characteristic. To be consistent with its application in engineering, the value of the SN ratio is intended to be large for favorable situations, and we use the following transformation for the smaller-the-better characteristic:

$$\mathrm{SN} = -10\log\left(\frac{1}{n}\sum_{i=1}^{n} y_i^2\right), \qquad (2.22)$$

which is also measured in decibels.

2.2.3 LARGER-THE-BETTER CHARACTERISTICS

Similar to the smaller-the-better characteristic, the following estimate of the expected loss $L = kE(1/y^2)$,

$$MSD = \frac{1}{n}\sum_{i=1}^{n}\left(\frac{1}{y} - 0\right)^2$$

$$= \frac{1}{n}\sum_{i=1}^{n}\frac{1}{y_i^2},$$

is used, and the corresponding SN ratio becomes

$$\mathrm{SN} = -10\log\left(\frac{1}{n}\sum_{i=1}^{n}\frac{1}{y_i^2}\right). \qquad (2.23)$$

Note that the target value of $1/y$ is 0 in the larger-the-better characteristic.

The SN equations in (2.22) and (2.23) are based on the loss function when there is a set of n characteristics. If we employ the loss-function approach for the nominal-is-best case, we can derive the following SN equation:

$$\mathrm{SN} = -10\log\left[\frac{1}{n}\sum_{i=1}^{n}(y_i - m)^2\right]. \tag{2.24}$$

This form of equation may be more desirable than (2.18) for three reasons. First, where y_i can take a negative or positive value, it is possible for S_m to be less than V, so that equation (2.18) cannot be used. Second, as $\bar{y}$ increases, SN in (2.18) increases. However, when $\bar{y}$ is greater than the target value m, the bigger $\bar{y}$ becomes, the bigger SN is, and at the same time, the more the y values are distanced from the target value m. Hence, where $\bar{y}$ is bigger than m, SN does not reflect desirable situations. Third, the SN values in (2.18) are not based on the concept of the loss function, and are not consistent with the loss function. Table 2.3 summarizes the average loss and SN ratio formulae for a set of n observations.

Table 2.3 Loss function and SN ratio for a set of n observations.

Type of characteristic	Average loss	SN ratio	
		Signal/Noise ratio approach	Loss function approach
Nominal-is-best	$k_1\left[\frac{1}{n}\sum_{i=1}^{n}(y_i - m)^2\right]$	$10\log\left[\frac{(1/n)(S_m - V)}{V}\right]$ $\approx 20\log\left(\frac{\bar{y}}{s}\right)$	$-10\log\left[\frac{1}{n}\sum_{i=1}^{n}(y_i - m)^2\right]$
Smaller-the-better	$k_1\left[\frac{1}{n}\sum_{i=1}^{n}y_i^2\right]$		$-10\log\left[\frac{1}{n}\sum_{i=1}^{n}y_i^2\right]$
Larger-the-better	$k_2\left[\frac{1}{n}\sum_{i=1}^{n}\frac{1}{y_i^2}\right]$		$-10\log\left[\frac{1}{n}\sum_{i=1}^{n}\frac{1}{y_i^2}\right]$

Example 2.1

Suppose we have a set of five observations:

32, 38, 36, 40, 37

Find the SN ratios for each of three types of quality characteristic.

Solution

(a) Nominal-is-best case:

$$\overline{y} = \frac{1}{n}\sum_{i=1}^{n} y_i = \frac{1}{5}(32+38+36+40+37) = 36.6,$$

$$S_m = \left(\sum_{i=1}^{n} y_i\right)^2 \Big/ n = \frac{1}{5}(32+38+36+40+37)^2 = 6697.8,$$

$$V = \left(\sum_{i=1}^{n} y_i - \overline{y}\right)^2 \Big/ (n-1) = \frac{1}{4}[(32+36.6)^2 + \cdots + (37-36.6)^2] = 8.8.$$

Using equation (2.18), we obtain

$$\begin{aligned} \text{SN} &= 10\log\left[\frac{(6697.8-8.8)/5}{8.8}\right] \\ &= 10\log(152.02) = 21.82 \text{ dB}. \end{aligned}$$

If we use equation (2.20), then

$$\begin{aligned} \text{SN} &= 20\log\left(\frac{\overline{y}}{s}\right) \\ &= 20\log\left(\frac{36.6}{\sqrt{8.8}}\right) = 20\log(12.34) = 21.82 \text{ dB}, \end{aligned}$$

which, to 2 decimal places, is equal to the result obtained by (2.18).

(b) Smaller-the-better case. From equation (2.22),

$$\begin{aligned} SN &= -10\log\left(\frac{1}{n}\sum_{i=1}^{n} y_i^{\,2}\right) \\ &= -10\log\left[\frac{1}{5}\sum_{i}(32^2+\cdots+37^2)\right] \\ &= -10\log(1346.6) = -31.29\,\text{dB}. \end{aligned}$$

(c) Larger-the-better case. From equation (2.23),

$$\begin{aligned} SN &= -10\log\left(\frac{1}{n}\sum_{i=1}^{n}\frac{1}{y_i^{\,2}}\right) \\ &= -10\log\left[\frac{1}{5}\sum\left(\frac{1}{32^2}+\cdots+\frac{1}{37^2}\right)\right] \\ &= -10\log(0.000759) = 31.20\,\text{dB}. \end{aligned}$$

2.3 FACTOR-FINDING METHODS FOR QUALITY PROBLEMS

In quality control (QC) circle activities, the techniques most often used in data analysis are the seven basic tools of QC, which are as follows:

1. Histograms;
2. Causes-and-effects diagrams;
3. Pareto diagrams;
4. Check sheets;
5. Graphs;
6. Scatter diagrams;
7. Stratification.

In addition to the above-mentioned, in order to cover QC activities broadly and analyze QC related data in depth, seven new tools of QC have been advanced. They are:

1. Relations diagrams;
2. Tree diagrams;
3. Matrix diagrams;
4. Affinity diagrams;
5. Process decision program chart (PDPC);
6. Arrow diagrams;
7. Matrix data analysis.

For detailed explanation of these methods, see Ishikawa (1983) and Tague (1995).

For any quality problem, whether it be one of low yield, poor quality (however measured) or high utility cost, the most important thing to do is to find the sources of trouble. The main sources could be design parameters or manufacturing conditions which are called factors. In order to find the factors relevant to quality problems, the following methods are the most often used in industrial practice.

1. Causes-and-effects diagrams, Pareto diagrams, scatter diagrams and stratification — which belong to the seven basic tools of QC;
2. Relations diagrams, tree diagrams, matrix diagrams, affinity diagrams and PDPC — which belong to the seven new tools of QC;
3. Statistical process control charts;
4. Quality function deployment;
5. Correlation and regression analysis.

In this section we will only briefly discuss correlation and simple regression analysis. Multiple regression analysis will be discussed in Section 2.4.

2.3.1 CORRELATION ANALYSIS

Suppose we have two random variables X and Y. The variances of X and Y, and the covariance of the two variables, are defined by:

$$\text{var}(X) = \sigma_x^2 = E[X - E(X)]^2,$$

$$\text{var}(Y) = \sigma_y^2 = E[Y - E(Y)]^2,$$

$$\text{cov}(X,Y) = \sigma_{xy} = E\{[X - E(X)][Y - E(Y)]\}.$$

If we have a set of observed data $(x_1, y_1), (x_2, y_2), \ldots, (x_n, y_n)$ for X and Y, we substitute $\bar{x}$ for $E(X)$ and $\bar{y}$ for $E(Y)$ to obtain the following estimates:

$$\hat{\sigma}_x^2 = \frac{\sum (x_i - \bar{x})^2}{n-1}, \tag{2.25}$$

$$\hat{\sigma}_y^2 = \frac{\sum (y_i - \bar{y})^2}{n-1}, \tag{2.26}$$

$$\hat{\sigma}_{xy} = \frac{\sum (x_i - \bar{x})(y_i - \bar{y})}{n-1}. \tag{2.27}$$

Suppose that we draw a scatter diagram of the data. If there is a strong tendency for Y to increase as X increases, as with height and weight, the estimated covariance in (2.27) is positive. On the other hand, if, for example, Y is the yield of a product and X is the percentage of impurity, Y will tend to decrease as X increases, and the estimated covariance will be negative.

The correlation coefficient of X and Y is defined by

$$\rho = \frac{\sigma_{xy}}{\sigma_x \sigma_y},$$

and its estimate from a sample, denoted by r, is given by

$$r = \frac{\hat{\sigma}_{xy}}{\hat{\sigma}_x \hat{\sigma}_y} = \frac{\sum (x_i - \bar{x})(y_i - \bar{y})}{\sqrt{\sum (x_i - \bar{x})^2 \sum (y_i - \bar{y})^2}}. \tag{2.28}$$

Note that r has the same sign as the covariance. The sample correlation coefficient has the following properties:

1. The range of r is $-1.0 \le r \le +1.0$.
2. The value r shows the degree of linear relationship between X and Y. If the points lie on a straight line, the correlation is ± 1.0, the sign of r being the same as the sign of the slope of the line.
3. The coefficient r is a scale-free measure of X and Y. That is, if X and Y are numerically transformed into W and Z by

$$W = aX + b, Z = cY + d,$$

where a, b, c, d are constants, and a and c have the same sign, the correlation coefficient of X and Y will be equal to that of W and Z.

When there is no correlation between X and Y, r could have a small positive or negative value. We often want to test whether there exists correlation between X and Y from the information given by r. For this case, the null hypothesis H_0 and the alternative hypothesis H_1 are

$$H_0 : \rho = 0$$

$$H_1 : \rho \neq 0.$$

The test statistic is $t_0 = \dfrac{r}{\sqrt{(1-r^2)/(n-2)}}$,

which is t-distributed with $n-2$ degrees of freedom under the null hypothesis. Hence, at the level of significance α, we accept H_0 if $|t_0| < t_{1-\alpha/2}(n-2)$, and reject H_0 if $|t_0| \ge t_{1-\alpha/2}(n-2)$. When H_0 is rejected, we say that correlation exists between X and Y, with a probability of $100\alpha\%$ that we have judged incorrectly.

When a correlation between X and Y exists, we can say that the variable X could be a factor which influences the variable Y.

2.3.2 SIMPLE REGRESSION ANALYSIS

When there exists correlation between X and Y, it is useful to fit a straight line, where the model is:

$$y_i = \beta_0 + \beta_1 x_i + \varepsilon_i, \quad i = 1, 2, \ldots, n. \tag{2.29}$$

Here, β_0 and β_1 are unknown regression coefficients to be estimated from the data, and ε_i represents the random noise in the system. We usually assume that they are identically and independently distributed as $N(0, \sigma^2)$.

Equation (2.29) is called the simple regression model, where X is the independent variable and Y is the dependent variable.

Suppose we estimate β_0 and β_1 by the **method of least squares**. Then the fitted simple regression equation is:

$$\hat{y}_i = \hat{\beta}_0 + \hat{\beta}_1 x_i, \tag{2.30}$$

where the **least-squares estimates** are obtained by the following formulas:

$$\hat{\beta}_0 = \bar{y} - \hat{\beta}_1 \bar{x},$$
$$\hat{\beta}_1 = \frac{\sum (x_i - \bar{x})(y_i - \bar{y})}{\sum (x_i - \bar{x})^2}. \tag{2.31}$$

The difference between y_i and $\hat{y}_i$ is called the ith **residual**, e_i. In fact, the method of least squares is to obtain $\hat{\beta}_0$ and $\hat{\beta}_1$, which are the values that minimize the sum of squares of the residuals:

$$SSE = \sum e_i^2 = \sum (y_i - \hat{y}_i)^2, \tag{2.32}$$

which is called the **residual sum of squares**.

In order to judge whether the straight-line equation in (2.29) is significant or not in explaining the relationship between X and Y, we normally use the following two methods: the coefficient of determination, R^2; and the F-test from the analysis of variance (ANOVA) table.

The coefficient of determination is the proportion of variation of y explained by x through the regression model:

$$R^2 = \frac{\sum(\hat{y}_i - \bar{y})^2}{\sum(y_i - \bar{y})^2}. \tag{2.33}$$

In the simple regression model, R^2 is simply the square of the correlation coefficient r. The larger R^2 is, the better the regression model is. The range of R^2 is [0,1].

Consider the following equality:

$$(y_i - \bar{y}) = (y_i - \hat{y}_i) + (\hat{y}_i - \bar{y}). \tag{2.34}$$

If we square both sides of (2.34) and sum over all i, we obtain

$$\sum(y_i - \bar{y})^2 = \sum(y_i - \hat{y}_i)^2 + \sum(\hat{y}_i - \bar{y})^2, \tag{2.35}$$

which can be expressed on words as follows:

Total sum of squares	=	Sum of squares due to residual errors	+	Sum of squares due to regression.
(SST)		(SSE)		(SSR)

These sums of squares may be computed more easily as follows:

$$\begin{aligned} SST &= \sum(y_i - \bar{y})^2 = \sum y_i^2 - n\bar{y}^2, \\ SSR &= \sum(y_i - \hat{y}_i)^2 = \frac{\left[\sum(x_i - \bar{x})(y_i - \bar{y})\right]^2}{\sum(x_i - \bar{x})^2}, \\ SSE &= SST - SSR. \end{aligned} \tag{2.36}$$

Based on this breakdown of the total sum of squares *SST*, we obtain the ANOVA table in Table 2.4. In this table *MSR* stands for the mean of squares due to regression, and is obtained by dividing *SSR* by the regression degrees of freedom, in this case 1. *MSE* means the mean of squares due to residual errors, and is obtained by dividing *SSE* by the residual degrees of freedom, in this case $n-2$.

Table 2.4 ANOVA table for simple regression.

Source of variation	Sum of squares	Degree of freedom	Mean square	F_0	$F(\alpha)$
Regression	SSR	1	MSR	MSR/MSE	$F_{1-\alpha}(1,n-2)$
Residual	SSE	$n-2$	MSE		
Total	SST	$n-1$			

By the regression theory, if $\beta_1=\beta_2=\cdots=\beta_k=0$, MSR/MSE follows the $F(k,n-k-1)$ probability distribution. Hence, if the test statistic $F_0=MSR/MSE$ is greater than the critical value $F_{1-\alpha}(1,n-2)$ at the α level of significance, then we conclude that the regression equation is significant. If this is the case, we may judge that the variable X is a factor which influences the variable Y.

2.4 MULTIPLE REGRESSION ANALYSIS

2.4.1 MODEL FITTING

Suppose we have k independent variables and the functional relationship with the dependent variable Y may be approximated by

$$y_j=\beta_0+\beta_1x_{1j}+\beta_2x_{2j}+\cdots+\beta_kx_{kj}+\varepsilon_j, \qquad j=1,2,\ldots,n. \tag{2.37}$$

Such a model is called a **multiple linear regression** model.

In matrix notation, we may write this model as

$$\mathbf{y}=\mathbf{X}\boldsymbol{\beta}+\boldsymbol{\varepsilon}, \tag{2.38}$$

where y is the n-dimensional vector of the observations, ε is the corresponding vector of errors, and $\boldsymbol{\beta}$ is a vector that consists of the $k+1$ coefficients. The matrix $\mathbf{X}$ is called the design matrix. It has n rows, one for each data point. Explicitly they are:

$$\mathbf{X}=\begin{pmatrix} 1 & x_{11} & x_{21} & \cdots & x_{k1} \\ 1 & x_{12} & x_{22} & \cdots & x_{k2} \\ \vdots & \vdots & \vdots & \vdots & \vdots \\ 1 & x_{1n} & x_{2n} & \cdots & x_{kn} \end{pmatrix},\quad \boldsymbol{\beta}=\begin{pmatrix} \beta_0 \\ \beta_1 \\ \beta_2 \\ \vdots \\ \beta_k \end{pmatrix},\quad \mathbf{y}=\begin{pmatrix} y_1 \\ y_2 \\ y_3 \\ \vdots \\ y_n \end{pmatrix},\quad \text{and } \boldsymbol{\varepsilon}=\begin{pmatrix} \varepsilon_1 \\ \varepsilon_2 \\ \varepsilon_3 \\ \vdots \\ \varepsilon_n \end{pmatrix}.$$

The least-squares estimate of the vector $\boldsymbol{\beta}$ is known to be

$$\hat{\boldsymbol{\beta}}=\begin{pmatrix} \hat{\beta}_0 \\ \hat{\beta}_1 \\ \hat{\beta}_2 \\ \vdots \\ \hat{\beta}_k \end{pmatrix}=(\mathbf{X}'\mathbf{X})^{-1}\mathbf{X}'\mathbf{y}, \tag{2.39}$$

and the vector of estimated values of the dependent variable and the vector of residuals are, respectively,

$$\hat{\mathbf{y}}=\begin{pmatrix} \hat{y}_1 \\ \hat{y}_2 \\ \hat{y}_3 \\ \vdots \\ \hat{y}_n \end{pmatrix}=\begin{pmatrix} \hat{\beta}_0+\hat{\beta}_1 x_{11}+\ldots+\hat{\beta}_k x_{k1} \\ \hat{\beta}_0+\hat{\beta}_1 x_{12}+\ldots+\hat{\beta}_k x_{k2} \\ \hat{\beta}_0+\hat{\beta}_1 x_{13}+\ldots+\hat{\beta}_k x_{k3} \\ \vdots \\ \hat{\beta}_0+\hat{\beta}_1 x_{1n}+\ldots+\hat{\beta}_k x_{kn} \end{pmatrix}=\mathbf{X}\hat{\boldsymbol{\beta}}, \tag{2.40}$$

$$\begin{aligned} \mathbf{e} &= \mathbf{y}-\hat{\mathbf{y}}=\mathbf{y}-\mathbf{X}\hat{\boldsymbol{\beta}} \\ &= \mathbf{y}-\mathbf{X}(\mathbf{X}'\mathbf{X})^{-1}\mathbf{X}'\mathbf{y} \\ &= (\mathbf{I}-\mathbf{X}(\mathbf{X}'\mathbf{X})^{-1}\mathbf{X}')\mathbf{y}, \end{aligned} \tag{2.41}$$

where $\mathbf{I}$ is the unit matrix, with 1s along the leading diagonal and 0s in all other positions. Note that the matrix $\mathbf{X'X}$ should not be singular in order for its inverse to exist. This implies that the columns (variables) in $\mathbf{X}$ must be linearly independent, which means that they must not have a linear relationship among themselves.

2.4.2 ANALYSIS OF VARIANCE

The total sum of squares of y in multiple regression is divided into two parts just as in equation (2.35) for simple regression. The way to compute SST and SSE is the same as in (2.36), but SSR is computed in a different way:

$$SSR = \sum_j (\hat{y}_j - \bar{y})^2 = \hat{\mathbf{y}}'\hat{\mathbf{y}} - n\bar{y}^2 = \hat{\boldsymbol{\beta}}'\mathbf{X}'\hat{\mathbf{y}} - n\bar{y}^2 . \quad (2.42)$$

The ANOVA table has the same form as Table 2.4, except that the degrees of freedom are different. The degrees of freedom of SSR and SSE are k and $n-k-1$, respectively. The ANOVA table for multiple regression is shown in Table 2.5. If $F_0 = MSR/MSE > F_{1-\alpha}(k, n-k-1)$, we can say that the fitted regression equation is significant at the α level of significance in explaining the relationship between Y and k independent variables.

Table 2.5 ANOVA table for multiple regression.

Source of variation	Sum of squares	Degree of freedom	Mean square	F_0	$F(\alpha)$
Regression	SSR	k	MSR	MSR/MSE	$F_{1-\alpha}(k,n-k-1)$
Residual	SSE	$n-k-1$	MSE		
Total	SST	$n-1$			

The coefficient of determination has the same form as (2.33):

$$R^2 = \frac{\sum(\hat{y}_i - \bar{y})^2}{\sum(y_i - \bar{y})^2} = \frac{SSR}{SST}. \tag{2.43}$$

The range of R^2 is again [0,1]. Values of R^2 that are close to 1 imply that most of the variability in y is explained by the regression model.

2.4.3 SELECTION OF VARIABLES

Suppose the fitted regression model is significant according to the F-test in the ANOVA table. This does not necessarily mean that every regressor significantly affects the dependent variable. Here, the regressor is the independent variable which is often called the predictor or the factor. Even though the model is significant, some of the regressors are important while others are not.

In most practical problems the analyst has a pool of candidate regressors that should include all the influential factors, but the actual regressors that should be used in the model should be a subset of all candidate regressors.

Finding an appropriate subset of regressors for the model is called the **variable selection** problem or **factor-finding problem.**

Building a regression model that includes only a subset of the available regressors involves two conflicting objectives.

1. To make the resultant equation useful for predictive purposes, we want our model to include as many regressors as possible.
2. Because of the costs involved in obtaining and maintaining information on a larger number of regressors, we want the equation to include as few regressors as possible.

The process of finding a model that is a compromise between these two objectives is called selecting the "best" regression equation. Unfortunately, there is no unique statistical procedure for doing this. The best-known methods are the following.

1. **All possible regression.** This procedure requires that the analyst fits all the regression equations involving one candidate

regressor, two candidate regressors, three candidate regressors, and so on. These equations are evaluated according to some suitable criterion such as R^2, and the "best" regression model is selected. However, as the number of candidate regressors increases, the computational burden rapidly increases, and it becomes impractical for problems involving more than seven or eight regressors.

2. **Forward selection.** This procedure begins with the model involving the intercept term only. Successive efforts are made to find an optimal subset by inserting regressors into the model one at a time until a suitable model is obtained.
3. **Backward selection.** This procedure begins with the full model that includes all k candidate regressors. Successive efforts are made to find an optimal subset by deleting regressors from the model one at a time until a suitable model is obtained. This procedure is just the opposite to forward selection.
4. **Stepwise regression.** This procedure is a popular combination of steps 2 and 3. It is a modification of forward selection in which at each step, all regressors previously entered into the model are reassessed, and if a regressor in the model is not significant, it may be deleted from the model. Note that a regressor entered at an earlier step may now be redundant because of the relationship between it and other regressors entered into the equation later.

2.4.4 CASE STUDY

This is a case study of an oil refinery company in Korea. One of the authors acted as consultant for several quality improvement teams (QITs) in this company, and the following is a part of one team's QIT report.

The company uses the $x - Rs$ control chart to control the color quality of SW-kerosene which is produced in the kerosene sweetening process U-1700. At that time the color quality was not satisfactory, and a QIT was formed to improve the quality. The quality is expressed by the Saybolt number, which is a larger-the-better characteristic. It should be at least 20 for customer satisfaction. At the beginning the team considered more than ten factors, but by means of a brainstorming discussion they reduced these to six factors which might influence the color quality. Of these factors, four are feeding conditions and two are operating conditions in the reactor, as shown in Figure 2.8.

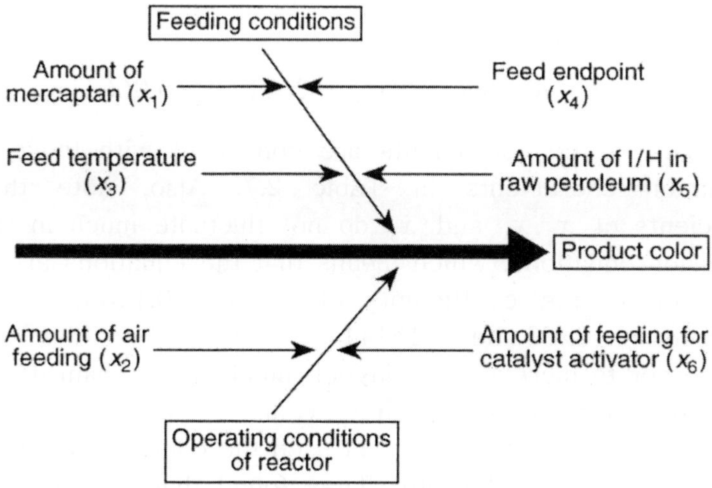

Figure 2.8 Causes-and-effects diagram for color quality.

In order to investigate the relationship between the product color y and the six factors $x_1,\ldots,x_6$, data from the previous two months was collected (see Table 2.6). The value in each piece of data represents the average of each variable in a day.

For data analysis, the team found the correlation matrix for a set of seven variables and the means and standard deviations of these variables, as shown in Table 2.7. Note that the average of y is about 21.9 and the standard deviation is 3.2, which is too large for desirable quality control. Observe that y is correlated with x_2, x_5 and x_6. However, since some strong correlation exists among the factors, it is difficult to say that x_2, x_5 and x_6 are the only important factors that affect the color quality y. In such situations the best way to find a suitable model for y is to use the stepwise regression approach. Table 2.8 shows the results of stepwise regression for the data set of Table 2.6.

From the results of stepwise regression in Table 2.8, the team decided to adopt the third equation with x_2, x_3 and x_5, since x_1 and x_6 which entered into the equation at the fourth and fifth steps are not easy to control, and the R^2 value of 0.68 is large enough for a regression model. Therefore, the final equation used was

$$\hat{y} = 55.65490 - 0.13588x_2 - 0.26217x_3 - 0.06633x_5. \quad (2.44)$$

The signs of the coefficients are consistent with those of the correlation coefficients in Table 2.7. Also, note that the coefficients of x_2, x_3 and x_5 do not fluctuate much in the five regression equations, which means that the equations are stable. For instance, the coefficients of x_2 are –0.14539, -0.14912, -0.13588, -0.13876 and –0.15444.

In order to increase the Saybolt number, the team decided to make the operating levels of x_2, x_3 and x_5 as low as possible. Before this team effort, they had used the conditions $x_2 = 140\%$, $x_3 = 50°\text{C}$, and $x_5 = 50\,\text{ppm}$. They found that they could make x_2, x_3 and x_5 as low as $x_2 = 120\%, x_3 = 42°\text{C}$ and $x_5 = 40\,\text{ppm}$. The predicted value of the Saybolt number under the new conditions is

$$\hat{y} = 55.65490 - 0.13588 \times 120 - 0.26217 \times 42 - 0.06633 \times 40 = 25.7\,.$$

The average Saybolt number before this effort was about 21.9. Hence, the team succeeded in improving the color quality by 3.8 on average.

Later the team confirmed this result by a series of confirmation tests under the new conditions. It obtained an average of 26.0 and a standard deviation of 1.9. The team was particularly happy with the small standard deviation, which meant that it had obtained more robust operating conditions for producing the kerosene product, and that the probability of obtaining product color with Saybolt number less than 20 was almost zero.

Table 2.6 SW-kerosene color data.

Data number	X_1 Feed mercaptan wt ppm	X_2 Air feeding %	X_3 Feed temperature °C	X_4 Feed E.P. °C	X_5 Catalyst activator wt ppm	X_6 I/H Petroleum vol %	y Saybolt number
1	111.6	130	46	247	62	100.0	21
2	91.5	135	50	251	70	94.2	21
3	90.8	135	50	245	70	94.1	21
4	80.8	175	43	247	80	16.3	13
5	82.8	145	46	248	80	22.0	20
6	78.6	145	46	255	80	22.0	20
7	93.2	120	43	254	80	22.0	21
8	93.7	120	38	253	80	16.4	23
9	95.9	120	38	253	80	10.9	21
10	97.6	115	42	256	70	10.8	27
11	100.0	115	42	254	70	10.8	26
12	100.0	115	50	260	50	10.0	26
13	100.0	115	51	264	40	0.9	26
14	100.0	119	47	268	40	0.9	27
15	98.7	119	46	259	40	1.0	28
16	100.0	119	44	268	40	1.4	27
17	94.0	135	45	261	50	2.8	23
18	138.7	160	53	268	70	96.9	17
19	157.0	143	56	264	36	95.1	20
20	160.0	150	55	265	36	94.4	18
21	160.0	137	47	261	65	94.4	21
22	119.0	128	59	274	40	86.2	21
23	116.0	139	53	273	52	13.8	19
24	118.6	139	52	265	52	13.8	17

Table 2.6 (Continued).

Data number	X_1 Feed mercaptan wt ppm	X_2 Air feeding %	X_3 Feed temperature °C	X_4 Feed E.P. °C	X_5 Catalyst activator wt ppm	X_6 I/H Petroleum vol %	y Saybolt number
25	117.7	139	50	270	40	13.8	17
26	117.5	150	50	273	40	13.8	21
27	114.3	173	40	259	40	1.9	18
28	133.2	115	58	270	60	9.6	19
29	129.5	107	53	274	40	9.6	23
30	128.4	128	59	262	40	9.6	21
31	121.4	128	57	268	40	11.4	21
32	150.0	110	51	269	40	2.7	23
33	150.8	110	50	276	40	3.0	23
34	150.0	110	49	275	40	3.0	24
35	106.9	110	49	271	40	3.0	26
36	124.3	120	53	271	40	3.0	20
37	139.2	105	52	274	40	2.2	24
38	139.7	119	53	270	40	1.2	22
39	140.4	119	52	266	40	1.2	22
40	140.4	119	48	276	40	1.2	23
41	138.9	119	50	270	40	1.3	26
42	139.5	119	51	271	40	1.8	23
43	105.9	133	52	260	40	4.9	20
44	110.0	133	52	269	40	3.9	23
45	101.6	133	53	273	40	3.9	22

Table 2.7 Correlation matrix, mean and standard deviation.

	x_1	x_2	x_3	x_4	x_5	x_6	y
x_1	1.0	- 0.17467**	0.52561**	0.63156**	- 0.55272--	0.15361	- 0.07459
x_2	- 0.17467	1.0	- 0.07785	- 0.34252*	0.25102†	0.39262**	- 0.74801**
x_3	0.53243**	- 0.07690	1.0	0.55343*	- 0.56221**	0.16607	0.18758
x_4	0.63156**	- 0.34252*	0.55672**	1.0	- 0.76345**	- 0.33973*	- 0.17077
x_5	- 0.55272**	0.25102†	- 0.56165**	- 0.76345**	1.0	0.28486†	- 0.26357†
x_6	0.15361	0.39262**	0.18396	- 0.33973*	0.28486†	1.0	- 0.35967*
y	- 0.07459	- 0.74801**	- 0.18681	0.17077	- 0.26357†	- 0.35967*	1.0

Variable	Mean	Standard deviation
x_1	117.29079	22.80504
x_2	128.26666	16.31419
x_3	49.42222	5.13672
x_4	264.00000	8.81373
x_5	50.95555	15.61558
x_6	23.04657	34.12817
y	21.88889	3.17104

** Significant at 1% level.
* Significant at 5% level.
† Significant at 10% level.

Table 2.8 Results of stepwise regression.

Step	Variable entered	Regression equation	Coefficient of determination (R2)
1	x_1	$\hat{y} = 40.53795 - 0.14539x_2$	0.560
2	x_2, x_3	$\hat{y} = 48.53840 - 0.14912x_2 - 0.15220x_3$	0.620
3	x_2, x_3, x_5	$\hat{y} = 55.65490 - 0.13588x_2 - 0.26217x_3 - 0.06633x_5$	0.688
4	x_2, x_3, x_5, x_1	$\hat{y} = 58.55692 - 0.13876x_2 - 0.21339x_3 - 0.08429x_5 - 0.03435x_1$	0.726
5	x_2, x_3, x_5, x_1, x_6	$\hat{y} = 65.62266 - 0.15442x_2 - 0.26746x_3 - 0.11384x_5 - 0.04627x_1 - 0.02256x_6$	0.761

2.5 PROCEDURE FOR QUALITY PROBLEM-SOLVING

A complete manufacturing process is either a continuous process, a batch process, or combination of both processes. It may consist of several units of continuous or batch processes. The basic goal of QC activities in the manufacturing processes is to check, analyze and control the mean and variation of quality in each unit of the whole manufacturing process.

The sources of quality variation are numerous. Usually the major sources are classified into six categories (we call them 5M1E, taking the initials of the six categories). They are **man, machine, method, material, measurement** and **environment.**

Figure 2.9 shows the general flow of quality problem-solving in a manufacturing process and the corresponding statistical methods which are often used for such purpose. The usual statistical methods used to check the variation quality are control charts, histograms, scatter diagrams, process capability indices and miscellaneous graphs. If some abnormality is suspected, then occurrence of quality problems is statistically tested for.

Often some types of statistical estimation for means and variances are necessary. When some problems of quality are detected, the present situation and the causes of trouble are investigated through data collection and analysis. For data analysis, statistical computation using statistical software is needed, since we normally handle large sets of past data which are related to the sources of trouble. The statistical methods which are often used for this analysis are scatter diagrams, correlation and regression analysis, multivariate analysis, reliability analysis, and so on.

Based on findings of sources of quality problems, some countermeasures to solve the problem are considered and implemented. If the problem is solved by these countermeasures, the flow goes to the usual checking process for the mean and variation of quality. If the problem is not solved, some type of experimental design should be adopted to find either the optimum operating conditions or another way of handling the situation.

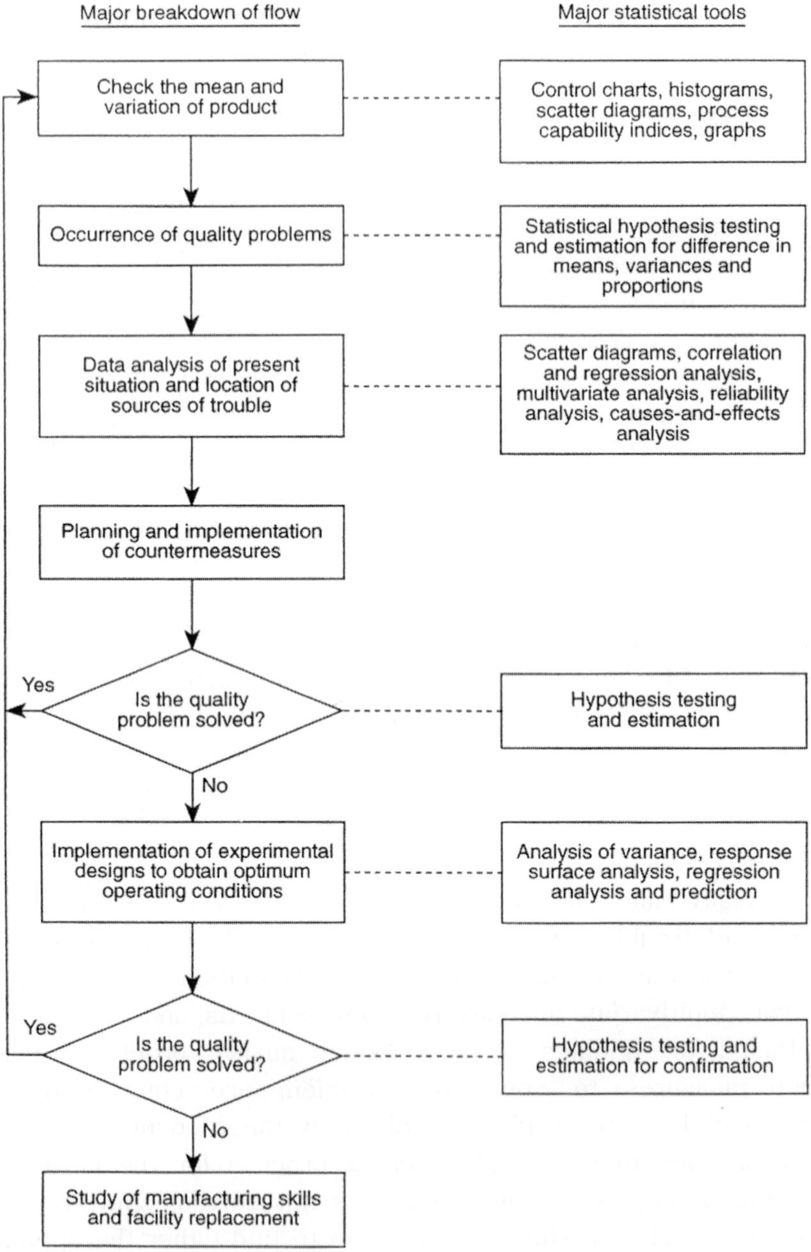

Figure 2.9 General flow of quality problem-solving.

Taguchi's robust design is a powerful tool for this stage. The most often used experimental designs in industry are perhaps factorial designs, fractional factorial designs, response surface designs, and mixture designs. For analysis of variance and prediction of responses, statistical computation is also needed at this stage. After such activity, one should check again whether the quality problem has been solved or not. If not, quite different countermeasures such as change of manufacturing skills or facility replacement should be studied.

If a QC circle wishes to solve a QC problem, the following nine steps are recommended:

1. Selection of project
2. Data collection to understand the present situation
3. Causes-and-effects analysis
4. Set-up goals
5. Planning and implementation of countermeasures
6. First-stage evaluation
7. Planning and implementation of experimental designs (if necessary)
8. Second-stage evaluation (if necessary)
9. Standardization and reports.

2.6 A STRATEGY FOR QUALITY IMPROVEMENT BY TEAM EFFORT

The key to the growth of a manufacturing company is the continuous effort for quality and productivity improvement. In order to uncover quality and productivity problems which are hidden in all departments, and solve those problems for company growth, the following four ingredients, which are shown graphically in Figure 2.10, are essential:

1. Senior management should have a long-term policy for quality management (QM), make it known to everybody, and show firm determination to implement it.

2. There must be sufficient education and training programs to maximize the ability of human resources.
3. The working atmosphere should be configured so as to enhance workers' pride in workmanship and self-confidence.
4. Teamwork should be improved by small-group activities, and barriers between departments should be removed to solve quality and productivity problems in which several departments have an interest.

Here we are most concerned with the last of these important ingredients for company growth. There are many quality and productivity problems in a company which cannot be tackled by an individual person, and which require team effort. Basically there are two types of team for quality problem-solving. One is the well-known QC circle whose members are mainly line workers. Another is the team organized by middle managers (engineers and researchers). Sometimes a mixed type is desirable, depending on the small-group organization for team effort.

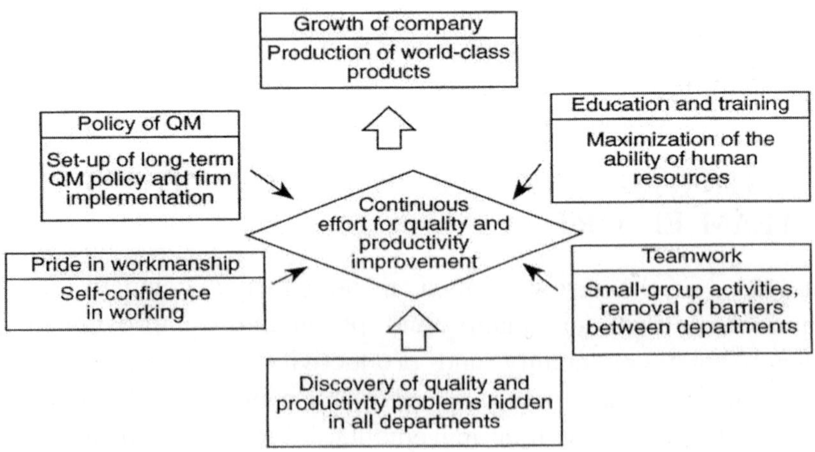

Figure 2.10 Continuous effort for quality and productivity improvement.

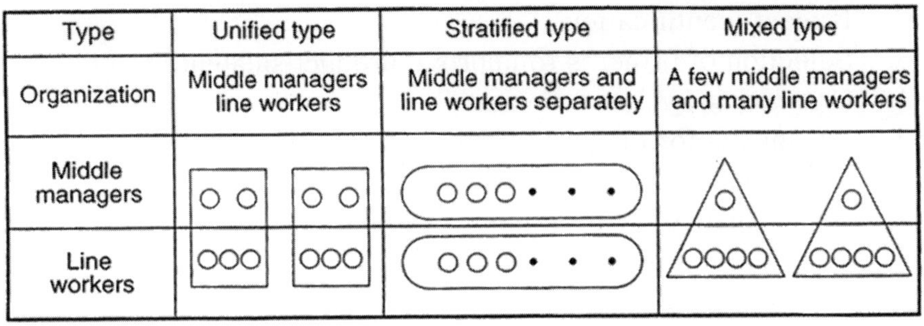

Figure 2.11 Organization of small groups for quality improvement.

In this section we will deal with only the teams organized by middle managers. Such a team is often called a quality improvement team (QIT), or, depending on the purpose of team, an SQC team, VE (value engineering) team, TPM (total productive maintenance) team or DOE (design of experiments) team.

Crosby (1979) and Juran (1981) have emphasized the importance of team effort for quality improvement programs. Crosby proposes the following 14 steps:

1. Management commitment
2. Formation of QIT
3. Quality measurement
4. Evaluation of quality cost
5. Quality awareness
6. Corrective action
7. Zero-defect planning
8. Supervisor training
9. Zero-defect day
10. Goal-setting
11. Error cause removal
12. Recognition
13. Quality council meetings
14. Do it over again.

Juran's ten steps are as follows.

1. Proof of the need

2. Project identification
3. Selection of projects to improve product salability
4. Organizing for improvements
5. Organizing for diagnosis: the diagnostic journey
6. Controllable errors
7. Use of diagnostic tools
8. Remedial journey
9. Motivation for quality
10. Holding the gains.

The authors have been involved as educational and consultative capacity in the quality programs of many companies over the last twenty years. From the authors' experience, the following strategy is a good way of planning and implementation for quality and productivity improvement effort by middle managers. It consists of seven steps.

1. Senior management's commitment for quality policy, and organization of QIT which will achieve quality target
2. Quality evaluation and selection of projects
3. Setting up goals for the team to achieve
4. Finding sources of trouble and determination of possible countermeasures
5. Implementation of countermeasures and preparation of team reports
6. Holding a report session for all managers, and recognition of team efforts for quality improvement results
7. Evaluation of the gains, standardization of the results and planning of future team efforts.

In order to plan, enable, evaluate and replan team activities, it is necessary to have a department whose major function is to initiate such activities. Such a department may be called the QM, SMT (scientific management techniques), PI (process innovation) or QIP (quality improvement program). Let us call it the QM department. Perhaps the best background for the staff of the QM department would be industrial engineering, statistics or management science. The most important thing to be borne in mind for the staff of the QM department is that they are supposed to help team members tackle quality problems by themselves. However, there are some areas in which QM department staff may be called upon to play more than a merely facilitative role: ordinary engineers, for example, often require help when they need to use statistical software, to use computers to collect and analyze data, and to prepare reports for a report session.

Exercises

2.1 The following 16 observations were taken from a process in which the thickness of an extremely thin metallic deposit at equally spaced intervals of time. It is desired to maintain the thickness of the deposit as close as possible to the target value M = 80. The USL and LSL of the thickness are USL = 100 and LSL = 60, respectively.

Data: 80, 75, 100, 61, 93, 85, 92, 86, 77, 82, 85, 102, 93, 90, 90, 88.

(a) Suppose the above data follow a normal distribution. Find the estimates of mean and variance.

(b) Calculate the process capability index for the thickness data.

(c) Evaluate the process capability. Is it good enough?

(d) Suppose that the loss A_0^* at the thickness of 60 or 100 is \$2.00. What is the estimate of expected loss? Note that the formula of the estimate of expected loss can be found in Table 2.1.

(e) Find the SN ratio for the data. Use the formula given in equation (2.18).

2.2 The following 25 observations are from a tire radial runout (mils) study. Low runout values are desirable.

Data: 0.51, 0.28, 0.65, 0.32, 0.47, 0.41, 0.23, 0.43, 0.67, 0.56, 0.53, 0.51, 0.35, 0.50, 0.24, 0.38, 0.45, 0.48, 0.49, 0.24, 0.53, 0.28, 0.22, 0.29, 0.78.

(a) Find the SN ratio for the data.
(b) Suppose that the upper specification limit is USL = 0.60, and the loss $A_0^* = \$10$. Find the average loss.

2.3 Suppose we have the following two sets of larger-the-better characteristics.

Data set 1: 4, 5, 6, 7, 8.

Data set 2: 1, 3, 6, 9, 12.

The average of data set 1 is 6, and the average of data set 2 is 6.2. The average of data set 2 is larger than that of data set 1. Does it mean that the SN ratio of data set 2 is larger than that of data set 1? Find the SN ratios, and compare them. If the SN ratio of data set 2 is not larger than that of data set 1, why does this result happen? Explain the reason.

2.4 Consider the follow data.

X	1.6	3.2	3.8	4.2	4.4	5.8	6.0	6.7	7.1	7.8	8.0	8.5
Y	5.6	7.9	8.0	8.2	8.1	9.2	9.5	9.4	9.6	9.9	9.8	10.5

(a) Find the sample correlation coefficient.
(b) Test the hypothesis that the correlation coefficient is zero. Use 5% level of significance.
(c) Find the simple regression equation. What is the coefficient of determination? Explain the meaning of this coefficient.
(d) Construct the ANOVA table for this simple regression. Is the regression equation significant at 5% level of significance?

2.5 For the case study in Section 2.4.4, there was a total of 45 observations. Suppose we have only first 30 observations in Table 2.6. Obtain the similar results as shown in Tables 2.7 and 2.8. Are the results much different?

2.6 Why is team effort important for quality improvement in an organization? Explain the reasons.

3

Fundamentals of Designing Experiments

3.1 FRAMEWORK OF EXPERIMENTAL DESIGN

3.1.1 OUTLINE OF EXPERIMENTAL DESIGN PROCEDURE

Experiments are carried out by researchers or engineers in all fields of study to compare the effects of several conditions or to discover something new. If an experiment is to be performed most efficiently, then a scientific approach to planning it must be considered. The statistical design of experiments is the process of planning experiments so that appropriate data will be collected, the minimum number of experiments will be performed to acquire the necessary technical information, and suitable statistical methods will be used to analyze the collected data.

The statistical approach to experimental design is necessary if we wish to draw meaningful conclusions from the data. Thus, there are two aspects to any experimental design: the design of the experiment and the statistical analysis of the collected data. They are closely related, since the method of statistical analysis depends on the design employed.

An outline of the recommended procedure for an experimental design is shown in Figure 3.1 and briefly explained below.

1. Statement of the experimental problem. It is necessary to develop a clear statement of the problem and its objectives.

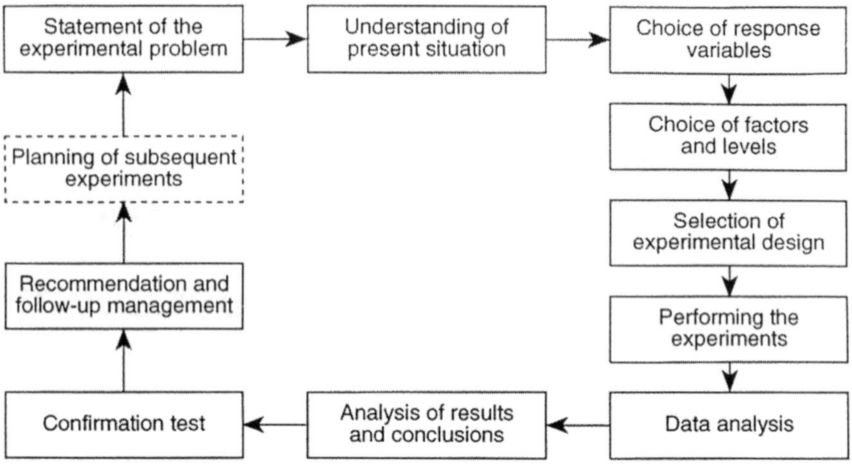

Figure 3.1 Outline of experimental design procedure.

A clear statement contributes to a better understanding of the phenomena under study and the form of final solution of the problem.

2. Understanding of present situation. It is important to collect as much related past data for the experimental problem as possible, and to understand the present situation. It is useful to collect related information from the literature and all concerned parties: engineering, quality assurance, manufacturing, marketing, operating personnel, and so on.
3. Choice of response variables. From the statement of the problem, an appropriate response variable should be selected. If necessary, multiple response variables may be chosen. Suppose that yield is the major response variable. Very often as the yield increases, the utility increases which should be minimized. For such a case, we should select not only yield but also utility. Thought must be given to how the response variables will be measured, and the probable accuracy of such measurements.
4. Choice of **factors** and **levels**. The experimenter must select the independent variables (factors) which affect the chosen

response variables. The values (levels) of the factors to be used in the experiment should be carefully selected. Usually between two and five levels are appropriate, and the range of levels should be as large as possible within the region of experimental interest to the experimenter.

5. Selection of experimental design. This step is the backbone in the experimental design procedure. The experimenter must decide the appropriate experimental design by considering the number of factors, levels, all possible level combinations, the cost of the experiment and the time available. If there are too many factors, a fractional factorial design using orthogonal arrays is usually recommended. The experimenter should also decide an appropriate number of replicates at each experimental condition (run) to guarantee desired statistical accuracy, and the order in which the experiments will be conducted (the method of randomization). A mathematical model for the experiment must also be proposed, so that a statistical analysis of the data may be performed.
6. Performing the experiments. This is the actual data-collection process. The experimenter should pay attention to the actual progress of the experiment to ensure that it is proceeding according to the plan, and also pay attention to maintaining as uniform an experimental environment as possible.
7. Data analysis. Statistical methods such as analysis of variance and statistical estimation should be employed in analyzing the data from the experiment. It is desirable to obtain all possible information from this data analysis for the stated experimental problem.
8. Analysis of results and conclusions. Once the data set has been analyzed, the experimenter must draw physical inferences from his statistical results, and evaluate their practical implications for the stated experimental problem. Then conclusions should be made for the stated problem.

9. Confirmation test. Before presenting the results to others and taking a practical course of action, the experimenter needs to carry out a confirmation test to evaluate the conclusions from the experiment.
10. Recommendations and follow-up management. The experimenter presents the results to others and takes any necessary action. In order to sustain the improvements suggested by the experiment, careful follow-up action is needed such as standardization of operating conditions and use of check sheets and control charts to evaluate the continuing implications of the experiment.
11. Planning of subsequent experiments. Very often a further round of experiments is recommended, as the experimental problem has not been completely solved. Experimentation is usually an iterative process, with one experiment solving the problem only partially and subsequent experiments hopefully dealing with outstanding questions.

3.1.2 CLASSIFICATION OF DATA TYPES

Experimental data are usually divided into two types: discrete (attribute) and continuous (variable). Taguchi (1987) further divided each of these two types into three categories, as explained in Table 3.1. There are three classes of discrete data: simple discrete, fixed marginal discrete and multi-discrete. There are also three classes of continuous data: simple continuous, multi-fractional continuous and multi-variable continuous. This classification is important in deciding on the number of replications that are necessary for the experiments, and then determining a suitable method of data analysis.

3.1.3 CLASSIFICATION OF FACTORS

The factors which are treated in experiments can be divided into two types: fixed and random. "Fixed" means that the factor levels are technically controllable and each level has some technical

Table 3.1 Classification of data types.

Type	Class	Example	Probability distribution
Discrete (attribute)	Simple discrete	Countable data, such as number of flaws and number of pinholes.	
	Fixed marginal discrete	Data on the number of individuals out of n individuals classified into several categories: for example, when one classifies the appearance of n objects as excellent, good, fair or poor.	Multinomial (binomial if there are two categories)
	Multi-discrete	The number of flaws is the simple discrete value. However, the numbers of large, medium and small flaws are counted; these numbers are multi-discrete data. This refers to the case where there are several grades and the number of units per grade is counted.	Poisson in each grade
Continuous (variable)	Simple continuous	Ordinary continuous values such as weight, hardness and tensile strength.	Normal
	Multi-fractional continuous	These values arise where the data of the individual categories are not the numbers of individual units, as with fixed marginal discrete values, but their percent-tage. For example, excellent 25.7%, good 20.9%, fair 39.2%, and poor 14.2%. Total must be 100%.	Approximately normal in each category, if n is large
	Multi-variable continuous	The total weight of apples in a tree is a simple continuous value. However, when apples, for example, in a tree are classified as first grade, 15.5kg; second grade, 20.9kg; and third grade, 5.5kg. These are multi-variable data.	Multivariate normal (normal in each grade)

meaning. Also the levels of a fixed factor can be reproduced and retested. However, "random" means that the factor levels are not technically controllable, and each level usually does not have any technical meaning. Also the levels of a random factor normally cannot be reproduced and retested.

Fixed factors can be subdivided into three classes:

1. Control factors: These are design or process variables which are controllable, and where we may be interested in finding the level which is, in some sense, best. The major variables in an experiment such as temperature, pressure and time are all control factors.
2. Indicative factors: These are factors that possess technical levels in the same way as control factors, but for which any notion of the "best" level may be meaningless. For example, in a tire-wear experiment, suppose that the position of tire (front right, front left, back right and back left) is a factor. Such factor is an indicative factor.
3. Signal factors: These are factors that influence the average value but not the variability of a response. They are called "target-control" factors.

Likewise, there are also three classes of **random factor**:

1. Block (group) factors: These are factors for which there are levels, but where there is no technical significance to such levels. Differences of geographic location, differences depending on the day, lot differences and differences between operators are block factors.
2. Supplementary factors: These are factors that include supplementary experimental or measured values, which are the result of recording the state of the experimental conditions. These factors are used as independent variables in the covariance analysis.
3. Noise (error) factors: These are factors that have an influence over a response but cannot be controlled in actual applications. They are of three kinds: inner noise, outer noise and between-product noise.

3.2 ONE-FACTOR-AT-A-TIME EXPERIMENT

Perhaps the most popular method of multi-factorial experimentation has been the one-factor-at-a-time experiment. This is the method in which the experimenter varies only one factor at a time, while keeping all the other factors fixed at a specific set of conditions. The popularity of this method is due to its simplicity. The method is, however, misleading and can lead to wrong results.

Let us take a simple example in which there are only two factors, say, temperature (factor A) and time (factor B), and the response is yield. Suppose a factory is interested in improving the yield of a chemical reactor. The experimenters wish to test four levels of A, and three levels of B as follows:

$$A_1 = 70°\text{C}, \quad A_2 = 75°\text{C}, \quad A_3 = 80°\text{C}, \quad A_4 = 85°\text{C}$$
$$B_1 = 40\,\text{min}, \quad B_2 = 44\,\text{min}, \quad B_3 = 48\,\text{min}.$$

The current operating levels of A and B are A_1 and B_1. First of all, the experimenters fix the temperature at A_1 and change the time from B_1 and B_2 and then B_3, obtaining the yields in Table 3.2. They find that B_3 is the optimal level of B, and the improvement in yield from the control is $+1.3\%$.

Next, they fix the time at B_1, the current operating level, and change the temperature from A_1 to A_2, then A_3, and finally A_4, obtaining the yields in Table 3.3. They find that A_3 is the optimal level of A, and the improvement of yield from the control is $+1.5\%$.

From the results of Tables 3.2 and 3.3, can we conclude that the optimal levels of A and B are A_3B_3? Summing the results we have obtained, can we say that the improvement of yield from A_1B_1 to A_3B_3 is 2.8%, giving a yield of 94.3%? The answers are "No". From this one-factor-at-a-time experiment, there is no guarantee whatsoever that A_3B_3 is the optimal condition, or that the improvement in yield is 2.8%. All we can confidently say about the different levels of B is that B_3 is optimal when A is fixed at A_1 (if A is fixed at A_3, B_2 may be optimal), and that difference from the control (B_3 over B_1) is +1.3% when A is fixed at A_1 (if A is fixed at A_3, the difference may be +1.0%).

Table 3.2 Influence of B while keeping A fixed at A_1.

	Temperature	Time	Yield (%)	Difference from the control (%)
Control	A_1	B_1	91.5	-
Change the levels of B	A_1	B_2	92.2	+0.7
	A_1	B_3	92.8	+1.3

Table 3.3 Influence of A while keeping B fixed at B_1.

	Temperature	Time	Yield (%)	Difference from the control (%)
Control	A_1	B_1	91.5	-
Change the levels of A	A_2	B_1	91.9	+0.4
	A_3	B_1	93.0	+1.5
	A_4	B_1	92.8	+1.3

Table 3.4 Yield data for two-factor factorial design.

	A_1	A_2	A_3	A_4	Total	Average
B_1	91.5	91.9	93.0	92.8	369.2	92.3
B_2	92.2	92.9	93.8	93.6	372.5	93.125
B_3	92.8	93.1	92.6	92.3	370.5	92.7
Total	276.5	277.5	279.4	278.7	112.5	
Average	92.167	92.633	93.133	92.9		92.708

If we employ the two-factor factorial design with no replication, there are 12 runs and the yield data may be as shown in Table 3.4. Note that the optimal condition of A and B in Table 3.4 is A_3B_2, not A_1B_3, and that the yield at A_3B_3 is not 94.3% but 92.6%. This means that the one-factor-at-a-time experiments led to wrong results in predicting the optimal condition of A and B, and in estimating the response value at the optimal condition.

When there are $k(k \geq 3)$ factors, if we use the one-factor-at-a-time experiment k times, it is very likely that we will end up with some incorrect results.

Hence, when there are k factors, it is desirable to employ the k-factor factorial design instead of k one-factor-at-a-time experiments. However, if k is greater than 2, the number of runs is often too large and it is necessary to use a fractional design. We will study fractional factorial design in Chapter 4.

If the yield response is sketched, Figure 3.2 is obtained. Observe that the connected lines are not parallel, which means that factors A and B have an interaction $A \times B$. If the connected line are parallel, we infer that factors A and B do not interact. If the difference in response between the levels of factor B is not the same at all levels of factor A, factors A and B are said to interact. In other words, when the effects of B are different at different levels of factor A, an interaction $A \times B$ exists.

If there exists an interaction between factors A and B, we have to experiment on all possible level combinations to find out the optimal conditions of A and B. Sequential experiments of two one-factor-at-a-time experiments cannot guarantee to produce the optimal levels of A and B, as shown in Tables 3.2–3.4.

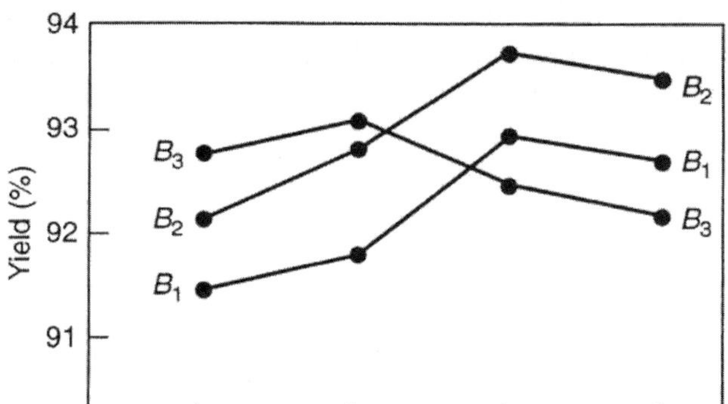

Figure 3.2 Sketch of yield response at 12 level combinations.

3.3 TWO-FACTOR FACTORIAL DESIGN

3.3.1 AN EXAMPLE

The simplest type of factorial design, and then one which is most often used in practice, is perhaps the two-factor factorial design with replicates. In this design there are l levels of factor A, m levels of factor B, and r replicates in each of lm level combinations. The order in which the lmr observations are taken is selected at random, so that this design is basically a completely randomized design.

As an example, a researcher is designing an adhesive that will be subjected to extreme variations in temperature when it is used. The design parameter that he wants to decide is the raw material for the adhesive, and he has three possible choices. When the adhesive is manufactured and is shipped to the customer, the researcher has no control over the temperature extremes that the adhesive will encounter, and he knows from experience that variations in temperature will impact the shear strength of the adhesive. However, temperature can be controlled in the product development laboratory for the purposes of a test.

The researcher decides to test all three materials at four temperature levels (−10°C, 5°C, 20°C, 35°C) as these temperature levels cover the temperature range of the product end-use environment. Two adhesives are tested at each combination of raw material and temperature, and all 24 tests are run in random order. The resulting observed shear-strength data is given in Table 3.5.

In this problem the researcher wants to answer the following questions:

1. What effects do material type and temperature have on the shear strength of the adhesive?
2. Is there any interaction between material type and temperature? For each material type, what temperature is best or worst for the shear strength?
3. Is there a choice of material that would give uniformly strong shear strength regardless of temperature?

The last question is particularly important. It may be possible to find a material that is not much affected by temperature. If this is so, the researcher can make the adhesive robust to temperature variation in the field. This is an example of using statistical experimental design for robust product design.

Table 3.5 Shear-strength ($kgcm^{-2}$) data for the adhesive design experiment.

Material type (factor *A*)	Temperature (factor *B*)				Sum	Mean
	-10°C	5°C	20°C	35°C		
1	35	38	45	34	305	38.1
	28	45	50	30		
2	26	54	53	20	308	38.5
	29	48	60	18		
3	18	26	32	25	210	26.3
	25	30	34	20		
Sum	161	241	274	147	823	34.3
Mean	26.8	40.2	45.7	24.5		

3.3.2 STATISTICAL MODEL

The design in Table 3.5 is a specific example of the general case of a two-factor factorial design with replicates. Proceeding to the general case, the observations may be described by the following linear model:

$$y_{ijk} = u + a_i + b_j + (ab)_{ij} + e_{ijk}, \tag{3.1}$$

$$i = 1, 2, \ldots, l,$$
$$j = 1, 2, \ldots, m,$$
$$k = 1, 2, \ldots, r,$$

where

μ = the overall mean effect,

a_i = the effect of the ith level of the row factor A,

b_i = the effect of the jth level of the column factor B,

$(ab)_{ij}$ = the effect of the interaction between A_i and B_j,

y_{ijk} = the observed response when factor A is at the ith level and the factor B is at the jth level for the kth replicate,

e_{ijk} = the random experimental error of y_{ijk}.

Table 3.6 Data arrangement for a two-factor design.

		Levels of factor B				Sum	Mean
		B_1	B_2	$\cdots$	B_m		
Levels of factor A	A_1	y_{111} $\vdots$ y_{11r}	y_{121} $\vdots$ y_{12r}	$\cdots$	y_{1m1} $\vdots$ y_{1mr}	$T_{1..}$	$\bar{y}_{1..}$
	A_2	y_{211} $\vdots$ y_{21r}	y_{221} $\vdots$ y_{22r}	$\cdots$	y_{2m1} $\vdots$ y_{2mr}	$T_{2..}$	$\bar{y}_{2..}$
	$\vdots$	$\vdots$	$\vdots$	$\cdots$	$\vdots$	$\cdots$	$\cdots$
	A_l	y_{l11} $\vdots$ y_{l1r}	y_{l21} $\vdots$ y_{l2r}	$\cdots$	y_{lm1} $\vdots$ y_{lmr}	$T_{l..}$	$\bar{y}_{l..}$
Sum		$T_{.1.}$	$T_{.2.}$		$T_{.m.}$	T	
Mean		$\bar{y}_{.1.}$	$\bar{y}_{.2.}$		$\bar{y}_{.m.}$		$\bar{\bar{y}}$

We assume that the error e_{ijk} is independently and identically normally distributed with mean 0 and variance σ^2, which we will abbreviate as i.i.d.

$N(0,\sigma^2)$. In addition, we assume that both factors are fixed, and that the treatment effects are defined as deviations from the overall mean, so that

$$\sum_i a_i = 0, \qquad \sum_j b_j = 0.$$

Similarly, the interaction effects are fixed and are defined such that

$$\sum_i (ab)_{ij} = 0 \text{ for all } j, \qquad \sum_j (ab)_{ij} = 0 \text{ for all } i.$$

There are lmr total observations in this design, since there are r replicated of the experiment. The observed data appears as in Table 3.6.

3.3.3 ANALYSIS OF VARIANCE AND HYPOTHESIS TESTING

In the two-factor factorial design, both row and column factors A and B are of equal interest. Specifically, we are interested in testing hypothesis about the equality of row factor effects, say

$$\begin{aligned} &H_0 : a_1 = a_2 = \ldots = a_l = 0 \ (A \text{ is not significant}) \\ &H_1 : \text{ at least one } a_i \neq 0 \ \ (A \text{ is significant}), \end{aligned} \tag{3.2a}$$

and the equality of column factor effects, say

$$\begin{aligned} &H_0 : \ b_1 = b_2 = \ldots = b_m = 0 \ (B \text{ is not significant}) \\ &H_1 : \ \text{ at least one } b_j \neq 0 \ \ (B \text{ is significant}). \end{aligned} \tag{3.2b}$$

We are also interested in determining whether row and column factors interact. Thus, we also wish to test.

$$\begin{aligned} &H_0 : \ (ab)_{ij} = 0 \text{ for all } i, j \ (A \times B \text{ is not significant}) \\ &H_1 : \ \text{ at least one } (ab)_{ij} \neq 0 \ \ (A \times B \text{ is significant}). \end{aligned} \tag{3.2c}$$

These hypotheses can be tested using a two-factor analysis of variance as explained below.

Let $T_{i..}$ denote the total of mr observations under the ith level of factor A, $T_{.j.}$ the total of lr observations under the jth level of factor B, $T_{ij.}$ the total for r observations in the ijth cell, and T the grand total of all the observations. Define $\bar{y}_{i..}, \bar{y}_{.j.}, \bar{y}_{ij.}$ and $\bar{\bar{y}}$ as the corresponding row, column, cell and grand means. We can express these definitions mathematically as follows:

$$T_{i..} = \sum_j \sum_k y_{ijk}, \quad \bar{y}_{i..} = \frac{T_{i..}}{mr}$$

$$T_{.j.} = \sum_i \sum_k y_{ijk}, \quad \bar{y}_{.j.} = \frac{T_{.j.}}{lr}$$

$$T_{ij.} = \sum_k y_{ijk}, \quad \bar{y}_{ij.} = \frac{T_{.j.}}{r}$$

$$T = \sum_i \sum_j \sum_k y_{ijk}, \quad \bar{\bar{y}} = \frac{T_{ijk}}{lmr}.$$

It is known that the least-squares estimates of u, a_i, b_j and $(ab)_{ij}$ in (3.1) are as follows:

$$\hat{\mu} = \bar{\bar{y}}, \quad \hat{a}_i = \bar{y}_{i..} - \bar{\bar{y}}, \quad \hat{b}_j = \bar{y}_{.j.} - \bar{\bar{y}}, \quad \widehat{(ab)}_{ij} = \bar{y}_{ij.} - \bar{y}_{i..} - \bar{y}_{.j.} + \bar{\bar{y}}. \tag{3.3}$$

If we substitute these estimates into (3.1), we can obtain the related data breakdown for the model (3.1) as

$$(y_{ijk} - \bar{\bar{y}}) = (\bar{y}_{i..} - \bar{\bar{y}}) + (\bar{y}_{ij.} - \bar{\bar{y}}) + (\bar{y}_{ij.} - \bar{y}_{i..} - \bar{y}_{.j.} + \bar{\bar{y}}) + (y_{ijk} - \bar{y}_{ij.}), \tag{3.4}$$

where the last component $y_{ijk} - \bar{y}_{ij.}$ is the residual for the observation y_{ijk}. If we square both sides and sum over all i, j and k, we can obtain the total corrected sum of squares for the left-hand side which can be divided into four sums of squares as follows:

$$\sum_i \sum_j \sum_k (y_{ijk} - \bar{\bar{y}})^2 = mr\sum_i (\bar{y}_{i..} - \bar{\bar{y}})^2 + lr\sum_j (\bar{y}_{.j.} - \bar{\bar{y}})^2$$
$$+ \sum_i \sum_j (\bar{y}_{ij.} - \bar{y}_{i..} - \bar{y}_{.j.} + \bar{\bar{y}})^2 + \sum_i \sum_j \sum_k (y_{ijk} - \bar{y}_{ij.})^2, \tag{3.5}$$

$$S_T = S_A + S_B + S_{A\times B} + S_E.$$

Notice that the total sums of squares (S_T) has been partitioned into a sum of squares due to factor $A(S_A)$, a sum of squares due to factor $B(S_B)$, a sum of squares due to the interaction between A an $B(S_{A\times B})$, and a sum of squares due to error (S_E).

Computational formulas for the sums of squares in (3.5) may be obtained more easily as in (3.6):

$$
\begin{aligned}
CT &= \text{correction term} = \frac{T^2}{lmr}, \\
S_T &= \sum_i \sum_j \sum_k (y_{ijk} - \bar{\bar{y}})^2 = \sum_i \sum_j \sum_k y_{ijk}^2 - CT, \\
S_A &= mr \sum_i (\bar{y}_{i..} - \bar{\bar{y}})^2 = \sum_i \frac{T_{i..}^2}{mr} - CT, \\
S_B &= lr \sum_j (\bar{y}_{.j.} - \bar{\bar{y}})^2 = \sum_j \frac{T_{.j.}^2}{lr} - CT, \\
S_{A\times B} &= r \sum_i \sum_j (\bar{y}_{ij.} - \bar{y}_{i..} - \bar{y}_{.j.} + \bar{\bar{y}})^2 = S_{AB} - S_A - S_B,
\end{aligned}
\tag{3.6}
$$

where

$$
\begin{aligned}
S_{AB} &= \sum_i \sum_j \frac{T_{ij.}^2}{r} - CT, \\
S_E &= \sum_i \sum_j \sum_k (y_{ijk} - \bar{y}_{ij.})^2 = S_T - S_{AB}.
\end{aligned}
$$

The S_{AB} in equation (3.6) is the sum of squares between the lm celled totals, which is necessary only for computation of S_{AB}.

The number of degree of freedom associated with each sum of squares is $l-1$ for S_A, $m-1$ for S_B, $(l-1)(m-1)$ for $S_{A\times B}$, $lm(r-1)$ for S_E, and $lmr-1$ for S_T. We denote these degrees of freedom by ϕ, with the appropriate subscript. Each sum of squares divided by its degrees of freedom gives a mean square. The expected values of the mean squares are as follows:

$$
\begin{aligned}
E(V_A) &= E\left(\frac{S_A}{l-1}\right) = \sigma^2 + mr\left(\frac{\sum_i a_i^2}{l-1}\right), \\
E(V_B) &= E\left(\frac{S_B}{m-1}\right) = \sigma^2 + lr\left(\frac{\sum_j b_j^2}{m-1}\right), \\
E(V_{A\times B}) &= E\left(\frac{S_{A\times B}}{(l-1)(m-1)}\right) = \sigma^2 + r\left(\frac{\sum_i\sum_j (ab)_{ij}^2}{(l-1)(m-1)}\right), \\
E(V_E) &= E\left(\frac{S_E}{lm(r-1)}\right) = \sigma^2 .
\end{aligned}
\tag{3.7}
$$

Notice that if the null hypotheses of no row-factor effects, no column effects, and no interaction are true, then V_A, V_B, $V_{A\times B}$, and V_E will all be approximately equal. However, if there are differences between row-factor effects, say, then V_A will be larger than V_E. Therefore, to test the significance of factor A, simply divide the mean square V_A by the error mean square V_E. A large value of this ratio implies that the null hypothesis in (3.2a) is not true, which means that factor A is significant. The same inference also holds for factor B and for interaction $A \times B$.

If we assume that the model (3.1) is adequate, the ratio V_A/V_E is distributed as F with $\phi_A = l-1$ numerator degrees of freedom and $\phi_E = lm(r-1)$ denominator degrees of freedom, and the critical region would be the upper tail of the F distribution. For a given level of significance, if $F_0 = V_A/V_E$ is greater than $F_{1-\alpha}(\phi_A, \phi_E)$, we say that A is significant. Similar hypothesis tests hold for each of B and $A \times B$. The test procedure is usually summarized in an analysis variance (ANOVA) table as shown in Table 3.7.

Note that in most statistical software the significance probability p is printed instead of $F(\alpha)$. If the value of p is less than 0.05, we say that factor A is significant at the 0.05 or 5% level.

Hence, the smaller the value of p is, the more significant factor A is. The same inference holds for factor B and interaction $A \times B$.

If the interaction $A \times B$ is negligible (not significant at a large value of α such as $\alpha = 0.2$) it is desirable to pool $A \times B$ and E, to make a new error E'. In this case the new sum of squares, degrees of freedom and mean square of E' should be as follows.

$$\begin{aligned} S_{E'} &= S_E + S_{A\times B}, \\ \phi_{E'} &= \phi_E + \phi_{A\times B}, \\ V_{E'} &= S_{E'}/\phi_{E'}. \end{aligned} \tag{3.8}$$

Table 3.8 provides guidelines on whether or not to pool $A \times B$ and E, based on the results of the F-test. Here $F(\alpha)$ means $F_{1-\alpha}(\phi_{A\times B}, \phi_E)$ for abbreviation.

From equation (3.7) we can obtain that

$$E(S_A) = (l-1)\sigma^2 + mr\sum_i a_i^2,$$

which tells us that there are about $l-1$ error variance in S_A. Therefore, the true effect of A, the pure variation, can be estimated from

$$S'_A = S_A - \phi_A V_E = S_A - (l-1)V_E. \tag{3.9}$$

Similarly, the pure variations of B, $A \times B$ and E are

$$\begin{aligned} S'_B &= S_B - \phi_B V_E = S_B - (m-1)V_E, \\ S'_{A\times B} &= S_{A\times B} - (l-1)(m-1)V_E, \\ S'_E &= S_T - S'_A - S'_B - S'_{A\times B} = S_E + (lm-1)V_E. \end{aligned} \tag{3.10}$$

The extent to which the pure variation S'_A accounts for the total variation S_T, expressed as a percentage, is represented by the symbol

$$\rho_A = (S'_A / S_T) \times 100\%, \tag{3.11}$$

which is called the contribution ratio of factor A. Similar contribution ratios can be defined for B, $A \times B$ and E, as shown in Table 3.9. One of the main uses of the contribution ratios lies in the field of tolerance design, which will be explained in Chapter 9.

Table 3.7 ANOVA table two-factor factorial experiments.

Source of variation	Sum of squares	Degree of freedom	Mean square	F_0	$F(\alpha)$
A	S_A	$\phi_A = l-1$	V_A	V_A/V_E	$F_{1-\alpha}(\phi_A, \phi_E)$
B	S_B	$\phi_B = m-1$	V_B	V_B/V_E	$F_{1-\alpha}(\phi_B, \phi_E)$
$A \times B$	$S_{A\times B}$	$\phi_{A\times B} = (l-1)(m-1)$	$V_{A\times B}$	$V_{A\times B}/V_E$	$F_{1-\alpha}(\phi_{A\times B}, \phi_E)$
E	S_E	$\phi_E = lm(r-1)$	V_E		
T	S_T	$\phi_T = lmr-1$			

Table 3.8 Guidelines on whether to pool $A\times B$ and E, based on the F-test.

$F_0 = V_{A\times B}/V_E$	Decision
$F_0 \le 1$	$A\times B$ is negligible and should be pooled with E
$1 \le F_0 \le F(0.2)$	$A\times B$ is almost negligible and could be pooled with E
$F(0.2) \le F_0 \le F(0.1)$	$A\times B$ is not very significant, but it is not desirable to pool $A\times B$ and E
$F(0.1) \le F_0 \le F(0.05)$	$A\times B$ is weakly significant, and it should not be pooled with E
$F(0.1) \le F_0 \le F(0.01)$	$A\times B$ is significant
$F_0 \ge F(0.01)$	$A\times B$ is very significant

Table 3.9 Pure variation and contribution ratio.

Source of variation	Pure variation	Contribution ratio
A	$S'_A = S_A - (l-1)V_E$	$\rho_A = (S'_A/S_T)\times 100\%$
B	$S'_B = S_B - (m-1)V_E$	$\rho_B = (S'_B/S_T)\times 100\%$
$A\times B$	$S'_{(A\times B)} = S_{A\times B} - (l-1)(m-1)V_E$	$\rho_{A\times B} = (S'_{(A\times B)}/S_T)\times 100\%$
E	$S'_E = S_E + (lm-1)V_E$	$\rho_E = (S'_E/S_T)\times 100\%$
T	S_T	100%

3.3.4 CONFIDENCE INTERVAL ESTIMATION

Let $u(A_i)$, $u(B_j)$ and $u(A_iB_j)$ denote the expected means of responses at A_i, B_j and A_iB_j, respectively. From equation (3.1) we obtain that

$$\begin{aligned} u(A_i) &= u + a_i, \\ u(B_j) &= u + b_j, \\ u(A_i, B_j) &= u + a_i + b_j + (ab)_{ij}, \end{aligned} \tag{3.12}$$

where it is known that the least-squares estimates of u, a_i, b_j and $(ab)_{ij}$ are

$$\hat{\mu} = \bar{\bar{y}},$$
$$\hat{a}_i = \bar{y}_{i\,..} - \bar{\bar{y}},$$
$$\hat{b}_j = \bar{y}_{.j\,.} - \bar{\bar{y}},$$
$$(\widehat{ab})_{ij} = \bar{y}_{ij.} - \bar{y}_{i\,..} - \bar{y}_{.j\,.} - \bar{\bar{y}}.$$

We are often interested in finding confidence intervals for the expected means in equation (3.12). Consider a confidence interval estimate of $u(A_i)$.

A point estimator of $u(A_i)$ would be

$$\hat{u}(A_i) = \hat{u} + \hat{a}_i = \bar{\bar{y}} + (\bar{y}_{i..} - \bar{\bar{y}}) = \bar{y}_{i..}.$$

Now, if we assume that the errors are normally distributed, each $\bar{y}_{i\,..}$ is i.i.d.

$N(u(A_i), \sigma^2 / mr)$. Using V_E as an estimator of σ^2, we would base the confidence interval on the t-distribution. Therefore, a $100(1-\alpha)\%$ confidence interval for $u(A_i)$ is

$$\bar{y}_{i..} \pm t_{1-\alpha/2}(\phi_E)\sqrt{\frac{V_E}{mr}}. \tag{3.13}$$

Similarly, $100(1-\alpha)\%$ confidence intervals on $u(B_j)$ and $u(A_iB_j)$ would be

$$\bar{y}_{.j\,.} \pm t_{1-\alpha/2}(\phi_E)\sqrt{\frac{V_E}{lr}},$$
$$\bar{y}_{ij.} \pm t_{1-\alpha/2}(\phi_E)\sqrt{\frac{V_E}{r}}. \tag{3.14}$$

However, if $A \times B$ is negligible, a point estimator for $u(A_iB_j)$ would be

$$\begin{aligned}\hat{u}(A_iB_j) &= \hat{u} + \hat{a}_i + \hat{b}_j \\ &= \bar{\bar{y}} + (\bar{y}_{i\,..} - \bar{\bar{y}}) + (\bar{y}_{.j.} - \bar{\bar{y}}) \\ &= \bar{y}_{i\,..} + \bar{y}_{.j.} - \bar{\bar{y}},\end{aligned}$$

and a $100(1-\alpha)\%$ confidence interval on $u(A_iB_j)$ for this case would be

$$\bar{y}_{i..} + \bar{y}_{.j.} - \bar{\bar{y}} \pm t_{1-\alpha/2}(\phi_E)\sqrt{V_{E'}/n_{e'}},$$

where $\phi_{E'}$ and $V_{E'}$ are obtained from equation (3.8). Here n_e denotes the effective number of replications which can be obtained from the relationship

$$\frac{1}{n_e} = \frac{1}{mr} + \frac{1}{lr} - \frac{1}{lmr} = \frac{l+m-1}{lmt}. \tag{3.15}$$

Note that mr, lr and mlr are the numbers of observations for obtaining $\bar{y}_{i..}, \bar{y}_{.j.}$ and $\bar{\bar{y}}$, respectively, in equation (3.15). Hence, n_e becomes

$$n_e = lmr/(l+m-1). \tag{3.16}$$

Table 3.10 Shear-strength data from Table 3.5.

	B_1	B_2	B_3	B_4	$T_{i..}$	$\bar{y}_{i..}$
A_1	35 28(63)	38 45(83)	45 50(95)	34 30(64)	305	38.1
A_2	26 29(55)	54 48(102)	53 60(113)	20 18(38)	308	38.5
A_3	18 25(43)	26 30(56)	32 34(66)	25 20(45)	210	26.3
$T_{.j.}$	161	241	274	147	T=823	
$\bar{y}_{.j.}$	26.8	40.2	45.7	24.5	$\bar{\bar{y}} = 34.3$	

3.3.5 EXAMPLE: THE ADHESIVE DESIGN EXPERIMENT

Table 3.10 presents the shear-strength data observed in the adhesive design experiment described earlier. The row and column totals are shown in the margins of the table, and the numbers in parentheses are the cell totals. The sums of squares are computed according to equation (3.6) as follows:

$$CT = \frac{T^2}{lmr} = \frac{823^2}{3 \times 4 \times 2} = 28222.04,$$

$$S_T = \sum_i \sum_j \sum_k y_{ijk}^2 - CT$$

$$= \frac{305^2 + 308^2 + 210^2}{4 \times 2} - 28222.04 = 776.58,$$

$$S_A = \sum_j \frac{T_{.j.}^2}{lr} - CT$$

$$= \frac{161^2 + 241^2 + 274^2 + 147^2}{3 \times 2} - 28222.04 = 1892.46,$$

$$S_{AB} = \sum_i \sum_j \frac{T_{ij.}^2}{r} - CT$$

$$= \frac{63^2 + 83^2 + \cdots + 147^2}{2} - 28222.04 = 3271.46,$$

$$S_{A \times B} = S_{AB} - S_A - S_B = 3271.46 - 771.58 - 1892.46 = 602.42,$$

$$S_E = S_T - S_{AB} = 3436.96 - 3271.46 = 165.5.$$

The pure variations and contribution ratios are computed using the formulas in Table 3.9 as follows:

$$V_E = \frac{S_E}{lm(r-1)} = \frac{165.5}{3 \times 4 \times 1} = 13.79,$$

$$S'_A = S_A - (l-1)V_E = 776.58 - 2 \times 13.79 = 749.00,$$

$$S'_B = S_B - (m-1)V_E = 1892.46 - 3 \times 13.79 = 1851.09,$$

$$S'_{(A \times B)} = S_{A \times B} - (l-1)(m-1)V_E = 602.42 - 2 \times 3 \times 13.79 = 519.68,$$

$$S'_E = S_E - (lm-1)V_E = 165.5 - 11 \times 13.79 = 317.19,$$

$$\rho_A = \frac{S'_A}{S_T} \times 100 = \frac{749.00}{3436.96} \times 100 = 21.8\%,$$

$$\rho_B = \frac{S'_B}{S_T} \times 100 = \frac{1851.09}{3436.96} \times 100 = 53.9\%,$$

$$\rho_{A\times B} = \frac{S'_{(A\times B)}}{S_T}\times 100 = \frac{519.68}{3436.96}\times 100 = 15.1\%,$$

$$\rho_E = \frac{S'_E}{S_T}\times 100 = \frac{317.19}{3436.96}\times 100 = 9.2\%.$$

The analysis of variance, including the contribution ratios, is shown in Table 3.11. We conclude that the main effects of material type (factor A) and temperature (factor B) are very significant, and also there is a very significant interaction between material types and temperature. Note that 21.8% of the total variation is attributable to the difference between the three types of material (A) used, 53.9% is attributable to the differences between temperature (B), 15.1% to the interaction, and only 9.2% is attributable to the error (E).

Table 3.11 ANOVA results for adhesive experiment.

Source of variation	Sum of squares	Degrees of freedom	Mean square	F_0	$F(0.01)$	$\rho(\%)$
A	776.58	2	388.29	28.2**	6.93	21.8
B	1892.46	3	630.82	45.7**	5.95	53.9
A×B	602.42	6	100.40	7.3**	4.82	15.1
E	165.50	12	13.79			9.2
T	3436.93	23				100.0

** Significant at 1% level

To assist in interpreting the results of this experiment, it is helpful to construct graph of the average responses for each treatment combination. This graph is shown in Figure 3.3. The significant interaction is indicated by the absence of parallel lines. In general, stronger shear strength is attained at temperature level $B_3(20°C)$, regardless of material type, and treatment combination A_2B_3 gives the greatest shear strength. At A_2B_3, a 95% confidence interval for $u(A_2B_3)$ from equation (3.14) would be

$$\bar{y}_{23.} \pm t_{0.975}(12)\sqrt{\frac{13.79}{2}} = 56.50 \pm 2.179 \times \sqrt{6.895}$$

$$= 56.50 \pm 5.71 \text{ kg cm}^{-2}.$$

However, the shear strength of A_2 (material type 2) rapidly changes as temperature changes, which means that the shear strength of A_2 is not robust to temperature changes. The shear strength is a larger-the-better characteristic, and the SN ratios of shear strength at A_1, A_2 and A_3 from equation (2.23) would be as follows:

$$\text{SN}(A_1) = -10\log\left(\frac{1}{n}\sum_i \frac{1}{y_i^2}\right)$$

$$= -10\log\left[\frac{1}{8}\left(\frac{1}{35^2} + \frac{1}{28^2} + \cdots + \frac{1}{30^2}\right)\right] = 31.14 \text{ dB},$$

$$\text{SN}(A_2) = 29.18 \text{ dB},$$

$$\text{SN}(A_3) = 27.28 \text{ dB}.$$

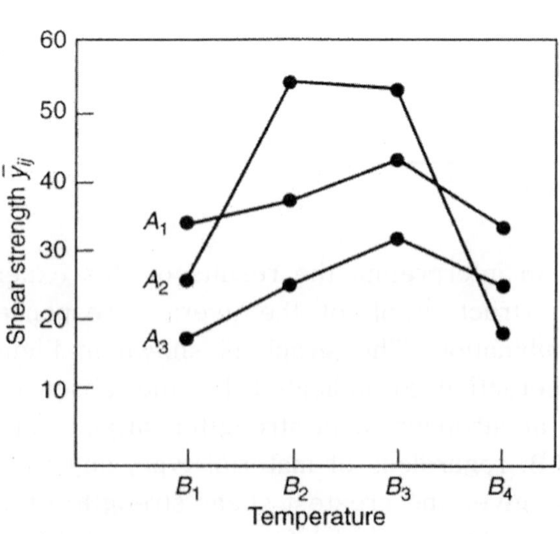

Figure 3.3 Plot of sample means at A_iB_j.

Notice that the SN ratio of A_1 is bigger than that of A_2, which means that A_1 (material type 1) is more robust to temperature change than A_2. Since the temperature range (B_1 through B_4) of the product end-use environment is inevitable, the researcher should recommend material type 1 instead of material type 2.

3.4 CLASSIFICATION OF EXPERIMENTAL DESIGNS

We have seen two designs in the previous sections: one-factor-at-a time experiment and two-factor factorial design. The one-factor-at-a-time experiment is in fact a series of single-factor design. The two-factor-factorial design we have considered is a two-factor factorial design with replications, and there is another two-factor factorial design without replicates, which has the structure of Table 3.4. There are numerous types of experimental design; these may be classified according to the allocation of treatment (factor) combinations and the degree of randomization of experiments as follows.

1. **Factorial design.** This is a design for investigating all possible treatment combinations which are formed from the factors under consideration. The order in which possible treatment combinations are selected is completely random. Single-factor, two-factor and three-factor factorial designs belong to this class, as do 2^k (k factors at 2 levels) and 3^k (k factors at 3 levels) factorial designs.
2. **Fractional factorial design.** This is a design for investigating a fraction of all possible treatment combination which are formed from the factors under investigation. Once again, the order in which the treatment combinations are chosen is completely random. Designs using tables of orthogonal arrays, Plackett-Burman designs, Latin square designs (when we assign factors to the rows and columns of a Latin square) and Graeco-Latin square designs are fractional factorial designs. This type of design is used when the cost of experiment is high and the experiment is time-consuming.

3. **Randomized complete block design, split-plot design and nested design.** All possible treatment combinations are tested in these designs, but some form of restriction is imposed on randomization. A design in which each block contains all possible treatments, and the only randomization of treatments is within the blocks, is called the randomized complete block design. For an explanation of split-plot design and nested design, see Montgomery (1991) and John (1971), for instance.
4. **Incomplete block design.** If every treatment is not present in every block in a randomized complete block design, it is an incomplete block design. This design is used when we may not be able to run all the treatments in each block because of a shortage of experimental apparatus or inadequate facilities. If each block contains the same number of treatments and they are arranged so that every pair of treatment occurs together in the same number of blocks, the design is said to be balanced.
5. **Response surface design and mixture design.** This is the design where the objective is to explore a regression model to find a functional relationship between the response variable and the factors (independent variables) involved, and to find the optimal conditions of the factors. Central composite designs, rotatable designs, simplex designs, mixture designs and evolutionary operation (EVOP) designs belong to this class. Mixture designs are used for experiments in which the various components are mixed in proportions constrained to sum to 1. In the mixture experiment, the factors are the components or ingredients whose optimal proportions or levels are of interest to establish. Note that, because of the constraint just mentioned, the factor levels are not independent.

Most experimental designs used in industrial practice are covered by the above classification. Since there are many designs, it is sometimes difficult to choose an appropriate one to use in practice. The flowchart in Figure 3.4 provides guidelines for the selection of experimental designs.

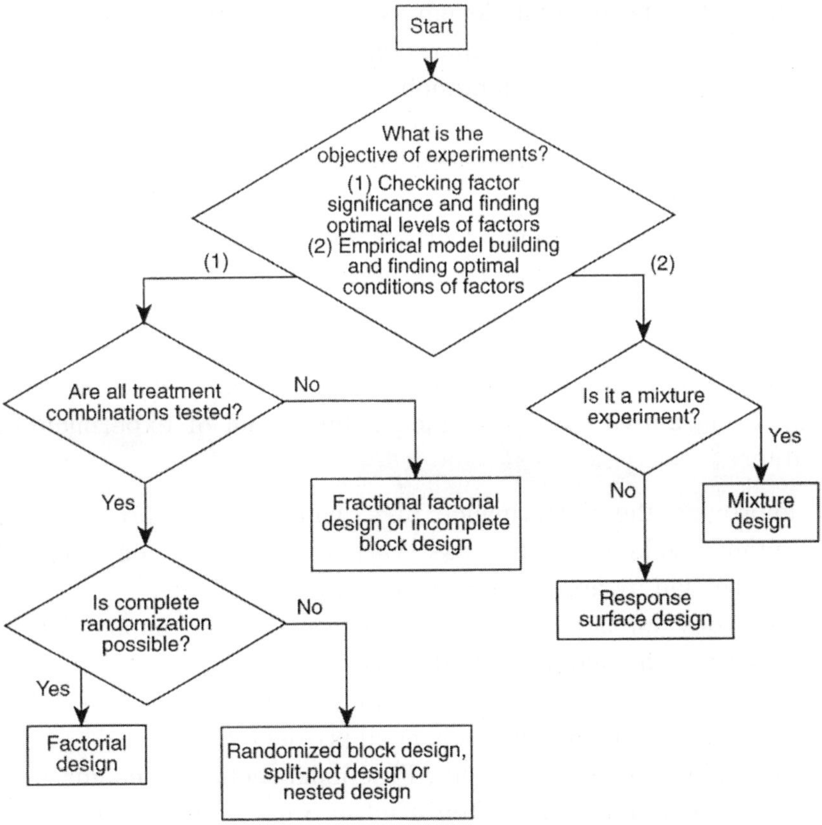

Figure 3.4 Flow chart for selection of experimental designs.

3.5 THE ROLE OF EXPERIMENTAL DESIGN

Earlier, in Section 3.1, we defined the design of experiments as the process of planning the experiment so that appropriate data is collected, the least number of experiments are performed to acquire the necessary technical information, and suitable statistical methods are used to analyze the data collected. Taguchi (1987) defined the design of experiments as a general technique for maximizing the efficiency of acquisition of technical information by experiment. From these definitions we can understand that the essence of experimental design is the scientific management of information acquisition by experiment.

We may view experimentation a part of the scientific iterative learning process and as one of the ways we learn about how systems or processes work. As Box, Hunter and Hunter (1978) noted, we learn through a series of activities in which we make conjectures about a process, perform experiments to generate data from the process, and then use the information from the experiment to establish new conjectures, which lead to new experiments, and so on.

The design of experiments plays a major role in many engineering activities. For instance, the design of experiments is used for:

1. Improving the performance of a manufacturing process. The optimal values of process variables can be economically determined by application of experimental designs.
2. The development of new processes. The application of experimental design methods early in process development can result in reduced development time, reduced variability about target requirements and enhanced process yields.
3. Screening important factors. Fractional factorial designs using orthogonal arrays are often used in order to screen important factors that impact product performance. This will help to enhance the efficiency of research activities.
4. Engineering design activities such as evaluation of material alterations, comparison of basic design configurations, and selection of design parameters so that the product is robust to a wide variety of field conditions.
5. Empirical model building to find out the functional relationship between the performance variable and the influence variables (factors or design/process parameters).

The use of experimental design in engineering activities can result in products that are easier to manufacture, have enhanced field performance and reliability, of lower costs, and shorter development times.

Park (1982; 1990; 1993) showed many examples which are useful for engineers and scientists who wish to employ experimental designs in their engineering and research activities.

3.6 HISTORY OF EXPERIMENTAL DESIGN AND ADVANCEMENT OF ROBUST DESIGN

The science of experimental design is a series of techniques used to enhance the efficiency of information acquisition and analysis. These techniques were first introduced in the 1920s by R.A. Fisher, working at the Agricultural Field Station at Rothamsted in England. He was concerned with arranging trials of fertilizers on plots to protect against the underlying effects of moisture, gradient, nature of soil, etc. Fisher founded the basic principles of factorial design and the associated data analysis known as "analysis of variance" (ANOVA) during his research in improving the yield of agricultural crops. These ideas are presented in Fisher (1935), the first book on experimental design.

In an industrial experiment, since there are many factors concerned and the number of possible experiments is limited, a fractional factorial design is often inevitable. The first systematic approach to choosing appropriate fractions was provided by Finney (1945). Plackett and Burman (1946) also contributed to this field, and Rao (1947) introduced further innovations with his fractional orthogonal arrays.

In the 1950s and 1960s experimental design was liberated from its agricultural roots and a variety of design concepts were suggested. The response surface methodology introduced by Box and Wilson (1951), and work by Box and Hunter (1957) led to more versatile modeling and statistical optimization. Box (1957) and Box and Draper (1969) introduced "evolutionary operation" for sequentially searching the design space for the optimal conditions. An optimal design theory was developed by Kiefer and Wolfowitz (1960), Kiefer (1984) and Fedorov (1971), among others. They set up the experimental design problem as a special form of the optimization problem based on the decision-theoretic approach. There are many excellent textbooks on this conventional experimental design: for example, Box and Draper (1987), Box, Hunter and Hunter (1978), John (1971) and Kempthorne (1952).

In the 1970s and 1980s it was noted that there are limitations when conventional experimental design techniques are applied to industrial experimentation. For instance, they are not adequate for handling noise factors, and for handling averages and dispersion simultaneously. Also data-analysis tools are not simple to understand, and the statistical F-test is often ineffective. Most critically, the existing fractional factorial layouts are quite limited and difficult to use. Robust design adds a new dimension to conventional experimental design. Taguchi developed the foundations of robust design, which Taguchi (1986) called parameter design in a narrow sense. The concept of robust design has many aspects, among them:

1. Finding a set of conditions for design variables which are robust to noise;
2. Achieving the smallest variation in a product's function about a desired target value;
3. Minimizing the number of experiments using orthogonal arrays and testing for confirmation.

The theory and applications of robust design have been advanced by many researchers. Among these are Taguchi and Wu (1979), Wu and Moore (1985), Kackar (1985), Ross (1986), Taguchi (1987), Leon, Shoemaker and Kackar (1987), Phadke (1989), Logothetis and Wynn (1989), Nair (1992), Park (1996), Wu and Hamad (2000), and many others.

The major differences between the Taguchi approach and the conventional experimental design approach are in philosophy, design and data analysis. The Taguchi approach is more engineering-oriented than science-oriented. Science explains nature while engineering utilizes nature. The Taguchi method is for quality improvement and cost reduction rather than discovering the causal relationships of how things happen.

The distinct idea of Taguchi's parameter design which is different from the conventional experimental design is that of designing for the simultaneous modeling of both mean and variability. Also orthogonal arrays are extensively used in parameter design for fractional factorials. In the conventional design orthogonal arrays are scarcely used.

Another difference is that, whereas classical statistical designs treat factors as fixed or random, Taguchi has classified them into control, noise, signal, adjustment, indicative and block factors. Each type of factor has an important meaning from an engineering point of view. For data analysis, Taguchi proposed several newly developed analytical techniques such as the use of SN ratios, the accumulation method and the scoring method for nonstandard data. Nonstandard responses such as binary (for instance, good or bad), ordered categorical (for instance, results are very poor, poor, good or excellent), or Poisson (number of defective items) frequently occurred in practical design situations. The classical analysis of design does not give much attention to these problems.

Exercises

3.1 What are the drawbacks of the one-factor-at-a-time experimental design? Give a two-way layout experiment which is similar to Tables 3.2-3.4, and which leads to incorrect results for optimization.

3.2 A two-factor factorial design was conducted to study the influence of pressure (A) and cycle time (B) on the occurrence of splay in an injection molding process. Factors A and B both have three levels; for each of the nine trials, 50 parts were made twice and the response observed was the number of parts (out of 50) in which splay occurred on the part surface. Two responses at each trial are given below in the table.

	B_1	B_2	B_3
A_1	12	15	10
	14	19	11
A_2	19	25	18
	17	22	15
A_3	23	14	10
	21	12	9

Perform analysis of this experiment which is similar to the adhesive design experiment.

3.3 A two-factor factorial design was constructed to test the effects of two different variables on the alertness of students in an early morning class. The variable levels tested and the responses are shown in the tables below. Three students are tested at each combination of two variable levels.

Variable	Level
Hours of sleep (A)	A_1=4, A_2=6, A_3=8
Ounces of coffee (B)	B_1=6, B_2=12

Test	A	B	Alertness (%)
1	1	1	56, 60, 45
2	1	2	67, 62, 71
3	2	1	65, 72, 75
4	2	2	75, 80, 79
5	3	1	69, 70, 79
6	3	2	87, 83, 75

(a) Perform analysis of this experiment which is similar to the adhesive design experiment.

(b) Let $A = x_1$, $B = x_2$ and y = alertness. Fit a multiple linear regression model and predict the alertness when $x_1 = 7$ and $x_2 = 10$.

3.4 In Section 3.5, the role of experimental design for engineering activities was listed. Recently experimental design is often used in service industry such as bank and insurance company. How can the design of experiments be used in service industry? Explain your ideas.

4

Orthogonal Array Experiments

4.1 STRUCTURE AND USE OF TWO-LEVEL ORTHOGONAL ARRAYS

4.1.1 STRUCTURE OF TWO-LEVEL ORTHOGONAL ARRAYS

Many designed experiments use matrices called orthogonal arrays for determining which combinations of factor levels to use for each experimental run and for analyzing the data. In the past, orthogonal arrays were known as "magic squares". Perhaps the effectiveness of orthogonal arrays in experimental design is magic! What is an orthogonal array?

An **orthogonal array** is a fractional factorial matrix which assures a balanced comparison of levels of any factor or interaction of factors. It is a matrix of numbers arranged in rows and columns where each row represents the level of the factors in each run, and each column represents a specific factor that can be changed from each run. The array is called orthogonal because all columns can be evaluated independently of one another.

Chapters 4 through 9 of this book deal with the methods of assignment (design) for experiments and calculations by orthogonal arrays, and with the associated data analysis. In fact, the use of orthogonal arrays is indispensable for robust design. Table 4.1 shows an orthogonal array, $L_8(2^7)$, which is simply written as L_8. The number to the left of each row is called the run number or experiment number, and runs from 1 to 8.

The vertical alignments are termed the columns of the orthogonal array, and every column consists of four 0 and four 1. The columns are generated as follows. Columns 1, 2 and 4 (basic marks a, b and c) are the fundamental columns which are given as (0 0 0 0 1 1 1 1), (0 0 1 1 0 0 1 1) and (0 1 0 1 0 1 0 1), respectively. All other columns are generated from these fundamental columns according to the following formula:

Number in the column of basic mark xy

$$= (\text{that of basic mark } x + \text{that of basic mark } y) \pmod 2.$$

Table 4.1 $L_8(2^7)$ orthogonal array.

Experiment number	Column number						
	1	2	3	4	5	6	7
1	0	0	0	0	0	0	0
2	0	0	0	1	1	1	1
3	0	1	1	0	0	1	1
4	0	1	1	1	1	0	0
5	1	0	1	0	1	0	1
6	1	0	1	1	0	1	0
7	1	1	0	0	1	1	0
8	1	1	0	1	0	0	1
Basic mark	a	b	ab	c	ac	bc	abc
Assignment							

For example, column 3 is generated from columns 1 and 2 in such a way that the number in column 3 is the sum of the numbers in columns 1 and 2. But if the sum is equal to 2, then it becomes 0, and if the sum is equal to 3, then it becomes 1.

$$\begin{matrix} \text{column 1} \\ (a) \\ \begin{bmatrix} 0\\0\\0\\0\\1\\1\\1\\1 \end{bmatrix} \end{matrix} + \begin{matrix} \text{column 2} \\ (b) \\ \begin{bmatrix} 0\\0\\1\\1\\0\\0\\1\\1 \end{bmatrix} \end{matrix} \pmod 2 = \begin{matrix} \text{column 3} \\ (ab) \\ \begin{bmatrix} 0\\0\\1\\1\\1\\1\\0\\0 \end{bmatrix} \end{matrix}$$

Similarly, column 5 is generated from columns 1 and 4, column 6 is generated from columns 2 and 4, and column 7 is generated from columns 1, 2 and 4.

Note that in Table 4.1, if we select any two columns and check the numerals 0 and 1, there are four possible combinations (0 0), (0 1), (1 0) and (1 1) in each row, and they appear exactly twice. We say that all possible combinations in each row for any two columns appear exactly with the same frequency; that is, these two columns are balanced, or orthogonal.

Taguchi (1987) prefers using (1 2) to using (0 1), and the corresponding L_8 orthogonal array is given in Table 4.2. In this book, we will use (0 1) instead of (1 2), since using (0 1) makes it easy to generate columns by binary arithmetic and there is no difficulty in applying (0 1) for factor levels. Usually (−1+1) notation is used for fractional factorial designs developed in western countries. The L_8 Taguchi design expressed in (−1+1) notation is shown in Table 4.3. Notice that in Table 4.3, column 3 is obtained by – (column 1)∗(column 2), which is indicated by – ***ab***.

Table 4.2 L_8 Taguchi design.

Experiment number	Column number						
	1	2	3	4	5	6	7
1	1	1	1	1	1	1	1
2	1	1	1	2	2	2	2
3	1	2	2	1	1	2	2
4	1	2	2	2	2	1	1
5	2	1	2	1	2	1	2
6	2	1	2	2	1	2	1
7	2	2	1	1	2	2	1
8	2	2	1	2	1	1	2
Basic mark	*a*	*b*	*ab*	*c*	*ac*	*bc*	*abc*

Table 4.3 L_8 equivalent design.

Experiment number	Column number						
	1	2	3	4	5	6	7
1	−1	−1	−1	−1	−1	−1	−1
2	−1	−1	−1	+1	+1	+1	+1
3	−1	+1	+1	−1	−1	+1	+1
4	−1	+1	+1	+1	+1	−1	−1
5	+1	−1	+1	−1	+1	−1	+1
6	+1	−1	+1	+1	−1	+1	−1
7	+1	+1	−1	−1	+1	+1	−1
8	+1	+1	−1	+1	−1	−1	+1
Basic mark	*a*	*b*	*−ab*	*c*	*−ac*	*−bc*	*−abc*

4.1.2 HOW TO USE TWO-LEVEL ORTHOGONAL ARRAYS

The general symbol for two-level standard orthogonal arrays is

$$L_{2^m}(2^{2^m-1}), \tag{4.1}$$

where

$$\begin{aligned} m &= \text{a positive integer which is greater than 1} \\ 2^m &= \text{the number of experimental runs} \\ 2 &= \text{the number of levels for each factor} \\ 2^m-1 &= \text{the number of column in orthogonal array.} \end{aligned}$$

The letter "L" comes from "Latin", the idea of using orthogonal arrays for experimental design having been associated with Latin square designs from the outset. The two-level standard orthogonal arrays most often used in practice are $L_4(2^3)$, $L_8(2^7)$, $L_{16}(2^{15})$ and $L_{32}(2^{31})$, which will be simply expressed as L_4, L_8, L_{16} and L_{32}. $L_4(2^3)$ is shown in Table 4.4, and $L_{16}(2^{15})$ can be found in Table 4.17.

We now present the rules for using two-level orthogonal arrays for experimental designs:

1. A factor can be arbitrarily allocated in a column.
2. For the basic marks, we let $a^2 = b^2 = c^2 = \cdots = 1$.
3. Each column has one degree of freedom, and a factor or an interaction is allocated to a single column. The columns to which no factor or interaction is allocated are error columns.
4. Suppose factor A has basic mark x and factor B has basic mark y. If we want to allocate interaction $A \times B$, we should allocate it to the column whose basic mark is xy. For example, if A is in basic mark ab, and B is in ac, then $A \times B$ should be allocated in basic mark $(ab)*(ac) = a^2bc = bc$.
5. The sum of squares for a factor or an interaction is obtained by the sum of squares of the column in which the factor or the interaction is allocated. The column sum of squares is computed as follows.

   Column sum of squares = $(T_1 - T_0)^2/$(total number of experiments), where T_i means the sum of the data for the ith level in the column.

6. The error sum of squares is the sum of sums of squares of error columns.

Table 4.4 L_4 orthogonal array.

Experiment number	Column number		
	1	2	3
1	0	0	0
2	0	1	1
3	1	0	1
4	1	1	0
Basic mark	*a*	*b*	*ab*
Assignment			

The above guidelines are explained by an example. Suppose, in a plastic molding experiment, that the quality characteristic is the tensile strength, and let the factors which are suspected to affect the tensile strength be as follows, with their two levels:

A: temperature: A_0 = 200°C, A_1 = 220°C
B: pressure: B_0 = 500 kgcm^{-2}, B_1 = 700 kgcm^{-2}
C: time: C_0 = 30 min, C_1 = 40 min
D: additive: D_0 = 3 %, D_1 = 5 %.

Since there are four factors, there are six possible two factor interactions. Our interest is confined to the interaction $A \times B$, which is the only one suspected to occur. Since there are four factors and one interaction, we only need five columns to allocate these sources of variation, and L_8 is enough for this purpose because it has seven columns. Here "source" means factor or interaction, and "variation" means sum of squares. The assignment of sources is shown in Table 4.5, and the eight experimental conditions are also shown in this table. Suppose the tensile strength data are as in Table 4.5. Then we can compute the sums of squares by using the computation scheme in Table 4.6. The last row in Table 4.6 shows the variation of each source. Since there are two columns for error, the error sum of squares is the sum of variations in columns 5 and 6. The variation of each column is computed by

$$[(\text{sum of data at level 1}) - (\text{sum of data at level 0})]^2/8 ,$$

which is calculated in Table 4.6. The ANOVA table is shown in Table 4.7, which shows that factors A and C are significant at the 10% level.

Table 4.5 Allocation of factors and interaction to L_8 for the tensile strength experiment.

Experiment number	Column number							Experimental condition	Tensile strength data ($kgcm^{-2}$)
	1	2	3	4	5	6	7		
1	0	0	0	0	0	0	0	$A_0B_0C_0D_0$	9
2	0	0	0	1	1	1	1	$A_0B_0C_1D_1$	12
3	0	1	1	0	0	1	1	$A_0B_1C_0D_1$	8
4	0	1	1	1	1	0	0	$A_0B_1C_1D_0$	15
5	1	0	1	0	1	0	1	$A_1B_0C_0D_1$	16
6	1	0	1	1	0	1	0	$A_1B_0C_1D_0$	20
7	1	1	0	0	1	1	0	$A_1B_1C_0D_0$	11
8	1	1	0	1	0	0	1	$A_1B_1C_1D_1$	13
Basic mark	a	b	ab	c	ac	bc	abc		$T = 104$
Assignment	A	B	$A\times B$	C	e	e	D		

Table 4.6 Computation of variation for L_8.

Column number	1		2		3		4		5		6		7	
Source	*A*		*B*		$A \times B$		*C*		*e*		*e*		*D*	
Level	0	1	0	1	0	1	0	1	0	1	0	1	0	1
Data	9	16	9	8	9	8	9	15	9	12	9	12	9	12
	12	20	12	15	12	15	8	15	8	15	15	8	15	8
	8	44	49	44	44	49	49	20	20	16	16	20	20	16
	15	13	20	13	13	20	11	13	13	11	13	11	11	13
(1) Level sum	44	60	57	47	45	59	44	60	50	54	53	51	55	49
(2) Sum of (1)	104		104		104		104		104		104		104	
(3) Difference of (1)	16		−10		14		16		4		−2		-6	
(4) $(3)^2$	256		100		196		256		16		4		36	
(5) (4)/8	32		12.5		24.5		32		2		0.5		4.5	

Table 4.7 ANOVA table for L_8 experiment.

Source of variation	Sum of squares	Degrees of freedom	Mean square	F_0	$F(0.1)$
A	32	1	32	25.6	8.53
B	12.5	1	12.5	10	8.53
$A \times B$	24.5	1	24.5	19.6	8.53
C	32	1	32	25.6	8.53
D	4.5	1	4.5	3.6	8.53
e (error)	2.5	2	1.25		
T	98	7			

Note that the design in Table 4.5 is a half-fraction of a 2^4 factorial design, which is often designated as a 2^{5-1} fractional factorial design. Also note that in this design A, B, and C are allocated in the columns of a, b, and c, and D is assigned into abc. Hence, the defining relation is $D = ABC$, and the defining contrast is

$$I = ABCD.$$

Since the word length of the defining contrast is 4, the resolution of this design is IV, and the main effects A, B, C and D are not confounded with two-factor interactions. The confounding pattern of this design is as follows.

$$A = BCD, B = ACD, C = ABC, D - ABC, AB = CD, AC = BD, AD = BC.$$

Figure 4.1 shows how eight experimental conditions in Table 4.5 are selected out of a total of 16 experimental conditions. Observe in this figure that among eight experiments exactly half are allocated into each level of each factor, and at each level of a factor exactly two experiments are performed at each level of another factor. This means the design is balanced or orthogonal, as mentioned before. Therefore, if there is a significant difference between A_0 and A_1, it really is the difference between A_0 and A_1, because at A_0 every other factor, say B, has an equal

		A_0		A_1	
		B_0	B_1	B_0	B_1
C_0	D_0	×			×
	D_1		×	×	
C_1	D_0		×	×	
	D_1	×			×

Figure 4.1 Eight selected experimental conditions out of 16.

number of experiments at B_0 and B_1, and at A_1 also every other factor, say B, has an equal number of experiments at B_0 and B_1. Hence, the effect of a factor allocated to a column can be estimated independently of all the other factor effects which are allocated to other columns.

4.1.3 ADVANTAGE OF ORTHOGONAL ARRAY DESIGN OVER ONE-FACTOR-AT-A-TIME DESIGN

Suppose we have seven factors with two levels each, and the factors are allocated in two layouts, as shown in Tables 4.8 and 4.9. The former table shows the L_8 orthogonal array design, and the latter shows a one-factor-at-a-time design.

The design represented in Table 4.9 has been used in industry for many years. With this method, the effect of A is investigated by varying the condition of A, from A_0 to A_1, while the levels of $B, C, \ldots, H$ are kept constant at $B_0C_0D_0F_0G_0H_0$. The effect of B is similarly investigated by varying the condition of B, from B_0 to

Table 4.8 L_8 orthogonal array design with seven factors.

Experiment number	Column number						
	1	2	3	4	5	6	7
1	0	0	0	0	0	0	0
2	0	0	0	1	1	1	1
3	0	1	1	0	0	1	1
4	0	1	1	1	1	0	0
5	1	0	1	0	1	0	1
6	1	0	1	1	0	1	0
7	1	1	0	0	1	1	0
8	1	1	0	1	0	0	1
Assignment	A	B	C	D	F	G	H

Table 4.9 One-factor-at-a-time design with seven factors.

Experiment number	Column number						
	1	2	3	4	5	6	7
1	0	0	0	0	0	0	0
2	1	0	0	0	0	0	0
3	1	1	0	0	0	0	0
4	1	1	1	0	0	0	0
5	1	1	1	1	0	0	0
6	1	1	1	1	1	0	0
7	1	1	1	1	1	1	0
8	1	1	1	1	1	1	1
Assignment	A	B	C	D	F	G	H

B_1, while levels $A_1C_0D_0F_0G_0H_0$ are kept constant, and so on. This method, called the one-factor-at-a-time method of experimentation, was introduced in Section 3.2. In this experiment, the level of only one factor is changed from one experiment to another. Hence, such an experiment is much easier to conduct than the experiments represented in the orthogonal array in Table 4.8. Note that both experiments in Tables 4.8 and 4.9 have the same number of experiments. Nevertheless, the layout by orthogonal array is much superior in several respects as follows.

1. The effect of each factor can be estimated much more reliably with smaller variance in the orthogonal array design than the one-factor-at-a-time design. For instance, the effect of A is estimated from two runs (difference between run 1 and run 2) in the latter design, but the effect of A is estimated from eight runs (difference between the average of runs 1-4 and the average of runs 5-8) in the former design.

2. The effect of each factor can be estimated much more accurately with no bias in the orthogonal array design compared to one-factor-at-a-time design. For instance, in the one-factor-at-a-time design, the effect of A is estimated only at the fixed condition $B_0C_0D_0F_0G_0H_0$ for the other factors. If the fixed condition changes, the effect of A may change and the effect could be biased. However, in the orthogonal array design, the effect of A is determined as the average effect while the conditions of other factors vary with equal frequency. For instance, at A_0, B_0 and B_1 are tested twice (which means that, at the conditions of A_0B_0 and A_0B_1, the experiment is conducted twice, respectively), and at A_1 also, B_0 and B_1 are tested twice. Such an average effect for the effect of A results in no bias and better reproducible results concerning factorial effects.
3. If the influence of A_0 and A_1 on experimental results is consistent while other factors vary, the effect of A is judged to be significant. On the other hand, if the difference between A_0 and A_1 varies greatly when other factors change, the effect of A tends to be insignificant. Such significance can only be evaluated through the orthogonal array design, not through the one-factor-at-a-time design. The property whereby factors maintain their effects under varied conditions is called robustness of factor effects. Such robustness can be well measured by orthogonal array designs.

4.1.4 HOW TO FIND THE OPTIMUM CONDITIONS

Having found the ANOVA table through the computation scheme shown in Tables 4.6 and 4.7, it is necessary to obtain the optimum levels of factors which maximize (or minimize or give a target value for) the response. The guidelines for finding the optimum conditions are as follows.

1. If there are significant interactions, construct a two-way table for each significant interaction, and find the optimum levels of two factors which are associated with the two-way table.

2. If there are significant factors which are not involved in the two-way tables in (1), construct a one-way table for each significant factor, and find the optimum level from this one-way table.
3. Combining the results in (1) and (2), find the optimum levels of factors which are involved in significant interactions or significant factors. If a factor has two levels for optimality in (1), the level which optimizes the one-way table for this factor is optimal. For instance, if $A \times B$ and $A \times C$ are significant, A_0 is optimal from the AB two-way table and A_1 is optimal from the AC two-way table, then the optimum level of A is decided from the one-way table of A.
4. If we want to find the optimum levels of insignificant factors whose optimum levels are not found in (3), the levels should be decided not by the response considered in the experiment, but by other considerations such as economy, operability or technicality.

Table 4.10 AB two-way table.

	A_0	A_0	Total
A_0	9+12=21	16+20=36	57
B_1	8+15=23	11+13=24	47
Total	44	60	104

Let us illustrate the above procedure by means of an example. Consider the experiment in Tables 4.5–4.7. The significant sources of variation are $A, B, A \times B$ and C. First of all, we have to construct the AB two-way table as shown in Table 4.10. The optimum levels of factors A and B are A_1B_0, since this gives the largest value of 36. Next, the optimum level of C, which is a significant factor but whose optimum level has not been decided, can be chosen from Table 4.6. The level sums of C_0 and C_1 are 44 and 60, respectively, hence C_1 is the optimum level. Therefore, the overall optimum levels of significant factors are $A_1B_0C_1$.

The estimated tensile strength at $A_1B_0C_1$ can be obtained as follows:

$$
\begin{aligned}
\hat{\mu}(A_1B_0C_1) &= est(\mu + a_1 + b_0 + (ab)_{10} + c_1) \\
&= est(\mu + a_1 + b_0 + (ab)_{10}) + est(\mu + c_1) - \hat{\mu} \\
&= \overline{A_1B_0} + \overline{C}_1 - \overline{T} \\
&= \frac{36}{2} - \frac{60}{4} - \frac{104}{8} \\
&= 18 + 15 - 13 = 20,
\end{aligned}
$$

where we use *est* instead of the "hat" notation for typographical convenience.

The optimum level of D was not considered above, since it was not significant. However, we can determine the optimum level of D from other considerations. Suppose we choose D_0. Then an estimated tensile strength at $A_1B_0C_1D_0$ would be

$$
\begin{aligned}
\hat{\mu}(A_1B_0C_1D_0) &= est(\mu + a_1 + b_0 + (ab)_{10} + c_1 + d_0) \\
&= est(\mu + a_1 + b_0 + (ab)_{10}) + est(\mu + c_1) + est(\mu + d_0) - 2\hat{\mu} \\
&= \overline{A_1B_0} + \overline{C}_1 - \overline{T} \\
&= \frac{36}{2} - \frac{60}{4} - \frac{55}{4} - 2 \times \frac{104}{8} = 20.75.
\end{aligned}
$$

This estimate is not much different from the previously estimated value of 20, since factor D does not significantly affect the tensile strength.

4.2 STRUCTURE AND USE OF THREE-LEVEL ORTHOGONAL ARRAYS

4.2.1 STRUCTURE AND USE OF THREE-LEVEL ORTHOGONAL ARRAYS

We have discussed the structure and use of two-level orthogonal arrays. Another group of orthogonal arrays which are frequently used is that of three-level standard orthogonal arrays. The general symbol for this group is

$$L_{3^m}(3^{(3^m-1)/2}), \tag{4.2}$$

where, imitating (4.1),

$$m = \text{a positive integer which is greater than 1,}$$

$$3^m = \text{the number of experimental runs,}$$

$$(3^m-1)/2 = \text{the number of columns in the orthogonal array.}$$

The three-level standard orthogonal arrays most often used are $L_9(3^4)$ and $L_{27}(3^{13})$, which are simply expressed as L_9 and L_{27}. L_9 is the simplest three-level orthogonal array, and is shown in Table 4.11.

Table 4.11 $L_9(3^4)$ orthogonal array.

Experiment number	Column number			
	1	2	3	4
1	0	0	0	0
2	0	1	1	1
3	0	2	2	2
4	1	0	1	2
5	1	1	2	0
6	1	2	0	1
7	2	0	2	1
8	2	1	0	2
9	2	2	1	0
Basic mark	a	b	ab	ab^2
Assignment				

The columns of L_9 are generated as follows. Columns 1 and 2 are fundamental columns which are given as (0 0 0 1 1 1 2 2 2) and (0 1 2 0 1 2 0 1 2), respectively. The other columns are generated by these fundamental columns according to the following formula:

Number in the column of basic mark xy

$$= (\text{that of basic mark } x + \text{that of basic mark } y) \pmod 3.$$

For example, column 3 is generated from columns 1 and 2 by summing the values in columns 1 and 2. But if the sum is equal to 3, then it becomes 0, and if the sum is equal to 4, then it becomes 1 and so on. The basic mark of column 4 began as a^2b, which means that column 4 is generated by calculating,

$$(2 \times (\text{number in column 1}) + (\text{number in column 2})) \pmod 3.$$

However, we use the notation ab^2 for the basic mark of column 4, since no problem occurs in using a^2b, and ab^2 interchangeably.

Note that in L_9, there are nine two-number combinations of 0s, 1s and 2s in any pair of columns, i.e. (00), (01), (02), (10), (11), (12), (20), (21) and (22). Since each combination occurs an equal number of times (in this case, once), the four columns of this array are, therefore, balanced.

4.2.2 HOW TO USE THREE-LEVEL ORTHOGONAL ARRAYS

We now present the rules for using three-level orthogonal arrays for experimental designs.

1. A factor can be arbitrarily allocated in a column.
2. For the basic marks, we let $a^3 = b^3 = c^3 = \ldots = 1$. In addition, the square of a basic mark is equivalent to the basic mark itself. For example $a^2b = (a^2b)^2 = a^4b^2 = aa^3b^2 = ab^2$.
3. Each column has two degrees of freedom. Since a two-factor interaction has four degrees of freedom, it should be allocated

to two columns. The columns to which no factor or interaction is allocated are error columns.

4. Suppose factor A has basic mark x, and B has basic mark y. If we want to allocate interaction $A \times B$, we should allocate it to the two columns whose basic marks are xy and xy^2.
5. The sum of squares for a factor is obtained by the sum of squares of the column in which the factor is allocated. The sum of squares for an interaction is the sum of sums of squares of the two columns in which the interaction is allocated. The error sum of squares is obtained similarly. The column sum of squares is computed as follows.

$$\text{Column sum of squares} = (3/N)(T^2{}_0 + T^2{}_1 + T^2{}_2) - T^2/N,$$

where

N = total number of observations taken,

T_i = sum of observations taken at level i in the column,

T = sum of all observations taken.

Suppose A and B are allocated to columns 4 and 7, respectively, in the $L_{27}(3^{13})$ orthogonal array which is shown in Table 4.12. Where should we allocate $A \times B$? Since the basic marks of columns 4 and 7 are $x = ab^2$ and $y = ac^2$, the interaction should be allocated to the columns where basic marks are

$$xy = (ab^2)(ac^2) = a^2b^2c^2 = (a^2b^2c^2)^2 = abc,$$
$$xy^2 = (ab^2)(ac^2)^2 = a^3b^2c^4 = b^2c = (b^2c)^2 = bc^2,$$

whose column numbers are 9 and 11, respectively. Since it is not easy to find the interaction columns as above, Table 4.13 is provided for this purpose. To find where to allocate the interaction between A (column 4) and B (column 7) in the above example, we look up the intersection of row 4 and column 7 in Table 4.13. There we will see that the column numbers for $A \times B$ are 9 and 11.

Table 4.12 $L_{27}(3^{13})$ orthogonal array.

Row number	Column number													Data
	1	2	3	4	5	6	7	8	9	10	11	12	13	
1	0	0	0	0	0	0	0	0	0	0	0	0	0	
2	0	0	0	0	1	1	1	1	1	1	1	1	1	
3	0	0	0	0	2	2	2	2	2	2	2	2	2	
4	0	1	1	1	0	0	0	1	1	1	2	2	2	
5	0	1	1	1	1	1	1	2	2	2	0	0	0	
6	0	1	1	1	2	2	2	0	0	0	1	1	1	
7	0	2	2	2	0	0	0	2	2	2	1	1	1	
8	0	2	2	2	1	1	1	0	0	0	2	2	2	
9	0	2	2	2	2	2	2	1	1	1	0	0	0	
10	1	0	1	2	0	1	2	0	1	2	0	1	2	
11	1	0	1	2	1	2	0	1	2	0	1	2	0	
12	1	0	1	2	2	0	1	2	0	1	2	0	1	
13	1	1	2	0	0	1	2	1	2	0	2	0	1	
14	1	1	2	0	1	2	0	2	0	1	0	1	2	
15	1	1	2	0	2	0	1	0	1	2	1	2	0	
16	1	2	0	1	0	1	2	2	0	1	1	2	0	
17	1	2	0	1	1	2	0	0	1	2	2	0	1	
18	1	2	0	1	2	0	1	1	2	0	0	1	2	
19	2	0	2	1	0	2	1	0	2	2	0	2	1	
20	2	0	2	1	1	0	2	1	0	0	1	0	2	
21	2	0	2	1	2	1	0	2	1	1	2	1	0	
22	2	1	0	2	0	2	1	1	0	0	2	1	0	
23	2	1	0	2	1	0	2	2	1	1	0	2	1	
24	2	1	0	2	2	1	0	0	2	2	1	0	2	
25	2	2	1	0	0	2	1	2	1	1	1	0	2	
26	2	2	1	0	1	0	2	0	2	2	2	1	0	
27	2	2	1	0	2	1	0	1	0	0	0	2	1	
Basic mark	a	b	ab	ab^2	c	ac	ac^2	bc	abc	ab^2c^2	bc^2	ab^2c	abc^2	
Assignment														

Table 4.13 Interaction table for $L_{27}(3^{13})$.

Column number	1	2	3	4	5	6	7	8	9	10	11	12	13
	(1)	3	2	2	6	5	5	9	9	8	12	11	11
		4	4	3	7	7	6	10	10	9	13	13	12
		(2)	1	1	8	9	10	5	6	7	5	6	7
			4	3	11	12	13	11	12	13	8	9	10
			(3)	1	9	10	8	7	5	6	6	7	5
				2	13	11	12	12	13	11	10	8	9
				(4)	10	8	9	6	7	5	7	5	6
					12	13	11	13	11	12	9	10	8
					(5)	1	1	2	3	4	2	4	3
						7	6	11	13	12	8	10	9
						(6)	1	4	2	3	3	2	4
							5	13	12	11	10	9	8
							(7)	3	4	2	4	3	2
								12	11	13	9	8	10
								(8)	1	1	2	3	4
									8	9	5	7	6
									(9)	1	4	2	3
										8	7	6	5
										(10)	3	4	2
											6	7	7
											(11)	1	1
												13	12
												(12)	1
													11

4.2.3 DESIGN AND ANALYSIS ON A THREE-LEVEL ORTHOGONAL ARRAY

The way to compute factor or interaction sums of squares mentioned in the above guidelines can be best explained by an example. Suppose the effects of three fertilizer ingredients on the yield of soy beans are under study. The fertilizer ingredients are nitrogen (N), phosphoric acid (P_2O_5) and potash (K_2O), and the response of interest is the average yield in kilograms (kg) per plot of soy beans. The factors are assigned as follows.

A : nitrogen: A_0 = 0.5kg, A_1 = 1.0kg, A_2 = 1.5kg.
B : phosphoric acid: B_0 = 0.3kg, B_1 = 0.6kg, B_2 = 0.9kg.
C : potash: C_0 = 0.4kg, C_1 = 0.7kg, C_2 = 1.0kg.

Table 4.14 Allocation of factors and data in soy-bean yield experiment.

Experiment number	Column number				Condition	Experimental data
	1	2	3	4		
1	0	0	0	0	$A_0B_0C_0$	8
2	0	1	1	1	$A_0B_1C_1$	12
3	0	2	2	2	$A_0B_2C_2$	9
4	1	0	1	2	$A_1B_0C_2$	11
5	1	1	2	0	$A_1B_1C_0$	12
6	1	2	0	1	$A_1B_2C_1$	15
7	2	0	2	1	$A_2B_0C_1$	21
8	2	1	0	2	$A_2B_1C_2$	18
9	2	2	1	0	$A_2B_2C_0$	20
Basic mark	a	b	ab	ab^2		
Assignment	A	B	e	C		T=126

In the orthogonal array of $L_9(3^4)$, the three-level factors A, B and C are assigned to columns 1, 2 and 4 respectively, as shown in Table 4.14. The yield data is also shown on the right-hand side of Table 4.14.

Note that this design is a 3^{3-1} fractional factorial design, since there is a total of 27 possible level combinations, of which only 9 combinations are tested. The conditions tested are shown in Figure 4.2.

	A_0			A_1			A_2		
	B_0	B_1	B_2	B_0	B_1	B_2	B_0	B_1	B_2
C_0	×				×				×
C_1		×				×	×		
C_2			×	×				×	

Figure 4.2 Nine selected experimental conditions out of 27.

Table 4.15 Computation of variation for L_9.

Column number		1			2			3			4	
Source		A			B			e			C	
Level	0	1	2	0	1	2	0	1	2	0	1	2
Data	8	11	21	8	12	9	8	12	9	8	12	9
	12	12	18	11	12	15	15	11	12	12	15	11
	9	15	20	21	18	20	18	20	21	20	21	18
(1) Level sum	29	38	59	40	42	44	41	43	42	40	48	38
(2) Sum of (1)		126			126			126			126	
(3) $(1)^2$	841	1444	3481	1600	1764	1936	1681	1849	1764	1600	2304	1444
(4) Sum of (3)		5766			5300			5294			5348	
(5) (4)/3		1922			1766.7			1764.7			1782.7	
(6) $CT=T^2/9$		1764			1764			1764			1764	
(7) (5)−(6)		158			2.7			0.7			18.7	

The way to compute the factor sum of squares, which is explained in the guidelines, can be implemented by using the computation scheme shown in Table 4.15. The last row in this table shows the variation of each source.

The ANOVA table for the experiment is constructed in Table 4.16, which shows that factors A and C are significant at the 5% level. However, the F_0 value of factor B is 4, which is not small enough to be ignored. In this situation, we may consider that factor B is also weakly significant. Usually, if the value of F_0 is larger than 2 in fractional factorial designs, we may say that the factor is weakly significant, and we should consider the level of the factor for optimization.

Table 4.16 Computation of variation for L_9.

Source of variation	Sum of squares	Degree of freedom	Mean square	F_0	F(0.05)
A	158	2	79	237	19.0
B	2.7	2	1.3	4	19.0
C	18.7	2	9.3	28	19.0
e	0.7	2	0.3		
T	180	8			

Since A, B and C are all significant, it is of interest to find optimum levels of A, B and C which maximize the average yield of soy beans. From Table 4.15, we find that A_2, B_2 and C_1 produce the largest average yield. An estimated value of the average yield can be obtained as

$$\begin{aligned}\hat{\mu}(A_2B_2C_1) &= est(\mu + a_2 + b_2 + c_1) \\ &= est(\mu + a_2) + est(\mu + b_2) + est(\mu + c_1) - 2\hat{\mu} \\ &= \overline{A}_2 + \overline{B}_2 + \overline{C}_1 - 2\overline{T} \\ &= \frac{59}{3} + \frac{44}{3} + \frac{48}{3} - 2 \times \frac{126}{9} \\ &= 22.3.\end{aligned}$$

Observe that the optimum condition $A_2B_2C_1$ was not tested in this experiment. However, this condition is predicted to produce the maximum average yield per plot of soy bean. Therefore it will be useful to carry out a confirmation test optimum condition.

4.3 LINEAR GRAPHS

4.3.1 LINEAR GRAPHS FOR TWO-LEVEL ORTHOGONAL ARRAYS

In Section 4.1, we explained how to allocate interactions by using the basic marks in orthogonal arrays. However, when there are too many interactions in L_{16} or L_{32}, using the basic marks to allocate interactions is not very convenient. **Linear graphs** are graphs which represent the interaction information and make it easy to assign factors and interactions to the various columns of an orthogonal array. Linear graphs are also useful tools for streamlining experimental layout, and for facilitating complicated assignment of factors with an unequal number of factor levels.

A linear graph is made up of dots, lines and numbers. A dot is used to indicate a main effect, and a line segment represents an interaction between the two main effects (dots) that it connects. A number assigned to a dot or a line indicates the column assignment. A factor can be assigned on a line. Figure 4.3 shows two linear graphs of L_8. In the first linear graph of Figure 4.3, the interaction between columns 1 and 2 goes into column 3, the interaction between columns 1 and 4 goes into column 5, and the interaction between columns 2 and 4 goes into column 6. Column 7 is an independent column available for a main effect.

To assign an experiment with five two-level factors, A,B,C,D,F and interactions $A\times B$ and $A\times C$, to the L_8 array, the graphical structure shown in Figure 4.4 is required. In this case, we can easily assign the factors and interactions to the column numbers on the first linear graph of Figure 4.3 by matching them graphically. The results are that A goes into column 1, B into 2,

C into 4, $A \times B$ into 3, $A \times C$ into 5, F into 7, an D into column 6. Alternatively, we may assign F to column 6 and D to column 7.

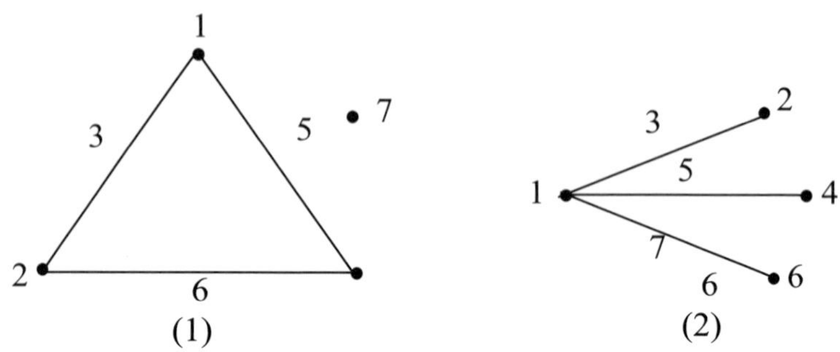

Figure 4.3 Two linear graphs of L_8.

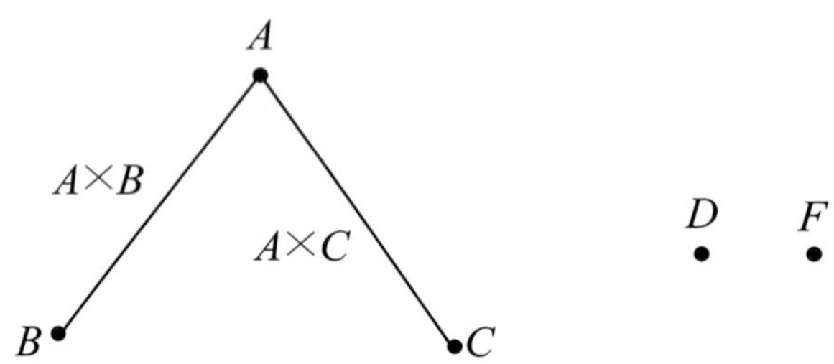

Figure 4.4 Graphical structure of an experiment.

4.3.2 LINEAR GRAPH OF L_{16}

One of the most commonly used orthogonal arrays is L_{16}, which is shown in Table 4.17. The rules for using this array for assigning factors and interactions by means of the basic marks were explained in Section 4.1. However, when there are several interactions to be assigned, it is more convenient to use linear graphs than the basic marks. The six linear graphs of L_{16} are shown in Figure 4.5.

Table 4.17 $L_{16}(2^{15})$ orthogonal array.

Row number	Column number															Data
	1	2	3	4	5	6	7	8	9	10	11	12	13	14	15	
1	0	0	0	0	0	0	0	0	0	0	0	0	0	0	0	
2	0	0	0	0	0	0	0	1	1	1	1	1	1	1	1	
3	0	0	0	1	1	1	1	0	0	0	0	1	1	1	1	
4	0	0	0	1	1	1	1	1	1	1	1	0	0	0	0	
5	0	1	1	0	0	1	1	0	0	1	1	0	0	1	1	
6	0	1	1	0	0	1	1	1	1	0	0	1	1	0	0	
7	0	1	1	1	1	0	0	0	0	1	1	1	1	0	0	
8	0	1	1	1	1	0	0	1	1	0	0	0	0	1	1	
9	1	0	1	0	1	0	1	0	1	0	1	0	1	0	1	
10	1	0	1	0	1	0	1	1	0	1	0	1	0	1	0	
11	1	0	1	1	0	1	0	0	1	0	1	1	0	1	0	
12	1	0	1	1	0	1	0	1	0	1	0	0	1	0	1	
13	1	1	0	0	1	1	0	0	1	1	0	0	1	1	0	
14	1	1	0	0	1	1	0	1	0	0	1	1	0	0	1	
15	1	1	0	1	0	0	1	0	1	1	0	1	0	0	1	
16	1	1	0	1	0	0	1	1	0	0	1	0	1	1	0	
Basic mark	*a*	*b*	*ab*	*c*	*ac*	*bc*	*abc*	*d*	*ad*	*bd*	*abd*	*cd*	*acd*	*bcd*	*abcd*	
Assignment																

Here are some examples of when to use the six types of linear graph in Figure 4.5.

Type 1: The interactions between factors are equally important.

Type 2: One factor is most important. The interactions between this factor and other factors are required. Two other interactions are needed in addition.

Type 3: Two separate groups of interactions are required.

Type 4: One factor is very important, and interactions between this factor and other factors are required.

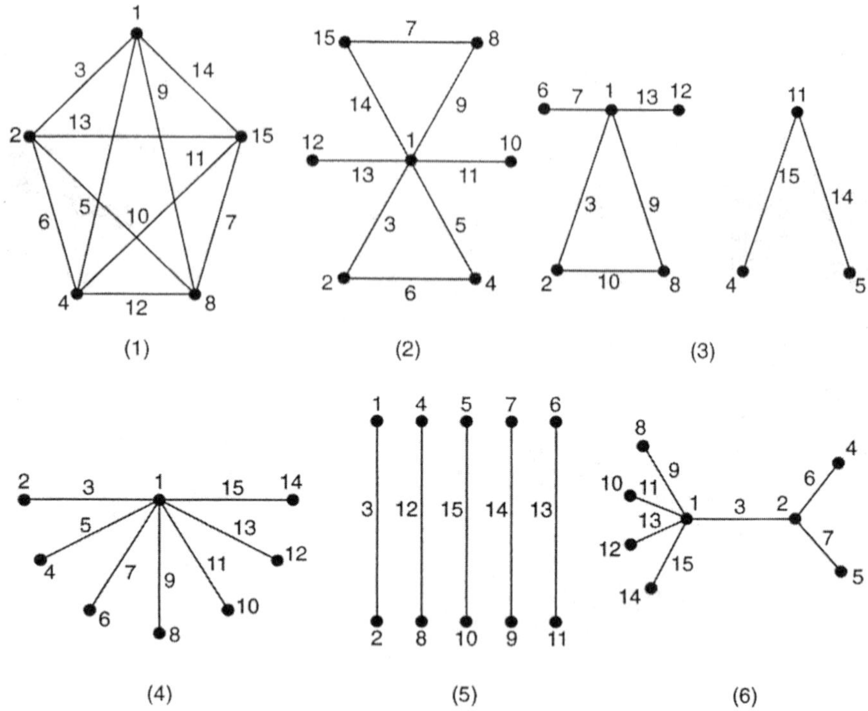

Figure 4.5 Six linear graphs of L_{16}.

Type 5: Five independent interactions may be assigned. Also five four-level factors may be assigned. Latin squares and Graeco-Latin squares may be prepared from this type.

Type 6: Two factors are important, and there are some interactions that are associated with these factors.

The way to use the six linear graphs of L_{16} may be summarized as follows:

Step 1: For a given problem, construct the required graphical structure which uses the dot and line conventions as explained in Figure 4.4.

Step 2: Choose, from the linear graphs of the orthogonal array being used, a linear graph which has all of the features of the required graphical structure.

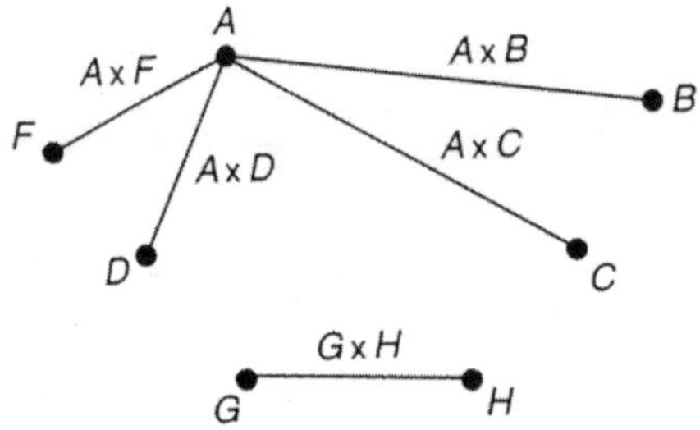

Figure 4.6 Required graphical structure.

Example 4.1

An experiment is to be designed which requires seven 2-level factors, A,B,C,D,F,G,H and five interaction effects, $A\times B$, $A\times C, A\times D, A\times F$ and $G\times H$. This example has 12 degrees of freedom and may be assigned to a linear graph of L_{16}. The required graphical structure of this example has the features shown in Fig 4.6.

The linear graphs (2), (3) and (6) contain all of the features of the required graphical structure. The assignments of factors and interactions on the linear graphs (2), (3) and (6) are shown in Figure 4.7, and the assignments on L_{16} are shown in Table 4.18.

4.3.3 LINEAR GRAPHS FOR THREE-LEVEL ORTHOGONAL ARRAYS

Each column of a three-level factor has two degrees of freedom associated with it, and the interaction between two 3-level columns has four degrees of freedom. Hence, to estimate the interaction we must assign the interaction to two columns. How to select two columns is summarized in the interaction table of Table 4.13. It is convenient to use the interaction table when we assign factors and interactions. However, it is perhaps easier to use the linear graphs of L_{27} in Figure 4.8 to assign them. We do not need the linear graphs of L_9, since we can assign three or four factors without interactions in L_9.

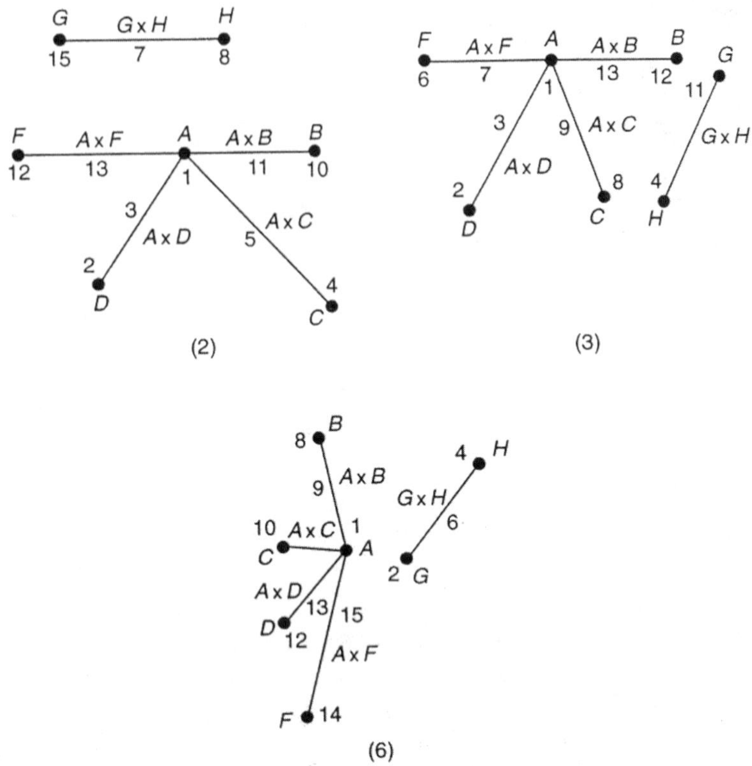

Figure 4.7 Assignments of factors and interactions on the linear graphs.

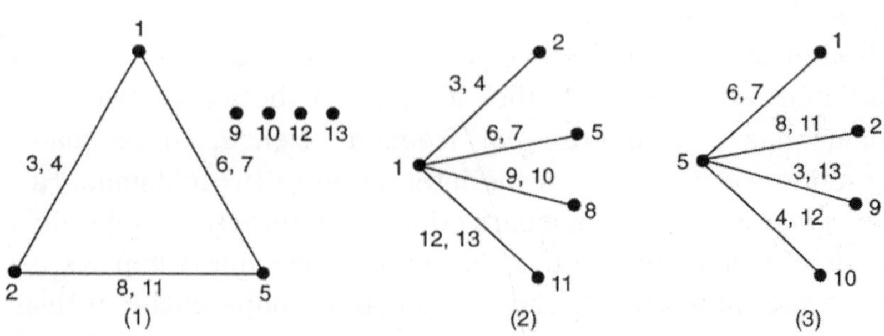

Figure 4.8 Linear graphs of L_{27}.

Table 4.18 Assignments of factors and interactions to L_{16} orthogonal array.

Column no.	1	2	3	4	5	6	7	8	9	10	11	12	13	14	15
Basic mark	*a*	*B*	*ab*	*c*	*ac*	*bc*	*Abc*	*D*	*ad*	*bd*	*abd*	*cd*	*acd*	*bcd*	*abcd*
(2)	A	D	$A\times D$	C	$A\times C$	e	$G\times H$	H	e	B	$A\times B$	F	$A\times F$	e	G
(3)	A	D	$A\times D$	H	e	F	$A\times F$	C	$A\times C$	e	G	B	$A\times B$	e	$G\times H$
(6)	A	G	e	H	e	$G\times H$	E	B	$A\times B$	C	$A\times C$	D	$A\times D$	F	$A\times F$

Example 4.2

An experiment is to be designed which requires six three-level factors, A, B, C, D, F, G and three interactions A×B, A×C and B×C. This example has 12 degrees of freedom for the six factors and 12 degrees of freedom for the three interactions. Hence, a total of 24 degrees of freedom may be assigned to a linear graph of L_{27}. The required graphical structure of this example has the features in Figure 4.9(1).

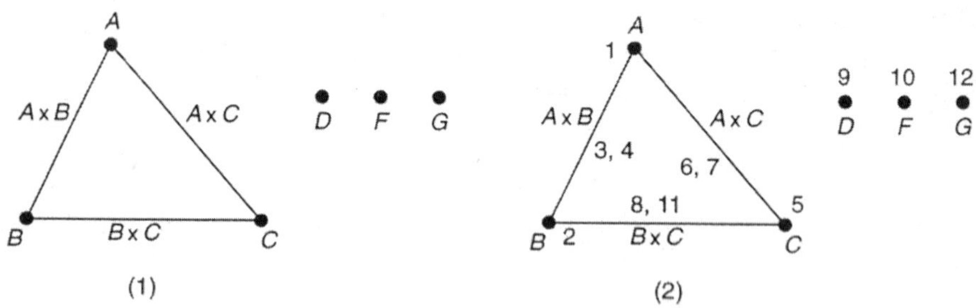

Figure 4.9 Required graphical structure and resulting assignment.

Linear graph (1) in Figure 4.8 satisfies all of the features of the required graphical structure. The resulting assignment of factors and interactions on the first linear graph of L_{27} is shown in Figure 4.9(2).

The linear graph of L_{81} (3^{40}) and $L_{27}(3^{22})$ are shown in Appendix G.

4.4 COLUMN-MERGING METHOD

The orthogonal arrays which have been studied so far have been used to assign only two-level factors or three-level factors. The number of levels should be the same for all factors, and moreover, factors with four or more levels were not discussed. This section addresses both of these apparent limitations with respect to the standard two-level or three-level orthogonal arrays.

The column merging method can be used to create a four-level column in a standard orthogonal array with all two-level columns, a nine-level column in a standard orthogonal array with all three-level columns, and a six-level column in an orthogonal array with a two-level and a three-level column.

4.4.1 CREATION OF A FOUR-LEVEL COLUMN

Since a four-level column has three degrees of freedom and a two-level column has one, the preparation of a four-level column requires three 2-level columns. Such an assignment is made using the following procedure.

1. Select any two columns and their interaction column, with basic marks x, y and xy, say. Merge these three columns into a four-level column. For example, in the L_8 array the interaction of columns 1 and 2 (basic marks a and b) lies in column 3 (basic mark ab). Thus, these three columns can be merged to form a four-level column.
2. Let the horizontal combinations of these two columns (basic marks x and y), 00, 01, 10 and 11, correspond to 0, 1, 2 and 3, respectively. A new four-level column is created, as shown in Table 4.19.
3. In the orthogonal array, replace the three columns with basic marks x, y and xy, with the new four-level column. Table 4.19 shows the new layout.

The above procedure can be simply implemented by using two-level linear graphs. From a linear graph for the orthogonal array being used, choose two dots and the connecting line segment. For example, columns 1, 2 and 3 are selected as shown in Figure 4.10 in an L_8 linear graph. Select any two of the three columns, for example, columns 1 and 2. Next, procedures 2 and 3 above are applied in the same way.

Suppose five factors are to be assigned to the L_8 array where A has four level, and B,C,D and F have two levels each. In the linear graph (Figure 4.10) of the L_8 array, columns 1, 2, and 3 are replaced by factor A with three degrees of freedom. In Table 4.19, factor A is assigned to the four-level column. It is observed that this column is orthogonal to all columns except columns 1, 2 and 3, which have been deleted. Looking at the two columns to which A and B are assigned, we note that for each level of A, each level of B occurs the same number of times, that is, only once in this instance.

Table 4.19 Assignments of a four-level factor to L8.

Experiment number	Column number							
	1	2	3	New column	4	5	6	7
1	0	0	0	0	0	0	0	0
2	0	0	0	0	1	1	1	1
3	0	1	1	1	0	0	1	1
4	0	1	1	1	1	1	0	0
5	1	0	1	2	0	1	0	1
6	1	0	1	2	1	0	1	0
7	1	1	0	3	0	1	1	0
8	1	1	0	3	1	0	0	1
Basic mark	*A*	*b*	*ab*		*c*	*ac*	*bc*	*abc*
Assignment				*A*	*B*	*C*	*D*	*F*

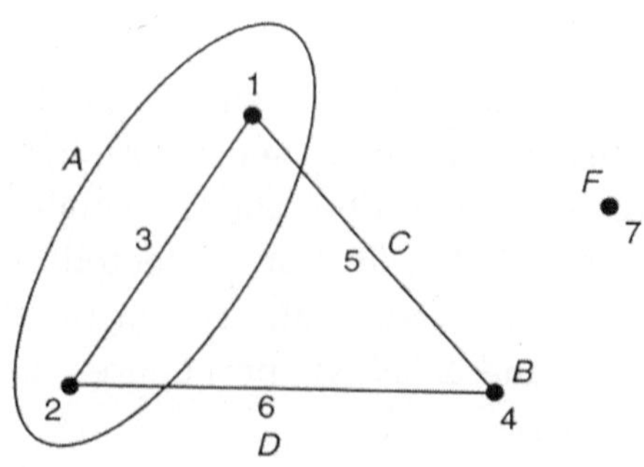

Figure 4.10 Assignment of a four-level factor to a linear graph of L_8.

4.4.2 CREATION OF AN EIGHT-LEVEL COLUMN

Preparation of an eight-level column also begins by considering degrees of freedom. Since an eight-level factor has seven degrees of freedom, it can be assigned to a two-level orthogonal array by replacing seven two-level columns, as in the following procedure.

1. Select any three "independent" columns and their interaction columns, with basic marks x, y, z, xy, xz, yz and xyz. (Here, independent x, y and z columns meant that one column should not be the interaction column between the other two columns.) Merge these seven columns into an eight-level column. For example, in the L_{16} array the interactions of columns 1, 2 and 4 (basic marks a, b and c) lie in columns 3, 5, 6 and 7 (basic marks ab, ac, bc and abc).
2. Let the horizontal combinations of these three columns (basic marks x, y and z), 000, 001, 010, 011, 100, 101, 110, 111, correspond to 0, 1, 2, 3, 4, 5, 6 and 7, respectively. A new eight-level column is created.
3. Replace the seven columns with basic marks x, y, z, xy, xz, yz and xyz with the new eight-level column.

The above procedure can be implemented by using linear graphs. From a linear graph for the orthogonal array being used, choose a closed triangle, plus a line to represent the interaction between the apex and the base as shown in Figure 4.11(1). Take the three columns at the vertices of the triangle. Next, procedures 2 and 3 above are applied in the same way.

Suppose seven factors are to be assigned to the L_{16} array where A has eight levels, and B, C, D, F, G and F have two levels each. Columns 1,2,4,3,5,6 and 7 are replaced by factor A with seven degrees of freedom as shown in Table 4.20. In a linear graph (Figure 4.11(2)) of the L_{16} array, columns 1 through 7 are replaced by a new column (factor A) with seven degrees of freedom. Note that in Table 4.20 this new column is orthogonal to all columns except columns 1, 2, 4, 3, 5, 6 and 7 which have been deleted.

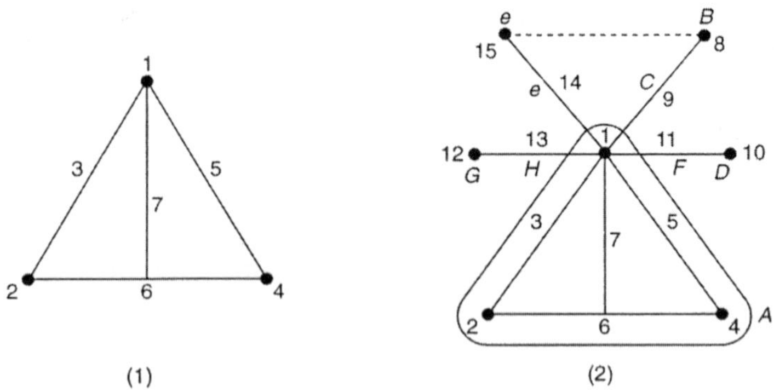

Figure 4.11 Assignment of an eight-level factor to a linear graph of L_{16}.

Table 4.20 Assignments of an eight-level factor to L16.

Experiment number	Column number											
	1	2	4 (3, 5, 6, 7)	New column	8	9	10	11	12	13	14	15
1	0	0	0	0	0	0	0	0	0	0	0	0
2	0	0	0	0	1	1	1	1	1	1	1	1
3	0	0	1	1	0	0	0	0	1	1	1	1
4	0	0	1	1	1	1	1	1	0	0	0	0
5	0	1	0	2	0	0	1	1	0	0	1	1
6	0	1	0	2	1	1	0	0	1	1	0	0
7	0	1	1	3	0	0	1	1	1	1	0	0
8	0	1	1	3	1	1	0	0	0	0	1	1
9	1	0	0	4	0	1	0	1	0	1	0	1
10	1	0	0	4	1	0	1	0	1	0	1	0
11	1	0	1	5	0	1	0	1	1	0	1	0
12	1	0	1	5	1	0	1	0	0	1	0	1
13	1	1	0	6	0	1	1	0	0	1	1	0
14	1	1	0	6	1	0	0	1	1	0	0	1
15	1	1	1	7	0	1	1	0	1	0	0	1
16	1	1	1	7	1	0	0	1	0	1	1	0
Basic mark	*A*	*b*	*c*		*d*	*ad*	*bd*	*abd*	*cd*	*acd*	*bcd*	*abcd*
Assignment				*A*	*B*	*C*	*D*	*F*	*G*	*H*	*e*	*e*

4.4.3 CREATION OF A NINE-LEVEL COLUMN

The column-merging procedure above generalizes to orthogonal arrays with columns with more than two levels. To form a nine-level column in a standard orthogonal array with three-level columns such as L_{27}, we follow a similar procedure to that above except that we must merge four columns, since a nine-level column should have eight degrees of freedom.

1. Select any two columns and their interaction columns, with basic marks x, y, xy and xy^2, say. Merge these four columns into a 9-level column. For example, select two columns 1 and 2, and their interaction columns 3 and 4.
2. Let the horizontal combinations of these two columns (basic marks x and y), 00, 01, 02, 10, 11, 12, 20, 21, and 22, correspond to 0, 1, 2,...,8, respectively. A new nine-level column is created as shown in Table 4.21.
3. In the L_{27} orthogonal array, replace the four columns with basic marks x, y, xy and xy^2 with the new nine-level column.

Table 4.21 Creation of a nine-level column.

Column with basic mark x	Column with basic mark y	Nine-level column
0	0	0
0	1	1
0	2	2
1	0	3
1	1	4
1	2	5
2	0	6
2	1	7
2	2	8

The above procedure can also be implemented by using linear graphs of L_{27}. From a linear graph of L_{27}, choose two dots and the connecting line segment to represent the four three-level columns

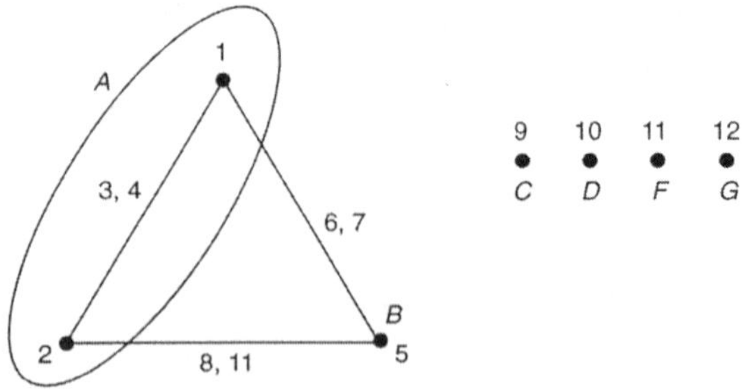

Figure 4.12 Assignment of a nine-level factor to a linear graph of L_{27}.

to be replaced by a nine-level factor. In Figure 4.12, let these columns be 1, 2, 3 and 4. Next, procedures 2 and 3 above are applied in the same way.

Suppose six factors are to be applied to the L_{27} array, where A has nine levels, B, C, D, F and G have 3 levels each. Factor A is assigned to the nine-level column made by columns 1 through 4, and the other factors are assigned to columns 5, 9, 10, 11 and 12, respectively, as shown in Figure 4.12.

4.4.4 CREATION OF A SIX-LEVEL COLUMN

The column-merging method can also be applied to an orthogonal array which has two-level columns and three-level columns such as $L_{18}(2^1 \times 3^7)$ and $L_{36}(2^1 \times 3^{12})$. These mixed-level orthogonal arrays are listed in Appendix G. To make a six-level column from $L_{18}(2^1 \times 3^7)$, select a two-level column and a three-level column and assign the levels as shown in Table 4.22.

Table 4.22 Creation of a six-level column.

Two-level column	Three-level column	Six-level column
0	0	0
0	1	1
0	2	2
1	0	3
1	1	4
1	2	5

4.4.5 CALCULATION OF SUM OF SQUARES FOR MULTI-LEVEL COLUMNS

When we have a set of experimental data for an orthogonal array with a multi-level column, the way to compute the sum of squares of the column is as follows. Suppose the column has t levels and a total of N observations with sum total T. Then

$$\text{Column sum of squares} = (t/N)({T^2}_0 + {T^2}_1 + \ldots + {T^2}_{t-1}) - T^2/N,$$

where T_i is the sum of the N/t observations at the ith level of the column. This column sum of squares has $t-1$ degrees of freedom.

4.5 CLASSIFICATION OF ORTHOGONAL ARRAYS

We have seen several orthogonal arrays which are commonly used in practice. Besides these arrays, there are many other arrays. Orthogonal arrays can be classified into standard orthogonal arrays, extended orthogonal arrays, mixed orthogonal arrays and column-merged orthogonal arrays. In this section we consider each of these in turn.

4.5.1 STANDARD ORTHOGONAL ARRAYS

Two-level orthogonal arrays, which were explained in Section 4.1, are standard ones whose general symbol is written as (4.1). Orthogonal arrays such as

$$L_4(2^3),\ L_8(2^7),\ L_{16}(2^{15}),\ L_{32}(2^{31}),\ L_{64}(2^{63})$$

are the major two-level standard orthogonal arrays.

Three-level orthogonal arrays, which were explained in Section 4.2, are standard ones whose general symbol is written as (4.2). Orthogonal arrays such as

$$L_9(3^4),\ L_{27}(3^{13}),\ L_{81}(3^{40})$$

are the major three-level standard orthogonal arrays. The arrays and their linear graphs of $L_4(2^3)$, $L_8(2^7)$, $L_{16}(2^{15})$, $L_9(3^4)$ and $L_{27}(3^{13})$, can be found in this chapter. The arrays and linear graphs of $L_{32}(2^{31})$, $L_{64}(2^{63})$ and $L_{81}(3^{40})$ are listed in Appendix G.

The four-level and five-level orthogonal arrays, $L_{64}(4^{21})$ and $L_{25}(5^6)$, are also classified as standard orthogonal arrays.

4.5.2 EXTENDED ORTHOGONAL ARRAYS

When there are too many factors to be assigned, and interactions can be ignored, we can extend the standard orthogonal arrays to have more columns to accommodate more factors. These arrays are called extended orthogonal arrays. Such arrays as

$$L_{12}(2^{11}),\ L_{27}(3^{22})$$

are typical extended orthogonal arrays. In fact, all Plackett-Burman designs such as $L_{12}(2^{11})$, $L_{20}(2^{19})$, and $L_{24}(2^{23})$ that do not belong to the standard orthogonal arrays may be considered as extended orthogonal arrays. $L_{36}(3^{13})$ is also classified as an extended orthogonal array.

4.5.3 MIXED ORTHOGONAL ARRAYS

Orthogonal arrays which contain two different levels of columns are called mixed orthogonal arrays. The ones most often used are

$$L_{18}(2^1\times3^7),\ L_{32}(2^1\times4^9),\ L_{36}(2^{11}\times3^{12})$$

$$L_{36}(2^3\times3^{13}),\ L_{50}(2^1\times5^{11}),\ L_{54}(2^1\times3^{25}).$$

These can be found in Appendix G. These arrays are used when there are many factors with two different levels, and interactions can be ignored.

4.5.4 COLUMN-MERGED ORTHOGONAL ARRAYS

The orthogonal arrays which can be constructed from the standard orthogonal arrays by using the column merging method are called column-merged orthogonal arrays. Example are

$$L_8(4^1\times 2^4),\ L_{16}(4^1\times 2^{12}),\ L_{16}(4^2\times 2^9),\ L_{16}(4^3\times 2^6)$$

$$L_{16}(4^4\times 2^3),\ L_{16}(4^5),\ L_{16}(8^1\times 2^8),\ L_{32}(4^1\times 2^{28})$$

$$L_{32}(4^2\times 2^{25}),\ L_{27}(9^1\times 3^9).$$

4.6 DUMMY-LEVEL TECHNIQUE

Suppose we have one two-level factor and three 3-level factors. If we want to use $L_9(3^4)$, we have to allocate the two-level factor into a three-level column. In order to put a two-level factor, A, in a three-level OA, A must be formally treated as a three-level factor. To create an extra level, one of the two actual levels is used twice. For instance,

$$A_0 = A_0, \quad A_1 = A_1, \quad A_2 = A_{0'}.$$

A_2 is formally treated as the third level of A, but actually it is the A_0 condition repeated. Here we have taken $A_2 = A_0$ and called it $A_{0'}$ to emphasize that this is a dummy level. If we put A in column 1, and three 3-level factors, $B, C,$ and D, in columns 2-4, the layout becomes as shown in Table 4.23. Note that in the first column of the original L_9 array, 2 is replaced by 0′, indicating that A_2 is the dummy level.

If we run nine experiments as in Table 4.23, the main effect A, with dummy level A_2, is calculated as follows:

$$S_A = \frac{(A_0 + A_{0'})^2}{6r} + \frac{A_1^2}{3r} - \frac{T^2}{9r} = \frac{(A_0 + A_{0'} - 2A_1)^2}{18r}, \tag{4.3}$$

where r is the number of repetitions for each experimental run, and A_i is the sum of observations taken at level i. The degrees of freedom of S_A should be 1, since the actual number of levels of A is two.

Table 4.23 Dummy-level assignment.

Experiment number	Column number			
	1	2	3	4
1	0	0	0	0
2	0	1	1	1
3	0	2	2	2
4	1	0	1	2
5	1	1	2	0
6	1	2	0	1
7	0′	0	2	1
8	0′	1	0	2
9	0′	2	1	0
Factor assignment	*A*	*B*	*C*	*D*

The dummy-level technique is to assign a factor with m levels to a column of n levels, where n is greater than m. Note that after we use the dummy level, the resulting array is still proportionally balanced and, hence, orthogonal. In Table 4.23, we could just as well have taken $A_2 = A_{1'}$. The choice between $A_2 = A_0$ and $A_2 = A_1$ depends on the following issues.

1. If we take $A_2 = A_0$, then the effect of A_0 will be estimated with twice as much precision in variance as the effect of A_1. Thus, the dummy level should be taken to be the one about which we want more precise information.
2. Economy, ease of experimentation and availability of experimental resources also play a role. Thus, if A_0 is cheaper, easier in experimentation, or less scarce than A_1, then we may choose $A_2 = A_0$.

One can apply the dummy-level technique for a three-level factor to be assigned to a two-level OA, If a three-level factor is to be assigned to a two-level OA, this may be done by providing a four-level column using the column-merging method, followed by a dummy treatment in the fourth level.

We can also apply the dummy-level technique to more than one factor in an experiment. Suppose there are two 2-level factors (A and B), and two 3-level factors (C and D). We can assign the four factors to the columns of the OA L_9 by taking dummy levels $A_2 = A_{0'}$ (or $A_2 = A_{1'}$) and $B_2 = B_{0'}$ (or $B_2 = B_{1'}$). Observe that the orthogonality is still preserved even when the dummy-level technique is applied to two or more factors.

We can further generalize the dummy-level technique, without loss of orthogonality, to assign an m-level factor to an n-level column, where m is less than n. For example, suppose we have a six-level factor, A, to the nine-level column in the column-merging OA $L_{27}(9^1 \times 3^9)$ by taking $A_6 = A_{0'}$, $A_7 = A_{1'}$ and $A_8 = A_{2'}$.

Example 4.3

A numerical example of dummy-level technique is illustrated for an emulsification experiment. The following factors were considered to be important.

A: temperature (°C)	$A_0 = 50$, $A_1 = 60$, $A_2 = 70$,
B: raw material concentration	$B_0 = 20$, $B_1 = 25$
C: treatment method	three levels C_0, C_1, C_2
D: type of additive	three levels D_0, D_1, D_2

Factor B was assigned to column 2 of L_9 by using the dummy-level technique as follows.

$$B_0 = B_0, \quad B_1 = B_1, \quad B_2 = B_0,$$

Factors A, C, D were assigned to columns 1, 3, and 4, respectively. For each of the nine experiments, two test pieces were similarly treated and measured under the same conditions. The purpose of the experiment was to find optimal factor conditions to prolong the emulsification time of a certain oil, and the data were in terms of emulsification time (minutes). The layout and results of the observations are given in Table 4.24.

Table 4.24 Emulsification-time data (minutes).

Experiment number	Column number and factor assignment				Data	Total
	1	2	3	4		
	A	*B*	*C*	*D*		
1	0	0	0	0	8 6	14
2	0	1	1	1	3 4	7
3	0	0′	2	2	1 1	2
4	1	0	1	2	4 7	11
5	1	1	2	0	6 5	11
6	1	0′	0	1	6 4	10
7	2	0	2	1	3 5	8
8	2	1	0	2	13 15	28
9	2	0’	1	0	11 13	24

There are 18 observations to be analyzed. The correction term and the total variation are

$$CT = T^2/18 = (115)^2/18 = 734.7,$$
$$S_T = 8^2 + 6^2 + 3^2 + \ldots + 13^2 - CT = 288.3(\phi = 17).$$

However, the total variation of primary units, 14, 7, 2, …, 24, is

$$S_{T_1} = [14^2 + 7^2 + 2^2 + \ldots + 24^2] - CT$$
$$= 272.8(\phi = 9).$$

In this problem with two repetitions, secondary error effects can be calculated. The secondary error variation, which is in fact the pure error sum of squares, is

$$S_{e_2} = S_T - S_{T_1} = 288.3 - 272.8 = 15.5(\phi = 9).$$

This variation can be also obtained from

$$S_{e_2} = [(8-6)^2 + (3-4)^2 + \ldots + (11-13)^2]/2 = 15.5.$$

In order to compute the sums of squares for *A*, *B*, *C* and *D*, we need to obtain the sums of observations at each level of each factor as shown in Table 4.25.

Table 4.25 Supplementary level totals.

Level	A	B	C	D
0	23	33 (B_0)	52	49
1	32	46 (B_1)	42	25
2	60	36 ($B_{0'}$)	21	41
Total	115	115	115	115

$$S_A = \frac{A_0^2 + A_1^2 + A_2^2}{6} - CT$$

$$= \frac{23^2 + 32^2 + 60^2}{6} - 734.7 = 124.1(\phi = 2),$$

$$S_B = \frac{(B_0 + B_{0'})^2}{12} + \frac{B_1^2}{6} - CT$$

$$= \frac{(33+36)^2}{12} + \frac{46^2}{6} - 734.7 = 14.7(\phi = 1),$$

$$S_C = \frac{C_0^2 + C_1^2 + C_2^2}{6} - CT$$

$$= \frac{52^2 + 42^2 + 21^2}{6} - 734.7 = 83.5(\phi = 2),$$

$$S_D = \frac{D_0^2 + D_1^2 + D_2^2}{6} - CT$$

$$= \frac{49^2 + 25^2 + 41^2}{6} - 734.7 = 49.8(\phi = 2).$$

Each column of a three-level OA has two degrees of freedom. In column 2, factor B has two levels and only one degree of freedom. Therefore, primary error e_1, with one degree of freedom, can be calculated.

$$S_{e_1} = \frac{(B_0 - B_{0'})^2}{12}$$

$$= \frac{(33-36)^2}{12} = 0.7(\phi = 1).$$

Table 4.26 ANOVA table for the numerical example.

Source of variation	Sum of squares	Degree of freedom	Mean square	F_0
A	124.1	2	62.05	38.3^{**}
B	14.7	2	14.7	9.1^{**}
C	83.5	2	41.75	25.8^{**}
D	49.8	2	24.9	15.4^{**}
E_1	0.7 ○	1	0.7	
T_1	272.8	8		
E_2	15.5	9	1.72	
(e)	(16.2)	(10)	(1.62)	
T	288.3	17		

We can also compute the primary error variation as follows.

$$S_{e_1} = S_{T_1} - [S_A + S_B + S_C + S_D]$$
$$= 272.8 - [124.1 + 14.7 + 83.5 + 49.8] = 0.7.$$

In Table 4.26, ** shows 1% significance. In addition, usually, * shows 5% significance and † shows 10% significance. From now on, we will use these notations to indicate statistical significance. Also, a "○" mark in the sum of squares column shows the pooled variations. The ANOVA table implies that e_1 is not significant, and is pooled together with e_2 to get (e). We can observe that all factors are very significant (at the 1% level).

Next, it is of interest to find the optimum and the confidence limits for the mean response at the optimum condition. The respective sample means of each factor level can be obtained as in Table 4.27.

Table 4.27 Sample mean (and number of data) at each factor level.

Level	A	B	C	D
0	3.83 (6)	5.75 (12)	8.67 (6)	8.17 (6)
1	5.33 (6)	7.67 (6)	7.00 (6)	4.17 (6)
2	10.00 (6)		3.50 (6)	6.83 (6)
Overall mean	6.39	6.39	6.39	6.39

Since all the factors are significant, the optimal condition which maximizes the emulsification time can be found in Table 4.27. This is $A_2B_1C_0D_0$, in which the estimate of the emulsification time mean is

$$\begin{aligned}\hat{\mu}(A_2B_1C_0D_0) &= \overline{A}_2 + \overline{B}_1 + \overline{C}_0 + \overline{D}_0 - 3\overline{T} \\ &= 10.00 + 7.67 + 8.67 + 8.17 - 3 \times 6.39 \\ &= 15.34. \end{aligned} \tag{4.4}$$

A 95% confidence interval for μ is

$$\begin{aligned} &15.34 \pm t_{0.975}(10)\sqrt{\frac{V_e}{n_e}} \\ &= 15.34 \pm (2.228)\sqrt{\frac{1.62}{2.0}} \\ &= 15.34 \pm 2.01, \end{aligned}$$

where the effective number of replications is n_e=2, since

$$\begin{aligned}\frac{1}{n_e} &= \frac{1}{6} + \frac{1}{6} + \frac{1}{6} + \frac{1}{6} - 3 \times \frac{1}{18} \\ &= \frac{1}{2}. \end{aligned} \tag{4.5}$$

The way to compute n_e was explained in equation (3.15). Note that the numbers 6, 6, 6, 6 and 18 in equation (4.5) are the numbers of data points to obtain the means $\overline{A}_2, \overline{B}_1, \overline{C}_0, \overline{D}_0$ and $\overline{T}$ in equation (4.4), respectively.

Exercises

4.1 A tile-production experiment was performed in a tile-manufacturing company. The object of the experiment was to use actual production scale conditions that had been arrived at in a small-scale experiment in the laboratory, and to shift to actual production once confirmation was obtained. The following four factors were related to the compounding of the raw materials; one of the levels was the material currently in use and the other level was the material predicted to be possibly better than the current one, taking into consideration both product quality and cost.

A : agalmatolite quantity	A_0 = 40%	A_1 = 50% (current)
B : feldspar quantity	B_0 = 2%	B_1 = 4% (current)
C : quantity of an additive	C_0 = 4%	C_1 = 2% (current)
D : quantity charged	D_0 = 1300kg	D_1 = 1100kg (current)

Since two-factor interactions $A \times B$ and $A \times C$ were of interest, the experimenters allocated four factors and two interactions, as shown in the table below. The table shows data on the number of defectives (tiles which did not become grade 1 products) out of 200 tiles produced under bad conditions near the edges of the kiln.

Experiment number	Column number and allocation of factors and interactions							Number of defectives in 200 tiles produced
	1	2	3	4	5	6	7	
	A	B	$A \times B$	C	$A \times C$	D	e	
1	0	0	0	0	0	0	0	68
2	0	0	0	1	1	1	1	54
3	0	1	1	0	0	1	1	37
4	0	1	1	1	1	0	0	47
5	1	0	1	0	1	0	1	50
6	1	0	1	1	0	1	0	95
7	1	1	0	0	1	1	0	68
8	1	1	0	1	0	0	1	116

(a) Perform an analysis of variance.

(b) Find the optimum conditions which produce the minimum number of defectives. Predict the number of defective sheets X when 200 tiles are produced under the optimum conditions.

4.2 The following experiment was carried out to find the best conditions of viscosity and molar ratio for the production of sodium silicate.

A : reaction temperature (°C) $A_0 = 170$, $A_1 = 180$, $A_2 = 190$, $A_3 = 300$

B : reaction time (h) $B_0 = 3$, $B_1 = 4$

C : agitation (pulley is changed) $C_0 = 100$, $C_1 = 120$

D : silica sand type $D_0 =$ type I, $D_2 =$ type II

F : evacuated water temperature (°C) $F_0 = 40$, $F_1 = 60$

Four interactions, $A \times B$, $B \times C$, $B \times D$ and $B \times F$ were suspected, and the following allocation was used (refer to OA L_{16} in Table 4.17). Note that since A has four levels, it needs three columns, as explained by the column-merging method.

Column number	1	2	3	4	5	6	7	8	9	10	11	12	13	14	15
Source allocation	A	A	A	B	$A \times B$	$A \times B$	$A \times B$	C	D	F	e	$B \times C$	$B \times D$	$B \times F$	e

The data obtained for viscosity and molar ratio for each experiment number in Table 4.17 are as follows.

Experiment number	Viscosity DATA	Molar ratio data
1	19	2.18
2	20	2.20
3	40	2.30
4	18	2.13
5	10	1.99
6	43	2.45
7	12	2.19
8	59	2.34
9	9	1.78
10	17	2.23
11	7	1.96
12	15	2.36
13	8	1.79
14	11	2.13
15	12	2.08
16	14	2.18

(a) Find the experimental conditions for each experiment number.

(b) Perform an analysis of variance. Find the optimum conditions for viscosity and molar ratio, respectively. Viscosity is a smaller-the-better characteristic, and molar ratio is a larger-the-better characteristic.

(c) Find the combinations of factors that will yield a viscosity of at most 40 and molar ratio of at least 2.20. If there are many combinations, find all of them.

4.3 Factors of three levels, A, B, C and D, were assigned as follows, rubber mixtures were produced, and the tensile strength ($kgcm^{-2}$) was measured.

Experiment number	Column number and factor assignment				300% tensile strength
	1 *A*	2 *B*	3 *C*	4 *D*	
1	0	0	0	0	560, 542
2	0	1	1	1	572, 578
3	0	2	2	2	643, 657
4	1	0	1	2	576, 587
5	1	1	2	0	578, 590
6	1	2	0	1	581, 596
7	2	0	2	1	635, 641
8	2	1	0	2	622, 619
9	2	2	1	0	611, 609

(a) Performed analysis of variance.

(b) Find the optimum conditions which maximize the tensile strength. Predict the tensile strength at the optimum conditions.

4.4 Using the L_{27} orthogonal array in Table 4.12, assign the following factors and interactions to find an appropriate experimental design for each of the following cases.

(a) Three-level factors A, B, C, D, F, G and interactions $A \times B$, $A \times C$, $B \times C$.
(b) Three-level factors, A, B, C, D, F, G, H and interactions $C \times D$, $C \times F$, $C \times G$.
(c) One nine-level factor A, and three-level factors, B, C, D, F, G, H, I.

5

Parameter Design for Continuous Data

5.1 STRUCTURE OF PARAMETER DESIGN

As mentioned in Chapter 1, parameter design is the essence of robust design; this chapter thus concerns probably the greatest contribution to quality engineering. Previous experimental approaches, since the invention of experimental design in the 1930s, did not consider noise factors as causes of quality variation. But if, for instance, a product is sensitive to ambient temperature variations, how can anyone control ambient temperature in a customer's environment? Note that ambient temperature is an external noise, and there are other noises such as internal noise and unit-to-unit noise as discussed in Section 1.2.

It is very difficult or very expensive to control or eliminate causes of variation which come from noise factors. Therefore, a different approach is required if variation of product quality is to be reduced. This approach was termed **parameter design** by Taguchi, and this design is used to improve quality without controlling or eliminating causes of variation, and to make the product robust against noise factors.

In general, parameter design has the following characteristics. First, parameter design classifies factors which affect quality characteristics into two groups: control factors and noise (or uncontrollable) factors. At the outset, it is quite useful to sketch a factor-characteristic relation diagram, as shown in Figure 5.1.

Note that "system" here means a process, a manufacturing line, an R&D environment, or a given experimental situation.

Second, parameter design generally uses two OAs: for the control factors, an OA which is called an "inner array" or "matrix of design variables"; and for the noise factors, another OA which is called an "outer array" or "matrix of uncontrollable factors". Table 5.1 shows an experimental layout with an inner array (here L_8) for control factors, and an outer array (here L_4) for noise factors. The design of such a structure is often called **direct product design**, since two separate OAs are used and observations are recorded for all combinations of the two arrays. Note that there are $8\times4=32$ observations in the parameter design in Table 5.1.

This parameter design strategy separates the control factors (A,B,C,D,F) from the noise factors (U,V,W) by using inner and outer arrays, respectively. Note that noise factors such as ambient temperature, operators and conditions of usage are assigned to the outer array to find the levels of control factors that do not have much variation in the results in spite of the noise factors being present. In this experimental arrangement there are 32 separate test conditions. For instance, y_{22} is observed at $A_0B_0C_0D_1F_1$ and $U_0V_1W_1$. If tests are very expensive to run, a thorough outer array may be precluded and only the single noise factor that is thought to be strongest may be used. If it is difficult to select a single strong noise factor, we may use two levels: the best and worst conditions of noise factors combined. Sometimes, it is not

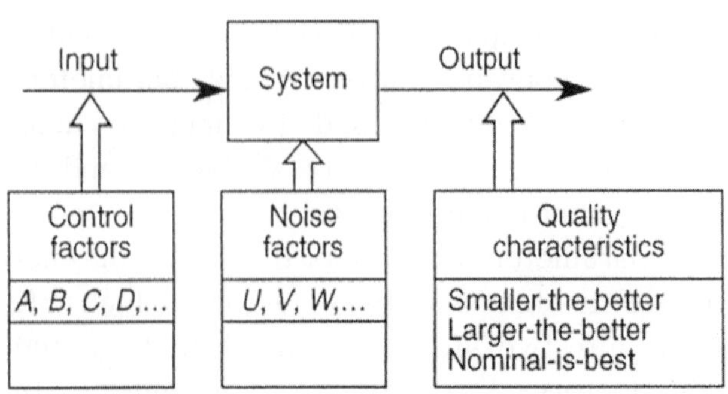

Figure 5.1 Factor–characteristic relation diagram.

Table 5.1 Basic structure of parameter design.

Type of array	Inner array (L8)							Outer array (L4)				
	Control factor assignment and column number							Row data				SN ratio
								Experiment no.				Noise factor assignment
								1	2	3	4	
Experiment number	*A*	*B*	*C*	*D*	*F*	*e*	*e*	0	0	1	1	U
	1	2	3	4	5	6	7	0	1	0	1	V
								0	1	1	0	W
1	0	0	0	0	0	0	0	y_{11}	y_{12}	y_{13}	y_{14}	SN_1
2	0	0	0	1	1	1	1	y_{21}	y_{22}	y_{23}	y_{24}	SN_2
3	0	1	1	0	0	1	1	y_{31}	y_{32}	y_{33}	y_{34}	SN_3
4	0	1	1	1	1	0	0	…				
5	1	0	1	0	1	0	1	…				
6	1	0	1	1	0	1	0	…				
7	1	1	0	0	1	1	0	…				
8	1	1	0	1	0	0	1	y_{81}	y_{82}	y_{83}	y_{84}	SN_8

easy to set up experimental conditions for noise factors. In this case we may take repeated observations by letting the conditions of the noise factors vary as they are naturally changed in practical situations.

Third, for the repeated observations, $y_{i1}, y_{i2}, y_{i3}, \ldots, y_{in}$, in each row i (ith experimental conditions of control factors), the signal-to-noise (SN) ratio is computed. As explained in Section 2.2, the SN ratio is a measure which consolidates several repetitions into one value, and which reflects the average response and the amount of variation present at the same time. The way to compute SN ratios is as follows (here n is the number of observation in each row).

(a) smaller-the-better characteristic

$$SN_i = -10\log\left(\frac{1}{n}\sum y_{ij}^2\right), \tag{5.1}$$

(b) larger-the-better characteristic

$$\mathrm{SN}_i = -10\log\left(\frac{1}{n}\sum_{j=1}^{n}\frac{1}{y_{ij}^2}\right), \tag{5.2}$$

(c) nominal-is-best characteristic

$$\mathrm{SN}_i = 10\log\left[\frac{1}{n}(S_{m(i)} - V_i)/V_i\right], \tag{5.3}$$

or approximately

$$\mathrm{SN}_i = 10\log\left[\frac{1}{n}(S_{m(i)})/V_i\right] = 10\log\left[\frac{(\bar{y}_i)^2}{V_i}\right], \tag{5.4}$$

where

$$V_i = \sum_{j=1}^{n}(y_{ij} - \bar{y})^2 \Big/ (n-1) = \text{ sample variance for the } i\text{th row,}$$

$$\mathrm{S}_{m(i)} = \frac{1}{n}(\sum_{j=1}^{n} y_{ij})^2 = n(\bar{y}_i)^2 = \text{ correction term for the } i\text{th row,}$$

$$\bar{y}_i = \sum_j y_{ij} \Big/ n = \text{sample mean for the } i\text{th row.}$$

For the nominal-is-best case the SN ratio is mostly influenced by the magnitude of V, not by $\bar{y}$. In order to find out which factor influences the change of mean response $\bar{y}$ most, we need to obtain the sensitivity, which is defined as follows for the set of n observations in the ith row.

$$\begin{aligned} S_{n(i)} &= 10\log\left[S_{m(i)}\right] \\ &= 10\log\left[n(\bar{y}_i^2)\right] \\ &= 20\log\left[\sqrt{n}\bar{y}_i\right]. \end{aligned} \tag{5.5}$$

5.2 STEPS OF PARAMETER DESIGN

Several steps are needed for parameter design and data analysis, differing slightly according to the type of quality characteristic as follows.

1. Smaller-the-better and larger-the-better characteristics:
   (a) Sketch the factor-characteristic relation diagram for a given experimental problem.
   (b) Select appropriate OAs for the inner array and the outer array, and assign control and noise factors to the arrays.
   (c) Conduct experiments, and compute the SN ratio for each row of the inner array.
   (d) Find significant control factors which affect SN ratios through ANOVA (or Pareto ANOVA, which will be explained in Section 5.3). Find the optimum condition for the significant control factors.
   (e) For the insignificant control factors, find the optimum condition based on other considerations such as economy, operability and easiness.
   (f) Estimate the mean response at the overall optimum condition, and conduct a confirmation test at the optimum condition to check that the obtained optimum condition produces desired responses.
2. Nominal-is-best characteristics:
   (a)-(b) Same as above.
   (c) Conduct experiments, and compute the SN ratio and sensitivity for each row of the inner array.
   (d) Find significant control factors which affect SN ratios through (Pareto) ANOVA. Also find significant control factors which affect sensitivities through (Pareto) ANOVA. Classify the control factors into the following three categories.
      (i) dispersion control factor: significant factors for SN ratios,
      (ii) mean adjustment factor: significant factors for sensitivities,

(iii) insignificant factor: other control factors.

If a factor is both a dispersion control factor and a mean adjustment factor, it is classified as a dispersion control factor.

(e) Find the optimum condition as follows.

(i) dispersion control factor: the level which maximizes the SN ratio,

(ii) mean adjustment factor: the level which gives the closest estimated response to the target response value,

(iii) insignificant factor: the level which is chosen based on other considerations such as economy, operability and simplicity.

(f) Same as above.

The three types of factors which were classified in the above step (d) may be best explained diagrammatically. Figure 5.2 shows three different types of factors (i), (ii) and (iii), and an additional factor which is both a dispersion control and a mean adjustment factor.

As discussed above in steps (d) and (e), lowest variation is the first priority in finding the optimum condition. The second priority is the mean response. As already mentioned, a condition which gives low variation against noise factors is a robust condition. This is the reason why parameter design is called robust design.

In parameter design there is no clear criterion for the number of observations taken at each experimental condition. For statistical analysis it is always true that the larger the number of observations is, the better. However, considering the economic and time constraints involved in the experiment, the following criteria are recommended for the number of observations: for continuous data, 2–10; for simple or multi-discrete data, 5–20; and for fixed marginal discrete data, 10–150.

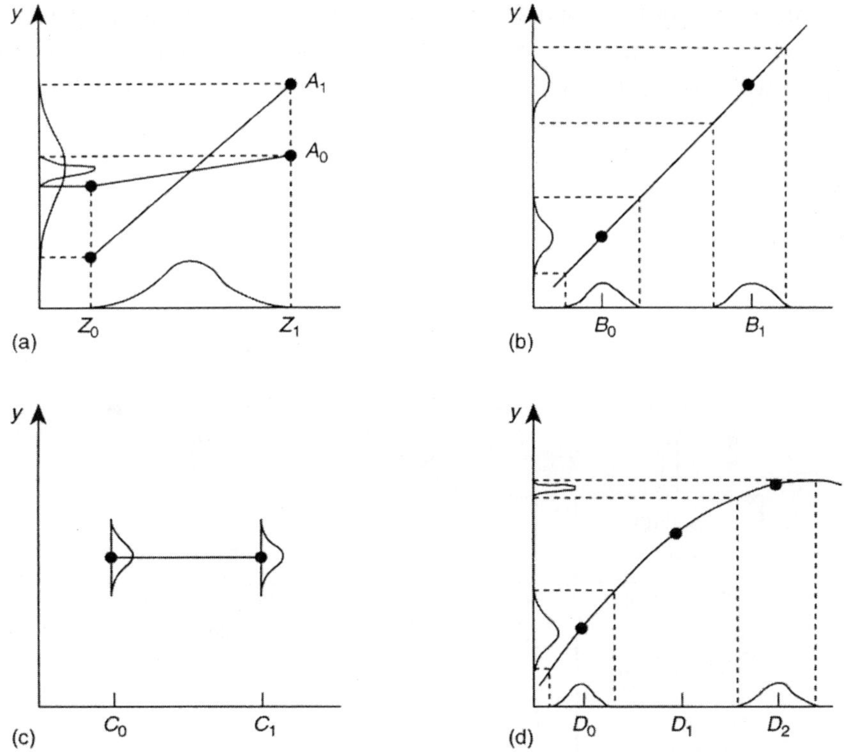

Figure 5.2 Types of control factor: (a) dispersion control factor A and noise factor Z; (b) mean adjustment factor B; (c) insignificant factor C; (d) dispersion control and mean adjustment factor D.

5.3 PARETO ANALYSIS OF VARIATION

The six steps in parameter design which were discussed in Section 5.2 are illustrated in this section with a real case study in a Korean electrical company. Here we will introduce a simplified ANOVA method which uses the Pareto principle. This method will be called **Pareto ANOVA.** This is a quick and easy method to analyze the results of parameter design which does not require an ANOVA table and, therefore, does not use F-tests.

An electrical company produces an electrical item called a ground fault circuit interrupter (GFCI), and the company was not happy with its high fraction defective for this product. It formed a task force to study the problem and decided to carry out a parameter design to reduce the fraction defective. The quality characteristic was the moving time, which is a smaller-the-better characteristic. The factor-characteristic relation diagram shown in Figure 5.3 was produced.

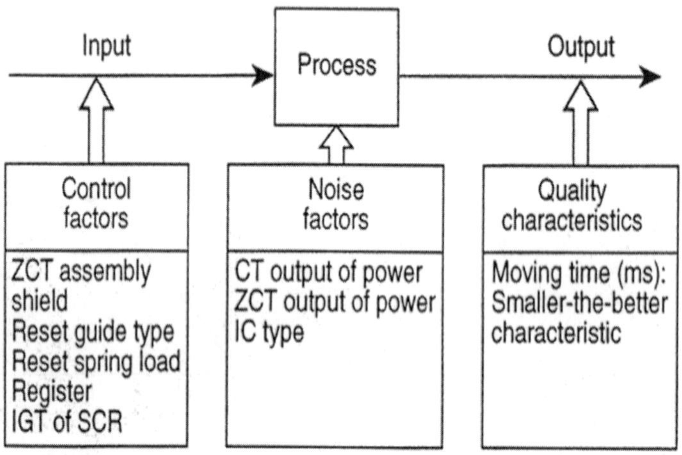

Figure 5.3 Factor-characteristic relation diagram for GFCI experiment.

For the five control factors in Figure 5.3, the task force decided to use two levels for each factor as follows, and assign the factors in L_8 as shown in Table 5.2:

A: use of ZCT assembly shield $(A_0 = \text{yes},\ A_1 = \text{no})$

B: reset guide type $(B_0 = \text{type I},\ B_1 = \text{type II})$

C: reset spring load $(C_0 = 640\text{g},\ C_1 = 740\text{g})$

D: use of resistor (2.7kΩ) $(D_0 = \text{yes},\ D_1 = \text{no})$

F: electrical gate turn-on (IGT, μA) of silicon control rectifier (SCR) $(F_0 = \text{below 20},\ F_1 = \text{above 20})$

For the noise factors, the task force decided to use two extreme conditions of the three noise factors combined as two levels of the combined noise factor N as follows.

$$N_0 : \text{CT}=16, \quad \text{ZCT}=16, \quad \text{IC}=\text{KA2803}$$

$$N_1 : \text{CT}=\text{above } 17, \quad \text{ZCT}=\text{above } 18, \quad \text{IC}=\text{GL7101}$$

At each level of the noise factor N, there are three observations for moving time. The current manufacturing condition for the control factors is $A_1B_0C_0D_0F_1$ in which there are also six observations, as shown in Table 5.2.

The task force suspected that there might be some interaction between the rest spring load (C) an the register (D), so they assigned $C \times D$ to a column as in Table 5.2. For this parameter design, the six steps in the previous section will be roughly sketched.

Table 5.2 Experimental data and SN ratios for moving time.

Array type	Inner array (L_8)							Outer array (one-way layout)						
Experiment number	Control factor assignment and column number							Raw data: Moving time (ms)						SN= SN+ 23
	A	B	C	D	F	e	C×D	Noise factor N						
	1	2	3	4	5	6	7	N_0			N_1			
1	0	0	0	0	0	0	0	7	14	16	11	17	10	0.73
2	0	0	0	1	1	1	1	8	14	13	12	16	12	0.90
3	0	1	1	0	0	1	1	13	11	13	8	12	16	1.13
4	0	1	1	1	1	0	0	11	13	13	10	14	13	1.12
5	1	0	1	0	1	0	1	17	14	14	13	15	12	-0.08
6	1	0	1	1	0	1	0	17	14	16	8	9	18	-0.05
7	1	1	0	0	1	1	0	13	10	13	12	13	13	1.14
8	1	1	0	1	0	0	1	8	11	9	13	10	12	2.46
	Current condition													0.54

5.3.1 COMPUTATION OF SN RATIOS

The first two steps have already been explained in Figure 5.3 and Table 5.2.

SN ratio for the ith row is computed using formula (5.1) since the quality characteristic is smaller-the-better. Note that there are six observations; hence, the SN ratio for the first row is

$$SN_1 = -10\log\left[\sum_j y_{1j}^2 \Big/ 6\right]$$
$$= -10\log\{[7^2 + 14^2 + \ldots + 10^2]/6\} = -22.27.$$

However, for computational simplicity, 23 is added to SN_1 and it becomes 0.73 dB as written in Table 5.2. It should be recalled that if the dispersion is small and the average is small, the SN ratio becomes large. Thus the condition which makes the SN ratio largest is optimal.

5.3.2 PARETO ANOVA AND DETERMINATION OF OPTIMUM COMDITION

For many engineers and researchers in industry, ANOVA and related F-tests are not easy to use. Pareto ANOVA is a quick and easy method for analyzing results. This method enables the significance of factors and interaction to be evaluated by Pareto-type analysis. It also allows the optimal levels of factors to be obtained. Table 5.3 shows the analysis for the GFCI experiment, from which we can see that the optimal levels of the control factors are $A_0B_1C_0D_1F_0$.

In Table 5.3 the square of the difference between the levels of A, for instance, is nothing but the numerator of S_A, since

$$S_A = \frac{(A_1 - A_0)^2}{8} = \frac{(3.47 - 3.88)^2}{8} = \frac{0.1681}{8}.$$

Also the contribution ratio of A, for instance, is computed as follows:

Table 5.3 Pareto ANOVA for two-level factors.

Factor and interaction		A	B	C	D	F	e	$C \times D$	Total
Sum at factor level	0	3.88	1.50	5.23	2.92	4.27	4.23	2.94	Sum of 0, 1 levels = 7.35
	1	3.47	5.85	2.12	4.43	3.08	3.12	4.41	
Difference at 2 levels		0.41	4.35	3.11	1.51	1.19	1.11	1.47	
Square of difference		0.1681	18.9225	9.6721	2.2801	1.4161	1.2321	2.1609	35.8519
Contribution ratio (%)		0.5	52.8	27.0	6.4	3.9	3.4	6.0	100.0
Pareto diagram		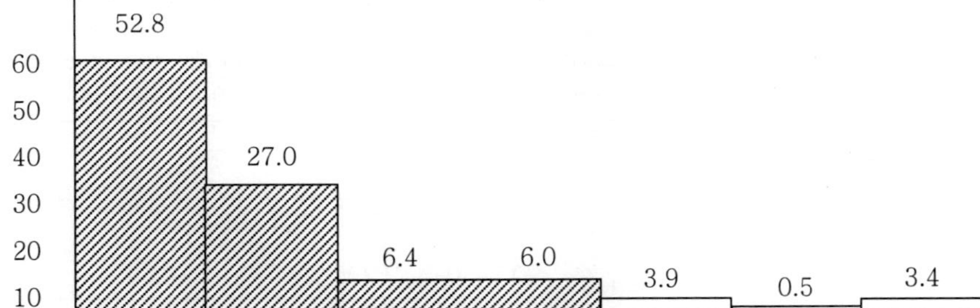							
Factor and interaction		B	C	D	$C \times D$	F	A	e	
Cumulative contribution ratio (%)		52.8	79.8	86.2	92.2	96.1	96.6	100.00	

Table 5.3 (Continued).

<table>
<tr><td>Check on significant interaction</td><td>CD 2-way table
<table><tr><td></td><td>C_0</td><td>C_1</td></tr><tr><td>D_0</td><td>0.73+ 1.14=1.87</td><td>1.13−0.08=1.05</td></tr><tr><td>D_1</td><td>0.90+ 2.46=3.36</td><td>1.12−0.05=1.07</td></tr></table>
C_0D_1 is optimal</td></tr>
<tr><td>Optimum combinations of significant factor levels</td><td>$B_1C_0D_1$</td></tr>
<tr><td>Remarks on optimum condition</td><td>The significant factors and interactions are from the left-hand side in the above Pareto diagram which cumulatively contribute about 90%. If interactions are significant, the optimal levels of the related factors are chosen from two-way tables. For other non-significant factors, the optimal levels can be chosen from other economic and technical considerations.</td></tr>
<tr><td>Overall optimum conditions for factors</td><td>A_0 : ZCT assembly shield, yes
B_1 : Reset guide type, type II
C_0 : Reset spring load, 640 g
D_1 : Register, no
F_0 : IGT of SCR, below 20</td></tr>
<tr><td>Estimate of error variance</td><td>V_e = (pooled variation of non-significant sources)/ (pooled degrees of freedom of non significant sources)
$= (S_a + S_F + S_e)/3$
$= (0.1681/8 + 1.4161/8 + 1.2321/8)/3 = 0.117$</td></tr>
<tr><td></td><td></td></tr>
</table>

$$(0.1681/35.8519)\times 100 = 0.5\%.$$

The factors and interactions of the shaded area in the Pareto diagram are considered to be significant sources of variation. Note that the significant factors and interactions are chosen from the left-hand side in the Pareto diagram, which cumulatively contribute about 90%. Since $C\times D$ is significant, we need to check the CD two-way table in order to find the optimal levels of C and D, and it turns out that C_0D_1 is optimal. Comparing the sums at B_0 and B_1 factor levels, we can see that B_1 is optimal. Hence, the optimal combination of significant factor levels is $B_1C_0D_1$.

If we choose A_0F_0 as non-significant factors for other economical and technical reasons, the overall optimal condition for all factors is $A_0B_1C_0D_1F_0$. At this condition the estimated SN ratio will be

$$\begin{aligned}\hat{\mu}(A_0B_1C_0D_1F_0) &= \overline{A}_0 + \overline{B}_1 + \overline{C_0D_1} + \overline{F}_0 - 3\overline{T} \\ &= \frac{3.88}{4} + \frac{5.84}{4} + \frac{3.36}{2} + \frac{4.27}{4} - 3\times\frac{7.35}{8} \qquad (5.6) \\ &= 2.42.\end{aligned}$$

For a smaller-the-better characteristic, the mean squared deviation, the loss function and the SN ratio may be written as

$$\begin{aligned} MSD_i &= \frac{1}{n}\sum_{j=1}^{n} y_{ij}^2, \\ L_i &= kMSD_i, \qquad (5.7) \\ SN_i &= -10\log[MSD_i]. \end{aligned}$$

Let the $MSDs$ of the current condition and the optimum condition be MSD_c and MSD_o, respectively. Since the increase in SN from the current condition to the optimum condition is $2.42 - 0.54 = 1.88$,

$$1.88 = -10\log MSD_o + 10\log MSD_c$$
$$= 10\log\left(\frac{MSD_c}{MSD_o}\right).$$

Hence,

$$\frac{MSD_c}{MSD_o} = 10^{0.188} = 1.54,$$

which means that, as compared with the current condition, the optimum condition produces the products with 35% less *MSD*. Also from equation (5.7), the loss decreases by 35%.

Suppose the GFCI is judged defective when the moving time is above 20ms (i.e. Δ_0 = 20), and the loss is A_0^* = \$10. Then the average loss per item for the current products becomes

$$L = \frac{A_0^*}{\Delta_0^2} MSD_c$$
$$= \frac{10}{(20)^2}\left[\frac{1}{6}(8^2 + 11^2 + 9^2 + 13^2 + 10^2 + 12^2)\right]$$
$$= 0.025 \times 175.67 = \$4.39.$$

However, at the optimum condition the loss is decreased by a factor of 1.54, and the loss becomes

$$L = \frac{4.39}{1.54} = \$2.85,$$

which means that for each item the effect of quality improvement is 4.39 − 2.85 = \$1.54. If the company produces 100,000 items of GFCI, the total quality improvement effect amounts to \$154,000 per year.

5.3.3 CLASSICAL ANOVA AND F-TESTS

We have used Pareto ANOVA instead of the classical ANOVA and related F-tests. In order to compare the results of Pareto ANOVA with those of the classical ANOVA, the results of the classical ANOVA are given below. The computation of sums of squares is as follows.

$$CT = \frac{(7.35)^2}{8} = 6.75,$$

$$S_T = 0.732^2 + 0.902 + \ldots + 2.46^2 - 6.75 = 4.48,$$

$$S_A = \frac{0.1681}{8} = 0.02, \qquad S_B = \frac{18.9225}{8} = 2.37,$$

$$S_C = \frac{9.6721}{8} = 1.21, \qquad S_D = \frac{2.2801}{8} = 0.28,$$

$$S_F = \frac{1.4161}{8} = 0.18, \qquad S_{C\times D} = \frac{2.1609}{8} = 0.27,$$

$$S_e = \frac{1.2312}{8} = 0.15.$$

Note that in Table 6.4 factors A and F are pooled into e to give the new error term (e). If the value of F_0 for a factor is greater than 2, it is often recommended that the factor should not be neglected. If this is the case in this situation, we can consider B, C, D and $C \times D$ to be significant, and this result is exactly the same as that of Table 5.3.

5.3.4 CONFIRMATION TEST

We obtained the estimated SN ratio in equation (5.6). If we subtract 23 (note that we added 23 to the SN ratio in Table 5.2) from this value, it becomes –20.58dB. A 95% confidence interval for the average SN ratio at the optimum condition is

$$-20.58 \pm t_{0.975}(3) = -20.58 \pm (3.182)\sqrt{\frac{V_e}{n_e}} \tag{5.8}$$

$$= -20.58 \pm 1.02,$$

Table 5.4 ANOVA table for moving time.

Source of variation	Sum of squares	Degree of freedom	Mean square	F_0
A	0.02	1		
B	2.37	1	2.37	20.3
C	1.21	1	1.21	10.3
D	0.28	1	0.28	2.4
F	0.18	1		
$C\times D$	0.27	1	0.27	2.3
e	0.15	1		
(e)	(0.35)	(3)	(0.117)	
T	4.48	7		

since the effective number of replication n_e is obtained from

$$\frac{1}{n_e}=\frac{1}{4}+\frac{1}{4}+\frac{1}{2}+\frac{1}{4}-\frac{3}{8}=\frac{7}{8}.$$

A confirmation test is conducted to check that the obtained optimum condition really produces the desired moving-time responses. The results are

12, 9, 8, 11, 10, 10, 9, 10, 11, 8.

The SN ratio for these observations is

$$\begin{aligned}\text{SN}&=-10\log\left[\frac{1}{10}(12^2+9^2+\ldots+8^2)\right]\\&=-19.89\text{dB}.\end{aligned}$$

Note that this value is contained within the 95% confidence interval obtained in (5.8), which means that the optimum condition is confirmed by a confirmation test.

5.3.5 PARETO ANOVA FOR THREE-LEVEL FACTORS

Table 5.3 showed the Pareto ANOVA for two-level factors. When we use tables of OAs such as L_9 and L_{27}, we need to analyze the sum of squares of three-level factors. For a three-level factor, say A, the sum of squares of SN ratios may be computed by the following formula:

$$\text{Sum of squares of } A = \sum_{1}^{3} \frac{3}{N}\left\{A_0^2 + A_1^2 + A_2^2\right\} - \frac{T^2}{N}$$
$$= \frac{1}{N}\left\{(A_0 - A_1)^2 + (A_0 - A_2)^2 + (A_1 - A_2)^2\right\},$$

where N is the total number of experiment, and A_i is the sum of SN ratios at the ith level of factor A. Hence, we can construct the computation scheme for Pareto analysis of variation for three-level factors shown in Table 5.5. Table 5.5 shows the simple case of L_9 for illustration, where four factors of three levels are allocated.

When the OA L_{27} is used and the interaction $A \times B$, for instance, is allocated to the appropriate two columns of the array, the squares of differences are computed for each of the two columns, and the sum of the two squares of differences is the sum for $A \times B$. If $A \times B$ is significant, we should check on $A \times B$ by means of a two-way table to obtain the optimum conditions of A and B.

5.3.6 PARETO ANOVA FOR MIXED-LEVEL FACTORS

When we use the tables of OAs for mixed levels such as L_{18} $(2^1 \times 3^7)$ and $L_{36}(2^{11} \times 3^{12})$, we need to analyze the sum of squares for factors of two and three levels. In such cases we should use the degrees of freedom of the sum of squares considered.

Suppose we use L_{18}, in which factor A has two levels and factors B, C and D have three levels. Then we can construct a table such as Table 5.6 for the computation scheme of Pareto ANOVA for two- and three-level factors.

Table 5.5 Pareto ANOVA for three-level factors.

Factors and interaction	A	B	C	D	Total
Sum at 0	A_0	B_0	C_0	D_0	
Factor 1	A_1	B_1	C_1	D_1	$T = A_0 + A_1 + A_2$
Level 2	A_2	B_2	C_2	D_2	
Sum of squares of differences (S)	$(A_0-A_1)^2+(A_0-A_2)^2 +(A_1-A_2)^2=S_A$	$(B_0-B_1)^2+(B_0-B_2)^2 +(B_1-B_2)^2=S_B$	$(C_0-C_1)^2+(C_0-C_2)^2 +(C_1-C_2)^2=S_C$	$(D_0-D_1)^2+(D_0-D_2)^2 +(D_1-D_2)^2=S_D$	$S_T = S_A + S_B + S_C + S_D$
Contribution ratio (%)	$(S_A/S_T)\times 100$	$(S_B/S_T)\times 100$	$(S_C/S_T)\times 100$	$(S_D/S_T)\times 100$	100%
Pareto diagram	(Continue as in Table 6.3.)				

Table 5.6 Pareto ANOVA for mixed level factors.

Factors	A	B	C	D	Total
Sum at 0	A_0	B_0	C_0	D_0	
Factor 1	A_1	B_1	C_1	D_1	$T = A_0 + A_1$
Level 2		B_2	C_2	D_2	
Sum of squares of differences (S)	$(A_0 - A_1)^2 = S_A$	$(B_0 - B_1)^2 + (B_0 - B_2)^2 + (B_1 - B_2)^2 = S_B$	$(C_0 - C_1)^2 + (C_0 - C_2)^2 + (C_1 - C_2)^2 = S_C$	$(D_0 - D_1)^2 + (D_0 - D_2)^2 + (D_1 - D_2)^2 = S_D$	
Degree of freedom (ϕ)	$\phi_A = 1$	$\phi_B = 2$	$\phi_C = 2$	$\phi_D = 2$	
S/ϕ	$S'_A = S_A / \phi_A$	$S'_B = S_B / \phi_B$	$S'_C = S_C / \phi_C$	$S'_D = S_D / \phi_D$	$S_T = S'_A + S'_B + S'_C + S'_D$
Contribution ratio (%)	$(S'_A / S_T) \times 100$	$(S'_B / S_T) \times 100$	$(S'_C / S_T) \times 100$	$(S'_D / S_T) \times 100$	100%
Pareto diagram	(Continue as in Table 6.3.)				

If there is a factor, say A, of four levels, the sum of squares of A, which has three degrees of freedom, may be computed as follows:

$$\text{Sum of squares of } A = \frac{4}{N}\sum_{1}^{3} A_i^2 - \frac{T^2}{N}$$

$$= \frac{1}{N}\{(A_0 - A_1)^2 + (A_0 - A_2)^2 + (A_0 - A_3)^2 + (A_1 - A_2)^2 + (A_1 - A_3)^2 + (A_2 - A_3)^2\}.$$

Hence, we can construct an analysis for mixed-level factors including four-level factors without much difficulty.

5.4 EXPERIMENTS INVOLVING LARGER-THE-BETTER CHARACTERISTICS

In this section an experiment concerned with a larger-the-better characteristic is presented. This is an experiment which was conducted in a chemical company. Suppose the adhesive strength of a plastic item is the larger-the better characteristic of interest, and the item is considered to be defective if the adhesive strength is less than 10kgcm^{-2}, i.e. the customer tolerance limit is $\triangle = 10\text{kgcm}^{-2}$. The loss at this limit is $A_0^* = \$94$. The control factors which might affect the adhesive strength during the manufacturing process are as follows, three levels for each factor being recommended for experiments.

Control factors	Levels		
	0	1	2
A: etching time	8min	10min	12min
B: etching temperature	60°C	70°C	80°C
C: mixing ratio	C_0	C_1	C_2
D: pre-treatment method	none	solvent	hot water

Because there is a limit on the number of experiments that can be conducted, L_9 is used. For noise factors, two conditions for user environments are taken as follows.

N_0 : Normal condition of use

N_1 : Bad condition of use.

Two plastic items are made and the adhesive strength for each item is tested after a certain time passes under N_0 and N_1. The design and data are given in Table 5.7. Since adhesive strength is a larger-the-better characteristic, the way to compute the SN ratio is

$$\mathrm{SN}_i = -10\log\left[\frac{1}{N}\sum_j \frac{1}{y_{ij}^2}\right].$$

For example, for the first experiment number, the SN ratio is

$$\mathrm{SN}_1 = -10\log\left[\frac{1}{2}\left(\frac{1}{6^2}+\frac{1}{5^2}\right)\right] = 14.7\mathrm{dB}.$$

Table 5.7 Design and data for adhesive strength experiment.

Experiment number	Control factor assignment and column number				Data (adhesive strength kgcm^{-2})		SN
	A	B	C	D	Noise factor		
	1	2	3	4	N_0	N_1	
1	0	0	0	0	6	5	14.7
2	0	1	1	1	10	8	18.9
3	0	2	2	2	12	10	20.7
4	1	0	1	2	8	6	16.6
5	1	1	2	0	11	7	18.4
6	1	2	0	1	13	9	20.4
7	2	0	2	1	23	18	26.0
8	2	1	0	2	24	23	27.4
9	2	2	1	0	20	16	24.9

Table 5.8 Pareto ANOVA for three-level factors.

Factor and interaction		A	B	C	D	Total
Sum at	0	54.3	57.3	62.5	58.0	
Factor	1	55.4	64.7	60.4	65.3	$T = 188.0$
Level	2	78.3	66.0	65.1	64.7	
Square of differences (S)		$(-1.1)^2 + (-24)^2 + (-22.9)^2 = 1101.62$	$(-7.4)^2 + (-8.7)^2 + (-1.3)^2 = 132.14$	$(2.1)^2 + (-2.6)^2 + (-4.7)^2 = 33.26$	$(-7.3)^2 + (-6.7)^2 + (0.6)^2 = 98.54$	$S_T = 1365.92$
Contribution ration (%)		80.7	9.7	2.4	7.2	100.0

Pareto diagram

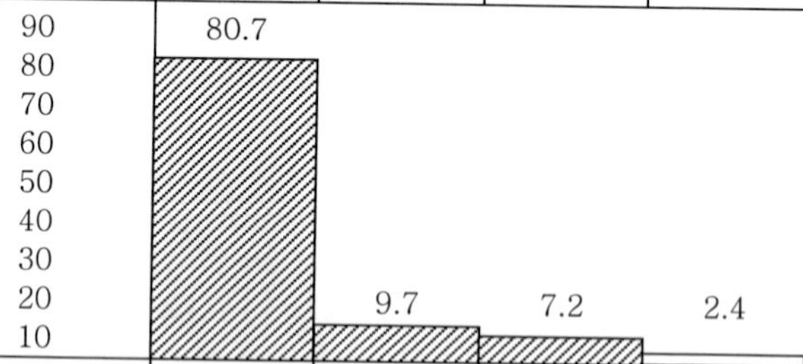

Factor and interaction	A	B	D	C
Cumulative contribution ratio (%)	80.7	90.4	97.6	100.0
Optimum combination of significant factor levels	$A_2B_2D_1$			
Remarks on optimum condition	The significant factors are chosen from the left-hand side in the above Pareto diagram which cumulatively contribute about 90%. For other non-significant factors, the optimal levels can be chosen by other economical and technical considerations.			
Overall optimum conditions for all factors	A_2: etching time, 12 minutes B_2: etching temperature, 80°C C_2: mixing ratio, C_2 D_1: pre-treatment method, solvent			
Estimate of error variance	$Ve = Sc/9 = 33.26/9 = 3.70$			

5.4.1 PARETO ANOVA

It is also useful to use the computation scheme for three-level factors for Pareto ANOVA. Table 5.8 shows the Pareto ANOVA for this experiment. The significant factors are A, B and D. Since the half-width of a 95% confidence interval at each level of each factor is

$$t_{0.975}(2)\sqrt{\frac{V_e}{n_e}} = 4.303\sqrt{\frac{1.85}{3}} = 3.38.$$

We can find 95% confidence intervals for SN ratio at each level of each significant factor as follows:

$A_0: 18.10 \pm 3.38$ $\quad B_0: 19.10 \pm 3.38$ $\quad D_0: 19.33 \pm 3.38$

$A_1: 18.47 \pm 3.38$ $\quad B_1: 21.57 \pm 3.38$ $\quad D_1: 21.77 \pm 3.38$

$A_2: 26.10 \pm 3.38$ $\quad B_2: 22.00 \pm 3.38$ $\quad D_2: 21.57 \pm 3.38$

If we sketch the intervals, Figure 5.4 is obtained.

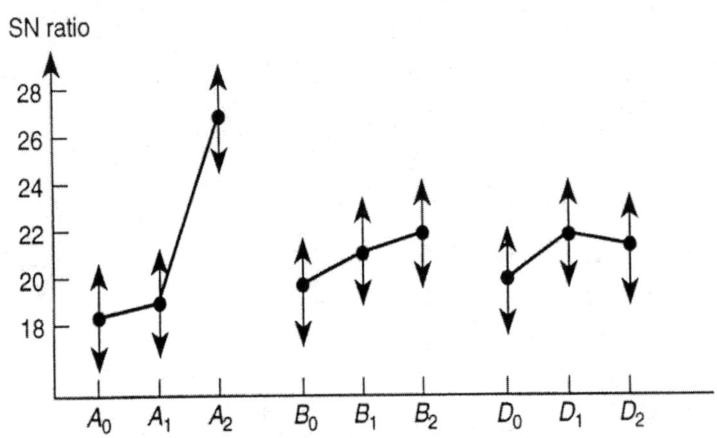

Figure 5.4 Ninety-five per cent confidence intervals at each level of each significant factor in the adhesive strength experiment.

5.4.2 ANALYSIS OF RESULTS

At the optimum condition $A_2B_2D_1$, the estimated SN ratio is

$$\hat{\mu}(A_2B_2D_1) = \overline{A}_2 + \overline{B}_2 + \overline{D}_1 - 2\overline{T}$$
$$= 26.10 + 22.00 + 21.77 - 2\times 188/9 \qquad (5.9)$$
$$= 28.09.$$

For a larger-the-better characteristic, the mean squared deviation, the loss and the SN ratio may be written as follows.

$$MSD = \frac{1}{n}\left(\frac{1}{y_1^2} + \frac{1}{y_2^2} + \ldots + \frac{1}{y_n^2}\right),$$
$$L = A_0^* \Delta_0^2 MSD,$$
$$\mathrm{SN} = -10\log[MSD].$$

Hence MSD at the optimum condition would be

$$MSD = 10^{-2.809} = 0.00155,$$

And since $\Delta_0 = 10$, and $A_0^* = 94$, the loss is,

$$L = 94\times 10^2 \times 0.00155 = \$14.57.$$

The prevailing condition before this experiment was $A_1B_1D_1$ for the factors A, B and D. At this condition the estimated SN ratio is

$$\hat{\mu}(A_1B_1D_1) = \overline{A}_1 + \overline{B}_1 + \overline{D}_1 - 2\overline{T}$$
$$= 18.47 + 21.57 + 21.77 - 2\times 20.89 = 20.03. \qquad (5.10)$$

And the MSD and the loss are

$$MSD = 10^{-2.003} = 0.00993,$$
$$L = 94\times 10^2 \times 0.00993 = \$93.34.$$

Therefore, the difference in loss per item is

93.34 - 14.57 = $78.77.

If the company produces 1 million items per year, the total saving from this experiment could be $78.77 million.

Note that when we estimate the SN ratio at the optimum condition, we may include the non-significant factor C in equations (5.9) and (5.10). However, since C is not significant, there is not much difference between C_0, C_1 and C_2. Hence, it is really not important whether we include the non-significant factor in the estimation of the SN ratio at the optimum condition. Note that this experiment requires a confirmation test, which is omitted here.

5.5 EXPERIMENTS INVOLVING NOMINAL-IS-BEST CHARACTERISTICS

A parameter design experiment involving a nominal-is-best characteristic is presented in this section. The R&D department of an automobile company is conducting an experiment to develop the ignition cable of an automobile, where the quality characteristic of interest is the core pulling force. The factor-characteristic relation diagram for this cable experiment is shown in Figure 5.5.

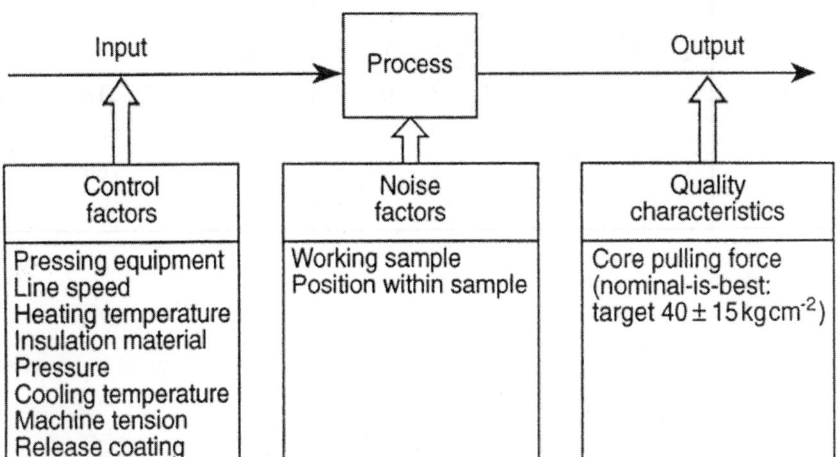

Figure 5.5 Factor-characteristic relation diagram for cable experiment.

Table 5.9 Core pulling force experiment and data on L_{18}.

Experiment number	Control factor assignment and column number								Data (core pulling force ($kgcm^{-2}$)) Noise factor assignment			
	A	B	C	D	F	G	H	I	S_0		S_1	
	1	2	3	4	5	6	7	8	P_0	P_1	P_0	P_1
1	0	0	0	0	0	0	0	0	30	40	38	49
2	0	0	1	1	1	1	1	1	10	15	25	25
3	0	0	2	2	2	2	2	2	49	53	53	55
4	0	1	0	0	1	1	2	2	62	58	52	68
5	0	1	1	1	2	2	0	0	30	50	49	62
6	0	1	2	2	0	0	1	1	10	25	29	36
7	0	2	0	1	0	2	1	2	58	42	41	50
8	0	2	1	2	1	0	2	0	28	29	32	31
9	0	2	2	0	2	1	0	1	110	74	94	115
10	1	0	0	2	2	1	1	0	76	86	66	103
11	1	0	1	0	0	2	2	1	52	37	54	59
12	1	0	2	1	1	0	0	2	55	79	62	98
13	1	1	0	1	2	0	2	1	5	35	16	42
14	1	1	1	2	0	1	0	2	52	96	79	91
15	1	1	2	0	1	2	1	0	50	70	56	65
16	1	2	0	2	1	2	0	1	15	20	18	21
17	1	2	1	0	2	0	1	2	51	62	59	70
18	1	2	2	1	0	1	2	0	77	83	66	74
Total									820	956	889	1114

Table 5.9 (Continued).

Experiment number	Total(T_i)	Mean($\bar{y}_i$)	Variance(V_i)	SN_i = $10\log[\bar{y}_i^2 / V_i]$	Sensitivity $S_{n(i)} = 10\log[T_i^2/4]$
1	157	39.25	60.92	14.0	37.90
2	75	18.75	56.25	7.8	31.49
3	210	52.50	6.33	26.4	40.41
4	240	60.00	45.33	19.0	41.58
5	191	47.75	174.92	11.0	39.60
6	100	25.00	120.67	6.9	33.98
7	191	47.75	62.92	15.6	39.60
8	120	30.00	3.33	24.3	35.56
9	393	98.25	341.58	14.5	45.87
10	333	83.25	254.25	14.3	44.42
11	202	50.50	89.67	14.5	40.09
12	294	73.50	368.33	11.6	43.34
13	98	24.50	289.67	2.6	33.80
14	318	79.50	387.0	12.1	44.03
15	241	60.25	80.25	16.5	41.61
16	74	18.50	7.00	16.9	31.37
17	242	60.50	61.67	17.7	41.64
18	300	75.00	50.00	20.5	43.52
Total	3779	944.75		266.2	709.81

For the eight control factors, the department decided to use two levels for the pressing equipment, and three levels for all the other factors. For the noise factors, a 2^2 factorial design is used. For this experiment L_{18} is used and the factors are assigned as in Table 5.9. The control factors are:

A : pressing equipment (A_0 = Type I, A_1 = Type II)
B : line speed (B_0 = low, B_1 = medium, B_2 = high)
C : heating temperature (C_0 = low, C_1 = medium, C_2 = high)
D : insulation material (D_0 = Type, D_1 = Type II, D_2 = Type III)
F : pressure (F_0 = low, F_1 = medium, F_2 = high)
G : cooling temperature (G_0 = low, G_1 = medium, G_2 = high)
H : machine tension (H_0 = low, H_1 = medium, H_2 = high)
I : release coating method (I_0 = method I, I_1 = II, I_2 = III)

The noise factors are

S : working sample (S_0 = sample 1, S_1 = sample 2)
P : position within sample (P_0 = position 1, P_1 = position 2).

5.5.1 ANALYSIS OF RATIOS

For the core pulling force data, we need to analyze SN ratios as well as sensitivities. First of all, the analysis of SN ratios is explained. The Pareto ANOVA of this experiment is shown in Table 5.10.

5.5.2 ANALYISIS OF SENSITIVITIES

For the sensitivities S_n in Table 5.9, if we also construct a Pareto ANOVA, we can obtain Table 5.11, which shows that the significant factors are A, C, D, F, G and I. These are the mean adjustment factors. From Table 5.10, since the dispersion control factors are B, D, G, H and I, we can classify the control factors as follows.

Dispersion control factors: B, D, G, H, I
Mean adjustment factors: A, C, F
Other control factors: none.

Note that D, G and I are both dispersion control factors and mean adjustment factors. This being case, they are classified as dispersion control factors.

We found in Table 5.10 that the optimal condition for the dispersion control factors was $B_2 D_2 G_2 H_2 I_2$. The optimal condition for the mean adjustment factors can be found by estimating the core pulling force from the equation

$$\hat{\mu}(A_i C_j F_k) = \overline{A}_i + \overline{C}_j + \overline{F}_k - 2\overline{T}.$$

Table 5.10 Pareto ANOVA of SN for the core pulling force experiment.

Factors		A	B	C	D	F	G	H	I	Total
Sum at factor level	0	139.5	88.6	82.4	96.2	83.6	77.1	80.1	100.6	
	1	126.7	68.1	87.4	69.1	96.1	88.2	78.8	63.2	$T = 266.2$
	2		109.5	96.4	100.9	86.5	100.9	107.3	102.4	
Squares of diff. (S)		163.8	2571.0	302.0	1767.7	256.8	850.9	1553.8	2938.6	10404.6
Degrees of freedom (ϕ)		1	2	2	2	2	2	2	2	
S/ϕ		163.8	1285.5	151.0	883.9	128.4	425.5	776.9	1469.3	5284.3
Contribution ratio (%)		3.1	24.3	2.9	16.7	2.4	8.1	14.7	27.8	100.0
Pareto diagram	40 30 20 10	27.8 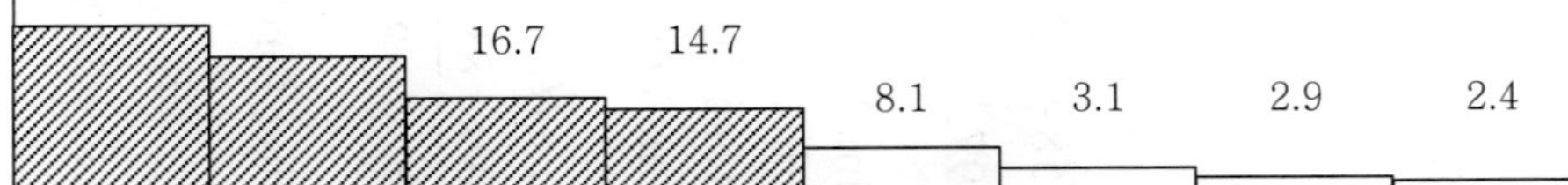	24.3	16.7	14.7	8.1	3.1	2.9	2.4	
Factor		I	B	D	H	G	A	C	F	
Cumulative contribution ratio (%)		27.8	52.1	68.8	83.5	91.6	94.7	97.6	100.0	
Optimum combination of significant factor levels		$B_2D_2G_2H_2I_2$ (The optimal level of each significant factor is the level which maximizes the sum of SN ratios.)								
Remarks on optimum conditions		The significant factors are chosen from the left-hand side in the above Pareto diagram which cumulatively contribute about 90%.								

Level	$\overline{A}$	$\overline{B}$	$\overline{C}$	$\overline{D}$	$\overline{F}$	$\overline{G}$	$\overline{H}$	$\overline{I}$
0	46.58	52.96	45.54	61.64	52.83	42.13	59.46	55.92
1	58.39	49.50	47.83	47.88	43.50	69.13	49.25	39.25
2		55.50	64.08	48.13	61.13	46.21	48.75	62.29
Average	52.49	52.49	52.49	52.49	52.49	52.49	52.49	52.49

When the levels of $A_iC_jF_k$ produce an estimated value close to 40, the target value, would be optimal. It turns out that at $A_0C_0F_0$

$$\begin{aligned}\hat{\mu}(A_0C_0F_0) &= \overline{A}_0 + \overline{C}_0 + \overline{F}_0 - 2\overline{T} \\ &= 46.58 + 45.54 + 52.83 + -2 \times 52.49 \\ &= 39.97.\end{aligned}$$

Therefore, $A_0C_0F_0$ is optimal.

If we combine the optimal levels for the dispersion control and mean adjustment factors, the overall optimum condition is $A_0B_2C_0D_2F_0G_2H_2I_2$. At this condition the estimated SN ratio is

$$\begin{aligned}\hat{\mu} &= \overline{A}_0 + \overline{B}_2 + \overline{C}_0 + \overline{D}_2 + \overline{F}_0 + \overline{G}_2 + \overline{H}_2 + \overline{I}_2 - 7\overline{T} \\ &= \frac{139.5}{9} + \frac{109.5}{6} + \frac{82.4}{6} + \frac{100.9}{6} + \frac{83.6}{6} + \frac{100.9}{6} + \frac{109.3}{6} + \frac{102.4}{6} - 7 \times \frac{266.2}{18} \\ &= 26.97\text{dB},\end{aligned}$$

and the estimated core pulling force

$$\begin{aligned}\hat{\mu} &= \overline{A}_0 + \overline{B}_2 + \overline{C}_0 + \overline{D}_2 + \overline{F}_0 + \overline{G}_2 + \overline{H}_2 + \overline{I}_2 - 7\overline{T} \\ &= 46.58 + 55.50 + 45.54 + 48.13 + 52.83 + 46.21 + 48.75 + 62.29 - 7 \times 52.49 \\ &= 38.40\text{kgcm}^{-2}\end{aligned}$$

Note that the estimated core pulling force is very close to 40 and is well within the limit 40 ± 15. Also the estimated SN ratio is very high, which means that the core pulling force at the optimum condition is robust to the noise factors, and has a small variation. The conclusion is that the experiment was quite successful.

Table 5.11 Pareto ANOVA of sensitivity for the core pulling force experiment.

Factors		A	B	C	D	F	G	H	I	Total
Sum at factor level	0	345.99	237.65	228.67	248.69	239.12	226.22	242.11	242.61	
	1	363.82	234.60	232.41	231.35	224.95	250.91	232.71	216.60	$T = 709.81$
	2		237.56	248.73	229.77	245.74	232.68	234.96	250.60	
Squares of diff. (S)		317.91	18.07	682.73	661.14	676.84	983.66	143.85	1896.36	5452.83
Degrees of freedom (ϕ)		1	2	2	2	2	2	2	2	
S/ϕ		317.91	9.04	341.37	330.57	338.42	491.83	71.93	948.18	2849.25
Contribution ratio (%)		11.2	0.3	12.0	11.6	11.9	17.3	2.5	33.2	100.0

Pareto diagram

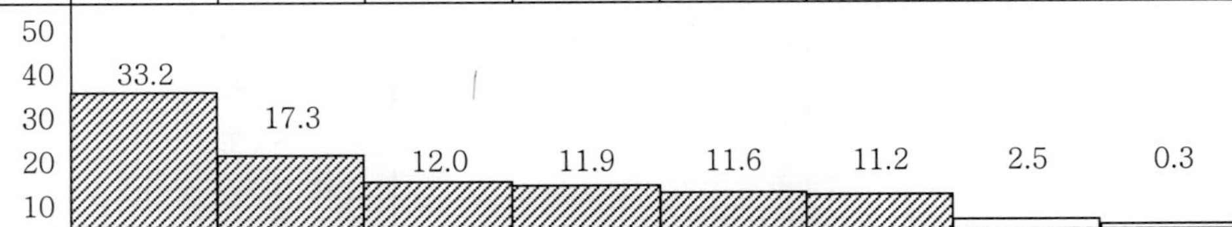

Factor	I	G	C	F	D	A	H	B
Cumulative contribution ratio (%)	33.2	50.5	62.5	74.4	86.0	97.2	99.7	100.0

Exercises

5.1 In order to reduce the amount of CO in the exhaust gas of automobiles, seven controllable factors $(A-H)$ are chosen and assigned to L_8 as follows. As for noise factors, three road conditions (R_1, R_2, R_3) are selected. At each road condition, and each controllable factor condition, a road test is performed and the amount of CO is measured in grams.

Experiment number	Control factor assignment and column number							Data (CO in exhaust gas, g)		
	A	*B*	*C*	*D*	*F*	*G*	*H*			
	1	2	3	4	5	6	7	R_1	R_2	R_3
1	0	0	0	0	0	0	0	1.04	1.20	1.54
2	0	0	0	1	1	1	1	1.42	1.76	2.10
3	0	1	1	0	0	1	1	1.01	1.23	1.52
4	0	1	1	1	1	0	0	1.50	1.87	2.25
5	1	0	1	0	1	0	1	1.28	1.34	2.05
6	1	0	1	1	0	1	0	1.14	1.26	1.88
7	1	1	0	0	1	1	0	1.33	1.42	2.10
8	1	1	0	1	0	0	1	1.33	1.52	2.13

(a) Find SN ratios for the amount of CO, which is a smaller-the-better characteristic.

(b) For the SN ratios data, perform Pareto ANOVA. Find significant factors and their optimal conditions.

(c) Perform an analysis of variance for the SN ratios. According to the F-tests, what are the significant factors? Compare your result with those you obtained in (b).

(d) For each of the non-significant factors, suppose that level 0 is optimal for economical and technical reasons. Find the overall optimum condition for all the control factors. Estimate the SN ratio at this condition.

(e) At the optimum condition, use the mean squared deviation to find the magnitude of CO dispersion by *MSD*. Suppose the upper specification is 5g, and the loss is \$15 when the amount of CO is 5g; what amount is lost at the optimum condition. The loss function is determined by

$$L = [15/5^2] \times MSD .$$

5.2 The lower specification of the tensile strength for an automobile part is 50 kgcm^{-2}. There are four control factors in the manufacturing process which influence the tensile strength, and their levels are as follows.

Control factors	0	1	2
A: heating temperature (°C)	500	600	700
B: material	I	II	III
C: heating time (min)	15	17	19
D: cooling time (min)	30	40	50

L_8 is used to assign the control factors, and at each experimental condition, 10 parts are made and the tensile strength is measured in kgcm^{-2}.

Experiment number	Control factor assignment and column number				Data	SN (dB)
	A	B	C	D		
	1	2	3	4		
1	0	0	0	0	46,50,...,42	33.3
2	0	1	1	1	52,51,...,54	34.4
3	0	2	2	2	52,52,...,56	34.9
4	1	0	1	2	52,59,...,52	34.7
5	1	1	2	0	50,54,...,55	34.2
6	1	2	0	1	60,55,...,50	34.7
7	2	0	2	1	50,56,...,50	34.6
8	2	1	0	2	60,58,...,63	35.2
9	2	2	1	0	56,56,...,49	34.8

The SN ratio (dB) for the first row is obtained by the following formula, since the data are for a larger-the-better characteristic.

$$SN = -10\log[MSD], MSD = \frac{1}{10}\left[\frac{1}{46^2} + \frac{1}{50^2} + ... + \frac{1}{42^2}\right].$$

(a) For the SN ratios, carry out Pareto ANOVA.

(b) Also carry out ANOVA and related F-tests. What are significant factors? Find the optimum condition for the significant factors. For the non-significant factors, select level 1 for the optimum condition. What is the overall optimum condition for all factors?

(c) Estimate the SN ratio at the overall optimum condition. Suppose the current condition is $A_2B_2C_2D_2$. What is the estimate of the SN ratio at the current condition? Compare them.

(d) The tolerance limit is 50 $kgcm^{-2}$, and the loss at this limit is $2. What are the losses at the current and optimum conditions? What is the difference? If the factory produces 100,000 parts per year, how much is gained by the change of operating condition? Note that the loss function is

$$L = A_0^* \Delta_0^2 MSD.$$

5.3 In the punching process for a product, the bias in the punching is a headache for the engineers. To tackle the bias, the engineers considered eight control factors and two noise factors, and assigned them in L_{18} as follows.

Control factors:	A(cutting equipment I)	2 levels
	B(cutting equipment II)	3 levels
	C(cutting equipment III)	3 levels
	D(transport speed)	3 levels
	F(RPM)	3 levels
	G(amount of cutting I)	3 levels
	H(amount of cutting II)	3 levels
	I(amount of oil flow for cutting)	3 levels
Noise factors:	U(type of washer)	2 levels
	V(machine number)	2 levels

The target punching distance is 10mm. Hence, the quality of interest is a nominal-is-best characteristic. The experimental data are as follows:

(a) Calculate T_i, $\bar{y}_i$, V_i, $\text{SN} = 10\log[y_i^2/V_i]$, and sensitivity $S_n = 10\log[T_i^2/4]$ for each experiment.

(b) Carry out both Pareto ANOVA and ANOVA for the SN ratios.

Experiment number	Control factor assignment and column number								Data			
									U_0		U_1	
	A	B	C	D	F	G	H	I				
	1	2	3	4	5	6	7	8	V_0	V_1	V_0	V_1
1	0	0	0	0	0	0	0	0	8	7	16	15
2	0	0	1	1	1	1	1	1	10	11	16	10
3	0	0	2	2	2	2	2	2	11	6	11	13
4	0	1	0	0	1	1	2	2	3	12	7	11
5	0	1	1	1	2	2	0	0	5	2	13	6
6	0	1	2	2	0	0	1	1	8	15	10	15
7	0	2	0	1	0	2	1	2	8	3	3	2
8	0	2	1	2	1	0	2	0	9	15	14	8
9	0	2	2	0	2	1	0	1	8	3	6	7
10	1	0	0	2	2	1	1	0	12	15	14	15
11	1	0	1	0	0	2	2	1	14	20	15	8
12	1	0	2	1	1	0	0	2	16	12	16	15
13	1	1	0	1	2	0	2	1	16	14	17	19
14	1	1	1	2	0	1	0	2	12	10	17	10
15	1	1	2	0	1	2	1	0	15	7	4	12
16	1	2	0	2	1	2	0	1	7	6	8	6
17	1	2	1	0	2	0	1	2	6	13	14	14
18	1	2	2	1	0	1	2	0	17	16	14	11

(c) Carry out both Pareto ANOVA and ANOVA for the sensitivities.

(d) Classify the control factors into dispersion control, mean adjustment and insignificant factors.

(e) If there are any mean adjustment factors, what is the optimum condition for these factors? How did you determine them?

(f) Find the optimum condition for the dispersion and mean adjustment factors. At the optimum condition, estimate the SN ratio and the punching distance.

(g) The current operating condition is $A_1B_1C_1D_1F_1G_1H_1I_1$. Estimate the punching distance at this condition. How great is the improvement in the punching distance?

(h) Suppose the specification is 10 ± 3, and the loss at the limit is $3. How much is the improvement in dollars if we change the current condition to the optimum condition?

[Hint: use the relationship $\text{SN} = 10\log(\hat{\mu}^2 / \hat{\sigma}^2)$. Substitute $\hat{\mu} = 10$ into the relationship and find $\hat{\sigma}^2$. Then use the loss function $L = (A_0^* / \Delta_0^2)\hat{\sigma}^2$].

6

Parameter Design for Discrete Data

In the previous chapters, all the experimental situations we were concerned with involved continuous data such as continuous flow rate, pressure, temperature and time. **Discrete** data, on the other hand, is in discontinuous form, which means experimental results can only take discrete values such as defective and non-defective or good and bad or 0 and 1. Discrete data is often called **attribute** data.

Two-class discrete data provides much less discrimination than continuous data. When a part is classified as good, no measure of how good is provided. Because of this reduced discrimination, many pieces of discrete data are required to provide the information equivalent to one piece of continuous data. Therefore, continuous data should be used whenever possible because it allows a greatly reduced sample size to be used. To improve the discriminatory power of discrete data, more classes may be used. As the number of classes increases, the data becomes semi-continuous in nature, which adds power of discrimination. For instance, if parts were inspected on appearance by a sensory test, then several classes may be used as follows.

Class number	Appearance
1	excellent
2	good
3	fair
4	poor

Note that this classification has more information than two-class discrete data restricted to good and poor only.

The design of experiments for discrete data is the same as that for continuous data. Any design method which can be used for continuous data can also be used for discrete data. The major differences in the experiments are the sample size required and the analysis of the results.

6.1 TWO-CLASS DISCRETE DATA: SN RATIO ANALYSIS

Suppose we have inspected n items, and the results are 0/1 data as follows:

$$y_i = \begin{cases} 0, & \text{if the item is defective} \\ 1, & \text{if the item is non-defective.} \end{cases}$$

A data set of this kind belongs to the category of fixed marginal discrete data in Table 3.1, where the number of categories (or classes) is two. Suppose the fraction non-defective is p'. Since y_i follows the Bernoulli distribution, the expected value and the variance of y_i are, respectively,

$$\mathrm{E}(y_i) = p',$$
$$\mathrm{var}(y_i) = p'(1-p').$$

If we obtain a set of data, $y_1, y_2, \ldots, y_n$, then the fraction non-defective becomes

$$p = \frac{y_1 + y_2 + \cdots + y_n}{n},$$

and the general mean variation is

$$S_m = \frac{T^2}{n} = \frac{(y_1 + y_2 + \cdots y_n)^2}{n} = \frac{1}{n}(np)^2 = np^2 .$$

In this situation, the total variation is defined as

$$S_T = \sum_{i=1}^{n} y_i^2 = \sum_{i=1}^{n} y_i = np,$$

and the error variation is

$$S_e = S_T - S_m = np - np^2 = np(1-p) .$$

In this case of 0/1 data, Taguchi recommended defining the SN ratio as in equation (2.20):

$$\begin{aligned} \mathrm{SN} &= 10\log\left(\frac{\bar{y}^2}{V}\right) = 10\log\left(\frac{\frac{1}{n}S_m}{V}\right) \\ &= 10\log\left(\frac{S_m}{nV}\right) = 10\log\left(\frac{S_m}{S_e}\right) \\ &= 10\log\left[\frac{np^2}{np(1-p)}\right] = 10\log\left[\frac{p^2}{p(1-p)}\right] \\ &= -10\log\left(\frac{1}{p} - 1\right). \end{aligned} \tag{6.1}$$

Note that as the fraction non-defective p increases, the SN value increases. Hence, the larger the SN value is, the better. In order to find the SN values for equation (6.1), Appendix E may be used.

The analysis of 0/1 data can be best explained by means of an example. An experiment was conducted to reduce the soldering defects in a semiconductor soldering process. If a single defect is found after soldering, the item is considered defective. Four control factors were considered as follows.

A : length of lead cutting	$A_0 = 1$ mm, $A_1 = 2$ mm, $A_2 = 3$ mm
B : preheating temperature	$B_0 = 100$°C, $B_1 = 110$°C, $B_2 = 120$°C

C: soldering temperature $C_0 = 245°C$, $C_1 = 250°C$, $C_2 = 255°C$
D: conveyor speed $D_0 = 4.5$ ft min^{-1}, $D_1 = 5.0$ ft min^{-1}
$D_2 = 5.5$ ft min^{-1}

The current operating condition is $A_1B_1C_1D_0$.

For this experiment the OA, L_9 was used, and at each experimental condition, 200 items soldered were inspected and the number of non-defective items counted. At the first experimental condition, 167 items were judged as non-defective, and the sample fraction non-defective is

$$p = \frac{167}{200} = 0.835.$$

Therefore, the SN ratio is

$$\begin{aligned} \text{SN} &= -10\log\left(\frac{1}{p} - 1\right) \\ &= -10\log\left(\frac{1}{0.835} - 1\right) = 7.042. \end{aligned}$$

The other SN ratios are obtained in the similar way.

The data analysis could be done by Pareto ANOVA, which was explained in Chapter 5. However, regular ANOVA is carried out here. The sums and means of SN ratios at each level of each factor are summarized below. The values in parentheses are the means of the SN ratios.

Level	Factor			
	A	B	C	D
0	28.160 (9.387)	23.116 (7.705)	24.297 (8.099)	27.032 (9.011)
1	24.563 (8.188)	39.119 (13.040)	28.053 (9.351)	31.438 (10.479)
2	28.756 (9.585)	19.244 (6.415)	29.129 (6.376)	23.009 (7.670)
Total	81.479 (9.053)	81.479 (9.053)	81.479 (9.053)	81.479 (9.053)

Table 6.1 Experimental data for soldering experiment.

Experiment number	Control factor assignment and column number				Data (number of non-defectives among 200 inspected items)	$\text{SN} = -10\log\left(\frac{1}{p}-1\right)$
	A 1	B 2	C 3	D 4		
1	0	0	0	0	167	7.042
2	0	1	1	1	194	15.097
3	0	2	2	2	160	6.021
4	1	0	1	2	158	5.754
5	1	1	2	0	190	12.788
6	1	2	0	1	160	6.021
7	2	0	2	1	183	10.320
8	2	1	0	2	186	11.234
9	2	2	1	0	168	7.202

The computation of variations for ANOVA can be done as follows, and the ANOVA table appears in Table 6.2.

$$CT = \frac{T^2}{n} = \frac{81.479^2}{9} = 737.65,$$

$$S_A = \frac{1}{3}(28.160^2 + 24.563^2 + 28.756^2) - 737.65 = 3.43,$$

$$S_B = \frac{1}{3}(23.116^2 + 39.119^2 + 19.244^2) - 737.65 = 74.01,$$

$$S_C = \frac{1}{3}(24.297^2 + 28.053^2 + 29.129^2) - 737.65 = 4.29,$$

$$S_D = \frac{1}{3}(27.032^2 + 31.438^2 + 23.009^2) - 737.65 = 11.85.$$

The results show that only factor B is significant. However, factor D cannot be ignored because of its high F value (normally if the value of F is greater than 2, then the factor may have some significance in the results of experiment by OAs). We can obtain the optimal levels of B and D as B_1D_1. For the non-significant factors, A and C, the optimal levels are chosen to be A_2C_1 for economical and technical reasons. Then the estimated SN ratio at the optimum condition is

$$\hat{\mu}(A_2B_1C_1D_1) = \overline{A}_2 + \overline{B}_1 + \overline{C}_1 + \overline{D}_1 - 3\overline{T}$$
$$= 9.585 + 13.040 + 9.351 + 10.479 - 3 \times 9.053 = 15.296.$$

Table 6.2 ANOVA table for soldering experiment.

Sources of variation	Sum of squares		Degrees of freedom	Mean square	F_0
A	3.43	O	2		
B	74.01		2	37.01	19.17*
C	4.29	O	2		
D	11.85		2	5.93	3.07
(e)	(7.73)		(4)	(1.93)	
T	93.58		8		

* indicates 5% significance.
O indicates pooled variations.

At the current operating condition, $A_1B_1C_1D_0$, the estimated SN ratio is

$$\hat{\mu}(A_1B_1C_1D_0) = \overline{A}_1 + \overline{B}_1 + \overline{C}_1 + \overline{D}_0 - 3\overline{T}$$
$$= 8.188 + 13.040 + 9.351 + 9.011 - 3 \times 9.053 = 12.431.$$

Hence, the increase in SN ratio is $15.296 - 12.431 = 2.865$, which is a measure of the amount of quality improvement. If this quality improvement is calculated in terms of fraction non-defective,

$$\text{SN of current condition} = -10 \log\left(\frac{1}{p} - 1\right) = 12.431 \rightarrow p = 0.946,$$

$$\text{SN of optimum condition} = -10 \log\left(\frac{1}{p} - 1\right) = 15.296 \rightarrow p = 0.971,$$

which means the fraction non-defective improves from 94.6% to 97.1%.

6.2 TWO-CLASS DISCRETE DATA: 0/1 DATA DIRECT ANALYSIS

In Section 6.1, SN ratios were obtained by equation (6.1) and analysis was performed using the SN ratios. Another approach to 0,1 data analysis is available, which may be called the 0/1 data direct analysis approach.

As before, the good parts (or non-defectives) and bad parts (or defectives) are assumed to be represented by 1s and 0s, respectively. In the soldering experiment in Table 6.1, the data for the first experiment number is really 167 ones and 33 zeros. If the 1s and 0s themselves are considered as the data, then, since the total number of defectives for the $N = 9$ experiments is $T = 1566$, the total sum of squares is

$$S_T = 1^2 + 1^2 + \cdots + 0^2 + 0^2 + \cdots + 0^2 - \frac{1566^2}{1800}$$
$$= 1566 - 1362.42 = 203.58.$$

Simplifying,

$$S_T = T - \frac{T^2}{N} = 1566 - \frac{1566^2}{1800} = 203.58.$$

The sum for the 0/1 data at each level of each factor is listed below.

Level	Factor			
	A	B	C	D
0	521	508	513	525
1	508	570	520	537
2	537	488	533	504
Total	1566	1566	1566	1566

The sums of squares for the factors are calculated as for continuous data:

$$S_A = \frac{1}{600}(521^2 + 508^2 + 537^2) - 1362.42 = 0.70.$$

Similar calculations give $S_B = 6.09$, $S_C = 0.34$ and $S_D = 0.93$.

The error sum of squares, S_e, comes from the repetitions in each experiment number. All 200 items were processed under the same condition of the control factors, yet not all are non-defectives or all defectives, because of some sources of errors. Mathematically, S_e can be calculated by finding the average of the 0/1 data for each experiment number, squaring the deviation of the 0/1 data from the average, and summing the squares of the deviations. However, an easier way to calculate S_e is to subtract the variation due to each accountable source from S_T:

$$\begin{aligned} S_e &= S_T - S_A - S_B - S_C - S_D \\ &= 203.58 - 0.70 - 6.09 - 0.34 - 0.93 \\ &= 195.52. \end{aligned}$$

The degrees of freedom associated with S_e are determined in a similar way; the accountable degrees of freedom are subtracted from those of the total variation.

A summary of the ANOVA results is shown in Table 6.3. From a practical point of view, non-significant sources of variation usually don't need to be pooled into error in a 0/1 classified data analysis. The degrees of freedom associated with e are usually very large, as in Table 6.3, and the error variance is changed very little when pooled. In Table 6.3 the pure variation (S') and the contribution ratio (ρ) are computed. The way to compute them was explained in Chapter 3. For example, S'_A and S'_e are obtained as follows.

$$\begin{aligned} S'_A &= S_A - \phi_A V_e = 0.70 - 2 \times 0.111 = 0.478, \\ S'_e &= S_e + (\phi_A + \phi_B + \phi_C + \phi_D) V_e \\ &= 195.52 + 8 \times 0.111 = 196.408, \end{aligned}$$

$$\rho_A = \frac{S'_A}{S_T} \times 100 = \frac{0.478}{203.58} \times 100 = 0.23\%,$$

$$\rho_e = \frac{S'_e}{S_T} \times 100 = \frac{196.408}{203.58} \times 100 = 96.48\%.$$

Note, also, in Table 6.3 that the percentage contributions of factors are very small, which is often the case in 0/1 data analysis. Usually the error variation is very large, and so is the percentage contribution of the error variation. Therefore, the percentage contributions of the factors are relatively small. This does not mean that the factors are not significant.

We can observe in Table 6.3 that factors *A*, *B* and *D* are significant, and C is not significant. This result is slightly different from that of Table 6.2 in Section 6.1, which can sometimes happen in statistical analysis since two different methods have been employed.

Table 6.3 ANOVA summary for soldering experiment.

Sources of variation	Sum of squares	Degrees of freedom	Mean square		Pure variation	Contribution ratio(%)
A	0.70	2	0.350	3.15*	0.478	0.23
B	6.09	2	3.045	127.43**	5.868	2.88
C	0.34	2	0.170	1.53	0.118	0.06
D	0.93	2	0.465	4.19*	0.708	0.35
E	195.52	1791	0.111		196.408	96.48
T	203.58	1799			203.58	100.00

**indicates 1% significance.

*indicates 5% significance.

The optimal levels of significant factors are determined from the sum at each level of each factor, that is, $A_2B_1D_1$. For the non-significant factor *C*, the optimal level is determined by economical or technical considerations. Suppose the optimal level for C is C_1 as before. Then the overall optimum condition for the control factors is, $A_2B_1C_1D_1$.

When the 0/1 data direct analysis approach is used, special care should be given to the estimation method for the fraction non-defective. Taguchi (1987) suggested using the logit transformation, which is called the **omega method**. This method is explained in the next section.

6.3 OMEGA METHOD FOR ESTIMATION FROM 0/1 DATA

Suppose we estimate the fraction non-defective at the optimum condition $A_2B_1C_1D_1$ under the assumption of additivity for fractions non-defective. Then

$$\begin{aligned}\hat{\mu}(A_2B_1C_1D_1) &= \overline{A}_2 + \overline{B}_1 + \overline{C}_1 + \overline{D}_1 - 3\overline{T} \\ &= \frac{537}{600} + \frac{570}{600} + \frac{520}{600} + \frac{537}{600} - 3 \times \frac{1566}{1800} \\ &= 0.895 + 0.950 + 0.867 + 0.895 + 3 \times 0.870 = 0.997 .\end{aligned} \tag{6.2}$$

This method does not have any guarantee that the estimated value should be between 0 and 1. To overcome this problem, Taguchi (1987) proposed the following method.

To compute $\hat{\mu}$ in equation (6.2) by means of the omega method, take the reciprocal minus 1 of each of the positive terms in equation (6.2), multiply together, and divide by the reciprocal minus 1 of the negative term to give the $\mathbf{\Omega}$ value which gives its name to the method:

$$\Omega = \frac{\left(\frac{1}{\overline{A}_2} - 1\right)\left(\frac{1}{\overline{B}_1} - 1\right)\left(\frac{1}{\overline{C}_1} - 1\right)\left(\frac{1}{\overline{D}_1} - 1\right)}{\left(\frac{1}{\overline{T}} - 1\right)^3} . \tag{6.3}$$

$\hat{\mu}$ is then derived by adding 1 to this value and taking the reciprocal:

$$\hat{\mu} = \frac{1}{1+\Omega}. \tag{6.4}$$

For the soldering experiment we analyzed in Sections 6.1 and 6.2, we obtain:

$$\Omega = \frac{\left(\frac{1}{0.895}-1\right)\left(\frac{1}{0.950}-1\right)\left(\frac{1}{0.867}-1\right)\left(\frac{1}{0.895}-1\right)}{\left(\frac{1}{0.890}-1\right)^3} = 0.033,$$

$$\hat{\mu} = \frac{1}{1+\Omega} = \frac{1}{1+0.033} = 0.968.$$

These calculations can also be performed by means of the following equation using decibel (dB) values:

$$\text{dB}(\hat{\mu}) = \text{dB}(\bar{A}_2) + \text{dB}(\bar{B}_1) + \text{dB}(\bar{C}_1) + \text{dB}(\bar{D}_1) - 3\text{dB}(\bar{T}). \tag{6.5}$$

The decibel values of $\bar{A}_2, \bar{B}_1, \bar{C}_1, \bar{D}_1$ and $\bar{T}$ can be obtained as follows from Appendix E.

$$\text{dB}(\bar{A}_2)(=89.5\%) = 9.306\text{dB},$$
$$\text{dB}(\bar{B}_1)(=95.0\%) = 12.788\text{dB},$$
$$\text{dB}(\bar{C}_1)(=86.7\%) = 8.142\text{dB},$$
$$\text{dB}(\bar{D}_1)(=89.5\%) = 9.306\text{dB},$$
$$\text{dB}(\bar{T})(=87.0\%) = 8.256\text{dB}.$$

From this, the decibel value of $\hat{\mu}$ is

$$9.306 + 12.788 + 8.142 + 9.306 - 3 \times 8.256 = 14.774\text{dB}.$$

From Appendix E, we obtain p (%) for 14.774 dB, which is

$$p = \hat{\mu}\left(A_2 B_1 C_1 D_1\right) = 96.8\% = 0.968.$$

This estimate of the fraction non-defective at the optimum condition is very close to 0.971, which was obtained in Section 6.2 by the SN ratio analysis approach.

Suppose we want to estimate the fraction non-defective at the optimum condition, $A_2B_1D_1$, of the significant factors only. Then the decibel value of $\hat{\mu}$ is

$$\begin{aligned}\mathrm{dB}(\hat{\mu}) &= \mathrm{dB}(\bar{A}_2) + \mathrm{dB}(\bar{B}_1) + \mathrm{dB}(\bar{D}_1) - 2\mathrm{dB}(\bar{T}) \\ &= 9.306 + 12.788 + 9.306 - 2.8256 = 14.888\,\mathrm{dB}.\end{aligned}$$

From Appendix E, we obtain p(%) for 14.888, which is

$$p = \hat{\mu}\left(A_2B_1C_1\right) = 96.9\% = 0.969.$$

The above two equations (6.3) and (6.5) are in fact identical, although (6.5) is easier to use than (6.3). The reason why these two equations are the same is as follows. The formula for the decibel value of a fraction $p\left(or\ \hat{\mu}\right)$ is defined as

$$-10\log\left(\frac{1}{p}-1\right).$$

Therefore, from equations (6.3) and (6.5),

$$\begin{aligned}-10\log\left(\frac{1}{\hat{\mu}}-1\right) &= -10\log\left(\frac{1}{\bar{A}_2}-1\right) - 10\log\left(\frac{1}{\bar{B}_1}-1\right) - 10\log\left(\frac{1}{\bar{C}_1}-1\right) \\ &\quad -10\log\left(\frac{1}{\bar{D}_1}-1\right) - 10\log\left(\frac{1}{\bar{T}}-1\right)^3 \\ &= -10\log\left(\frac{\left(\frac{1}{\bar{A}_2}-1\right)\left(\frac{1}{\bar{B}_1}-1\right)\left(\frac{1}{\bar{C}_1}-1\right)\left(\frac{1}{\bar{D}_1}-1\right)}{\left(\frac{1}{\bar{T}}-1\right)^3}\right) \\ &= -10\log\Omega,\end{aligned}$$

which results in equation (6.4). Observe that as Ω changes in the range (0, ∞), the estimated value $\hat{\mu}$ changes only in the range (0,1).

We have examined two approaches to handling 0/1 data. The advantage of the SN ratio analysis approach is the simplicity of analysis. On the other hand, the advantage of the direct analysis approach is the statistical reliability. The former approach has only eight degrees of freedom for total variation in ANOVA, but the latter approach has 1799 degrees of freedom. That is, the latter approach handles more information, which gives greater statistical reliability. However, since the omega method should be used in the latter approach, the former approach needs less computation.

6.4 MULTI-CLASS DISCRETE DATA: SCORING METHOD

If there are more than two classes, say t classes, in fixed marginal discrete data, such as excellent, good, fair and poor, the weights $0,1,2\cdots, k-1$ or $1,2,3\cdots, k$ are given to the respective classes. Suppose the number of items which belong to each class is $n_1, n_2, ..., n_k$, with weights $w_1, w_2, ..., w_k$ and the total is n, where $n_1 + n_2 + \cdots + n_k = n$. Then the SN ratio is defined as

$$\mathrm{SN} = -10\log\left(\frac{\sum_i w_i^2 n_i}{n}\right), \tag{6.6}$$

and the usual ANOVA is then carried out using these SN ratios.

If $n_1 = n$ and all other n_i are zero, then the weights 0, 1, $2\cdots$, $k-1$ should not be used. In this case the numerator inside the log in (6.6) becomes zero, and the log function is not defined. In this situation the weights 1, 2, $3\cdots$, k should be used.

This method of analysis is called the scoring method, since we use the weights as scores to compute SN ratios. The quantity $\sum_i w_i^2 n_i$ in equation (6.6) is obtained in the following way. Let

$$y_{ij} = \begin{cases} 1 & \text{if an item belongs to the } j\text{th item of the } i\text{th class} \\ 0 & \text{otherwise.} \end{cases}$$

Then the sum of the square of $w_i y_{ij}$ for all the items becomes

$$\sum_{i=1}^{k}\sum_{j=1}^{n}(w_i y_{ij})^2 = \sum_i w_i^2 \sum_j y_{ij}^2 = \sum_i w_i^2 n_i .$$

For example, suppose 150 items are produced, and among them 80, 55 and 15 are classified as high-quality, medium-quality and low-quality, respectively. If the weights assigned are 0, 1 and 2, respectively, then the SN ratio is computed by the scoring method as follows:

$$\begin{aligned} \text{SN} &= -10\log\left(\frac{\sum_i w_i^2 n_i}{n}\right) \\ &= -10\log\left(\frac{0^2 \times 80 + 1^2 \times 55 + 2^2 \times 15}{150}\right) \\ &= -10\log\left(7.67 \times 10^{-1}\right) = 1.15\,\text{dB}. \end{aligned}$$

Observe that if the portion of high quality increases or the portion of low quality decreases, the SN ratio increases.

An example is given to illustrate the scoring method. The most critical quality problem in the manufacturing of cast aluminum machine parts is the occurrence of flow-holes. In an effort to decrease the occurrence of flow-holes, four control factors and two interactions are investigated with the following levels:

A: material	A_0 = current, A_1 = new
B: type of molding	B_0 = current, B_1 = new
C: hardening	C_0 = current, C_1 = new
D: melting temperature	D_0 = current, D_1 = current + 30°C

Two interactions, $A \times B$ and $A \times D$, are to be investigated. The L_8 OA is used for the experimental design, as in Table 6.4. At each experimental condition 20 parts are made and checked for flow-holes, and the results are classified as follows.

Table 6.4 Molding experiment data.

Experiment number	Control factor assignment, column number and basic marks							Data(n=20) Weights: excel=0 good=1 fair=2 poor=3				SN=−log $[\sum w_i^2 n_i/n]$
	A 1 a	B 2 b	A×B 3 ab	C 4 c	e 5 ac	D 6 bc	A×D 7 abc	excel	good	fair	poor	
1	0	0	0	0	0	0	0	15	3	1	1	0.97
2	0	0	0	1	1	1	1	14	3	2	1	00.00
3	0	1	1	0	0	1	1	16	2	2	0	3.01
4	0	1	1	1	1	0	0	19	1	0	0	13.01
5	1	0	1	0	1	0	1	17	1	0	2	0.22
6	1	0	1	1	0	1	0	17	2	1	0	5.23
7	1	1	0	0	1	1	0	10	2	4	4	-4.31
8	1	1	0	1	0	0	1	16	3	1	0	4.56

excellent: no flow-holes
good: a few flow-holes
fair: several flow-holes
poor: many flow-holes.

For experiment 1 the SN ratio is

$$\mathrm{SN} = -10\log\left(\frac{0^2 \times 15 + 1^2 \times 3 + 2^2 \times 1 + 3^2 \times 1}{150}\right)$$

$$= -10\log(0.8) = 0.97\,\mathrm{dB}.$$

In order to compute variations of sources in ANOVA, the sum of SN ratios at each level of each factor is shown in Table 6.5, and the source variations are computed accordingly.

Table 6.5 Sum of SN ratios at each level.

Source		A	B	$A\times B$	C	e	D	$A\times D$	Total
Level	0	16.99	6.42	1.22	−0.11	13.77	18.76	14.90	22.69
	1	5.70	16.27	21.47	22.80	8.92	3.93	7.79	

$$S_A = (A_1 - A_0)^2/8 = (5.70 - 16.99)^2/8 = 15.93,$$
$$S_B = (B_1 - B_0)^2/8 = (16.27 - 6.42)^2/8 = 12.13,$$
$$S_{A\times B} = (21.47 - 1.22)^2/8 = 51.26,$$
$$S_C = (C_1 - C_0)^2/8 = (22.80 - (-0.11))^2/8 = 65.61,$$
$$S_D = (D_1 - D_0)^2/8 = (3.93 - 18.76)^2/8 = 27.49,$$
$$S_{A\times D} = (7.79 - 14.90)^2/8 = 6.32,$$
$$S_e = (8.92 - 13.77)^2/8 = 2.94.$$

The ANOVA table for this experiment is shown in Table 6.6. In this table, $A\times D$ is pooled into the error term. The F_0 values for A, B, C, D and $A\times B$ are greater than 2, which may be judged to be significant. Since $A\times B$ is significant, the following AB two-way table is constructed.

	A_0	A_1
B_0	0.97	5.45
B_1	16.02	0.25

Observe that A_0B_1 is optimal for the factors A and B. For C and D, we can see from Table 6.5 that C_0D_1 is optimal. Hence, the overall optimum condition for the control factors is $A_0B_1C_1D_0$. The current molding condition is $A_0B_0C_0D_0$ happens to be the first experimental condition. At this condition the experimental SN ratio is 0.97. How great is the difference in SN ratio between the optimum and the current conditions? To answer this question, we can estimate the SN ratio at the optimum condition.

$$\hat{\mu}(A_0B_1C_1D_0) = \overline{A_0B_1} + \overline{C}_1 + \overline{D}_0 - 2\overline{T}$$
$$= \frac{16.02}{2} + \frac{22.08}{4} + \frac{18.76}{4} - 2\times\frac{22.69}{8}$$
$$= 12.73\text{dB}.$$

Therefore, the increase in the SN ratio from the current condition to the optimum condition is 12.73 - 0.97 = 11.76, which is very large.

Table 6.6 ANOVA table for molding experiment.

Source of variation	Sum of squares		Degrees of freedom	Mean square	F_0
A	15.93		1	15.93	3.44
B	12.13		1	12.13	2.62
C	65.61		1	65.61	14.17
D	27.49		1	27.49	5.94
$A \times B$	51.26		1	51.26	11.07
$A \times D$	6.32	O	1		
E	2.94	O	1		
(e)	(9.26)		2	(4.63)	
T	181.68		7		

O indicates pooled variation.

If we substitute SN =12.73 into equation (6.6) to find out the practical meaning of the SN value,

$$12.73 = -10\log\left(\frac{\sum_i w_i^2 n_i}{20}\right),$$

we find that

$$\sum_i w_i^2 n_i = 20 \times 10^{-1.273}$$

which is equal to

$$0^2 \times n_1 + 1^2 \times n_2 + 2^2 \times n_3 + 3^2 \times n_4 = 1.066$$

or approximately

$$n_2 + 4n_3 + 9n_4 = 1 .$$

Hence, the only solution for this equation is $n_1 = 19$, $n_2 = 1$, $n_3 = n_4 = 0$. We expect that under the optimum condition, of 20 items inspected, 19 would be excellent, one good, none fair and none poor.

Note that the optimum condition happens to be the same condition as in experiment 4 in Table 6.4, with SN ratio 13.01 dB. However, the optimum condition does not have to be one of the experimental conditions implemented in the experiment. In this experiment, there are a total of $2^4 = 16$ possible experimental conditions, and any one could be the optimum condition. Note that eight conditions out of 16 are tested in this experiment.

6.5 MULTI-CLASS DISCRETE DATA: ACCUMULATING ANALYSIS

For the analysis of categorized or ranked data — say, excellent, good, fair and poor — or for the cases when the characteristic values, although fundamentally continuous, have been divided into a number of classes depending on their magnitude, Taguchi (1987) recommends accumulating analysis. **Accumulating analysis** takes frequency into account and is used when the class number has some practical meaning. The molding result classes are given in Table 6.7.

Table 6.7 Rating of molding result.

Molding result	Class number
Excellent	1
Good	2
Fair	3
Poor	4

Table 6.8 Frequency and cumulative frequency.

Experiment number	Class				Cumulative class			
	1	2	3	4	I	II	III	IV
1	15	3	1	1	15	18	19	20
2	14	3	2	1	14	17	19	20
3	16	2	2	0	16	18	20	20
4	19	1	0	0	19	20	20	20
5	17	1	0	2	17	18	18	20
6	17	2	1	0	17	19	20	20
7	10	2	4	4	10	12	16	20
8	16	3	1	0	16	19	20	20
Total	124	17	11	8	124	141	152	160

The steps to follow for the accumulating analysis are best described by means of an example. The same example, the molding experiment in Table 6.4, will be analyzed in order to compare the accumulating analysis with the scoring method. The molding results for 20 items in each experiment are summarized again in Table 6.8.

The first step in accumulation analysis is to create a cumulative frequency table by summing the frequencies from left to right for each experiment. The cumulative values are calculated as follows:

$$\text{class I} = \text{class 1}$$

$$\text{class II} = \text{class 1} + \text{class 2}$$

$$\text{class III} = \text{class 1} + \text{class 2} + \text{class 3}$$

$$\text{class IV} = \text{class 1} + \text{class 2} + \text{class 3} + \text{class 4}.$$

Table 6.8 shows the cumulative frequencies for the four cumulative classes. One should notice that the last cumulative class shows no variation from experiment to experiment. For this reason, this last class is ignored in accumulation analysis.

The second step is to calculate a weight for each cumulative class, which is a function of the cumulative frequency of occurrence in that class. The formula for the weight of class i is

$$W_i = \frac{1}{p_i(1-p_i)}, \tag{6.7}$$

where p_i is the cumulative proportion of occurrences in class i. Therefore, the weights are

$$W_1 = \left[\frac{1}{\frac{T_1}{N}\left(1-\frac{T_1}{N}\right)}\right] = \left[\frac{1}{\frac{124}{160}\left(1-\frac{124}{160}\right)}\right] = 5.73,$$

$$W_2 = \left[\frac{1}{\frac{141}{160}\left(1-\frac{141}{160}\right)}\right] = 9.56,$$

$$W_3 = \left[\frac{1}{\frac{152}{160}\left(1-\frac{152}{160}\right)}\right] = 21.05.$$

The reason why weight formula (6.7) is used is as follows. In estimation of population proportion, the sample proportion $p = r/n$ (the number of "successes" out of n trials is r) has the estimated variance

$$\text{var}(p) = p(1-p)/n$$

from the theory of the binomial distribution. If this variance is large, the degree of reliability of this estimate is not good. Therefore, if we want to use the degree of reliability for the class weight, the reciprocal of this variance $n/p(1-p)$ should be used. Since n is a fixed constant, it is not necessary to use n, so that only $1/p(1-p)$ is used for the class weight.

The third step is to calculate the total sum of squares,

$$S_T = \sum_{i=1}^{k-1} S_{T_i} W_i ,$$

where k is the number of classes and $k-1$ is the number of classes analyzed. Similarly to the two-class discrete data analysis in Section 6.2,

$$S_{T_i} = T_i - \frac{T_i^2}{N},$$

where T_i is the cumulative frequency of occurrences in class i. Then, in this example,

$$S_T = S_{T_1} W_1 + S_{T_2} W_2 + S_{T_3} W_3 . \tag{6.8}$$

Notice that the sum of squares for each class is multiplied by the class weight to obtain a weighted sum of squares for each class. The weighted sums of squares are then added to obtain the total sum of squares.

The first term on the right-hand side of (6.8) becomes

$$\begin{aligned} S_{T_i} W_1 &= \left[T_i - \frac{T_i^2}{N} \right] \left[\frac{1}{T_i - \frac{T_i^2}{N}} \right] \\ &= T_1 \left[\frac{N - T_1}{N} \right] \left[\frac{N^2}{T_1 (N - T_1)} \right] \\ &= N . \end{aligned}$$

Similarly, $S_{T_2} W_2 = S_{T_3} W_3 = N$. Therefore, $S_T = 3N = 3 \times 160 = 480$ in this experiment. In general, the total sum of squares in accumulation analysis is

$$S_T = (k-1)N . \tag{6.9}$$

The fourth step is to calculate the OA column sums of squares, and carry out ANOVA. The sums of squares for each class are computed in the same way as for two-class discrete 0, 1 data.

Table 6.9 Cumulative frequency at each level of each column.

Column		*A*			*B*			*A* × *B*			*C*		
Cumulative class		I	II	III	I	II	III	I	II	III	I	II	III
Level	0	64	73	78	63	72	76	55	66	74	58	66	73
	1	60	68	74	61	69	76	69	75	78	66	75	79
Total		124	141	152	124	141	152	124	141	152	124	141	152

Column		*E*			*D*			*A* × *D*		
Cumulative class		I	II	III	I	II	III	I	II	III
Level	0	64	74	79	67	75	77	61	69	75
	1	60	67	73	57	66	75	63	72	77
Total		124	141	152	124	141	152	124	141	152

Then the column sums of squares are multiplied by the class weight and summed to obtain the complete sums of squares. For instance, for factor A (first column),

$$S_A = \sum_{i=1}^{k-1} S_{A_i} W_i ,$$

$$S_{A_i} = \frac{1}{N}(A_{1i} - A_{0i})^2 ,$$

where S_{A_i} is the sum of squares due to factor A in class i, and A_{0}, is the accumulative occurrences for the zero level of factor A in class i. Referring to Tables 6.4, 6.8 and 6.9,

$$S_{A_1} = \frac{1}{160}[(17+17+10+16)-(15+14+16+19)]^2 = \frac{[60-64]^2}{160} = 0.100,$$

$$S_{A_2} = \frac{1}{160}[(18+19+12+19)-(18+17+18+20)]^2 = \frac{[68-73]^2}{160} = 0.156,$$

$$S_{A_3} = \frac{1}{160}[(18+20+16+20)-(19+19+20+20)]^2 = \frac{[74-78]^2}{160} = 0.100,$$

$$S_A = (0.100)(5.73)+(0.156)(9.56)+(0.100)(21.05) = 4.169.$$

The remaining columns are calculated in a similar fashion, and the results are shown in Table 6.10.

The degrees of freedom are obtained in similar fashion to the other methods. However, there is a distinct difference. Each discrete class allows a comparison of the 0 and 1 levels of each OA column, and hence, each class analyzed provides one degree of freedom. Therefore, the degrees of freedom for factor A is

$$\phi_A = (k-1)\times 1 = 3\times 1 = 3$$

and the total degrees of freedom is

$$\phi_T = (k-1)(N-1) = 3\times 159 = 477.$$

In Table 6.10 $S_{e'}$ was obtained by subtracting S_A, S_B, $S_{A\times B}$, S_C, S_D, $S_{A\times D}$ and S_e from S_T, and e' was used for the F-test. In this instance, we observe that $A\times B$, C and D are very significant and the conclusions from the accumulating analysis are nearly the same as for the scoring method in Section 6.4.

Table 6.10 ANOVA summary for molding experiment by accumulating analysis.

Source of variation	Sum of squares	Degrees of freedom	Mean squares	F_0
A	4.169	3	1.390	1.47
B	0.681	3	0.227	0.24
$A\times B$	13.965	3	4.655	4.93^{**}
C	11.868	3	3.956	4.19^{*}
D	8.947	3	2.982	3.16^{*}
$A\times D$	1.207	3	0.402	0.43
e	8.237	3	2.746	2.91^{*}
e'	430.926	456	0.945	
T	480	477		

** indicates 1% significance.
* indicates 5% significance.

The last step is to find the optimum condition for the control factors. Since $A \times B$, C and D are significant, Table 6.11 and Figure 6.1 are constructed to find the optimal levels. The values in parentheses in Table 6.11 show the percentage frequencies.

Table 6.11 Frequency, percentage and cumulative percentage.

Factor level	Frequency and percentage					Cumulative percentage			
	excel	good	fair	poor	total	I	II	III	IV
A_0B_0	29 (72.5)	6 (15.0)	3 (7.5)	2 (5.0)	40	72.5	87.5	95.0	100
A_0B_1	35 (87.5)	3 (7.5)	2 (5.0)	0 (0.0)	40	87.5	95.0	100.0	100
A_1B_0	34 (85.0)	3 (7.5)	1 (2.5)	2 (5.0)	40	85.0	92.5	95.0	100
A_1B_1	26 (65.0)	5 (12.5)	5 (12,5)	4 (10.0)	40	65.0	77.5	90.0	100
C_0	58 (72.5)	8 (10.0)	7 (8.75)	7 (8.75)	80	72.5	82.5	91.25	100
C_1	66 (82.5)	9 (11.25)	4 (5.0)	1 (1.25)	80	82.5	93.75	98.75	100
D_0	67 (83.75)	8 (10.0)	2 (2.5)	3 (2.5)	80	83.75	93.75	96.25	100
D_1	57 (71.25)	9 (11.25)	9 (11.25)	5 (6.2)	80	71.25	82.5	93.75	100

Figure 6.1 tells us that A_0B_1, C_1 and D_0 are the optimal factor levels. At this optimum condition, the estimated cumulative percentages (or proportions) belonging to class I and class II appear to be obtained as follows.

$$\hat{\mu}_I(A_0B_1C_1D_0) = \overline{A_0B_1} + \overline{C}_1 + \overline{D}_0 - 2\overline{T}$$
$$= 0.875 + 0.825 + 0.8375 - 2 \times 124/160 = 0.9875,$$
$$\hat{\mu}_{II}(A_0B_1C_1D_0) = 0.950 + 0.9375 + 0.9375 - 2 \times 141/160 = 1.0625.$$

Obviously, this computation is invalid, since the result violates the proportion constraint [0,1]. As described in Section 6.3, the omega method is appropriate for estimating such proportions. Adapting the method of (6.5), and using Appendix E,

$$\begin{aligned}\mathrm{dB}(\hat{\mu}_{\mathrm{I}}) &= \mathrm{dB}(\overline{\mathrm{A}_0\mathrm{B}_1}) + \mathrm{dB}(\overline{\mathrm{C}}_1) + \mathrm{dB}(\overline{\mathrm{D}}_0) - 2\mathrm{dB}(\overline{\mathrm{T}}) \\ &= \mathrm{dB}(0.875) + \mathrm{dB}(0.825) + \mathrm{dB}(0.8375) - 2\mathrm{dB}(0.775) \\ &= 8.451\mathrm{dB} + 6.734\mathrm{dB} + 7.137\mathrm{dB} - 2 \times 5.371\mathrm{dB} \\ &= 11.580\mathrm{dB}.\end{aligned}$$

Therefore,

$$\hat{\mu}_I = 0.935,$$

and similarly, for class II,

$$\hat{\mu}_{II} = 0.988.$$

The method of accumulating analysis has some disadvantages. Firstly, if one of the end classes accounts for a large proportion of the frequency, the accumulating analysis does not have much to offer. Accumulating analysis works best when the distribution of frequency of occurrence is spread over many classes. A second disadvantage is the fact that the frequencies of the cumulative categories are not independent, as demonstrated by Hamada and Wu (1990). It was shown in (6.9) that always

$$S_T = (k-1)N = \text{constant}.$$

Then, for example in the two-factor main effects setting,

$$S_e = \text{constant} - S_A - S_B,$$

and consequently, when testing for significance of A, the distribution of the F_0 statistic for A depends on factor B. This is obviously not desirable. Another disadvantage is that the observations are not normally distributed.

Despite these disadvantages, accumulating analysis is still a powerful tool which is often used in practice for multi-class

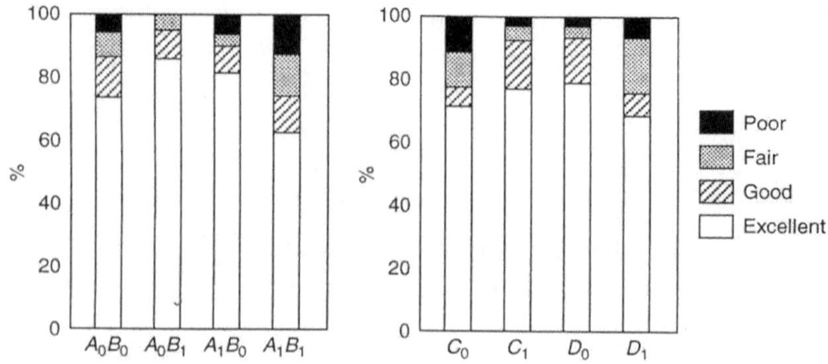

Figure 6.1 Bar chart for $A \times B$, C and D effects.

discrete data. For this type of data two methods of analysis, the scoring method and accumulating analysis have been examined. It seems that the scoring method works fine if we want to find the significant factors and their optimal conditions. Also the SN ratios obtained by (6.6) are approximately normally distributed. However, an important disadvantage is the arbitrariness of the weights $w_1, w_2, \ldots, w_k$. If different weights are given, the results could be different to some extent. The choice of either method relies on the preferences of the analyst.

Exercises

6.1 An experiment took place with the aim of increasing reliability for an automatic vending machine in the process of product design. Five control factors were selected and assigned to L_9, and the experimental results were recorded as follows. The result is the number of successful trials out of 100.

A: part A:	A_0 = part I, A_1 = part II
B: part B:	B_0 = domestic part, B_1 = foreign part
C: circuit:	C_0 = circuit I, C_1 = circuit II, C_2 = circuit III
D: electric power:	D_0 = 70% of standard, D_1 = standard, D_2 = 130% of standard
F: anti-dust method:	F_0 = old type, F_1 = new type

Experiment number	Control factor assignment and column number				Results (good trials out of 100 trials)
	(AB) 1	C 2	D 3	F 4	
1	00	0	0	0	57
2	00	1	1	1	84
3	00	2	2	0'	60
4	01	0	1	0'	58
5	01	1	2	0	88
6	01	2	0	1	55
7	10	0	2	1	83
8	10	1	0	0'	78
9	10	2	1	0	56

Note that A and B are assigned to column 1 using the combination design, and F is assigned to column 4 and the dummy-level technique is used for factor F.

(a) Carry out ANOVA by finding SN ratios for the two-class discrete data.

(b) Carry out ANOVA by the approach of 0/1 data direct analysis.

(c) Find the optimum condition.

6.2 To develop a new coffee mix, an experiment takes place in a coffee making company. The control factors considered are as follows:

A: mixing ratio	A_0, A_1 and A_2
B: process temperature	B_0, B_1 and B_2
C: heating method	C_0, C_1 and C_2
D: home of coffee plant	D_0, D_1 and D_2
F: drying time	F_0, F_1 and F_2
G: operating method	G_0 and G_1.

Factor C is assigned to column 1 of the OA $L_{18}(2^1 \times 3^7)$, and factors, A, B, C, D and F are assigned to columns 2, 3, 4, 5 and 6, respectively. For each experiment, ten evaluators taste the coffee and rank its taste as good, normal and bad as follows.

(a) Assign the weights 1, 2 and 3 to good, normal and bad, respectively, and carry out ANOVA by the scoring method. What is the optimum condition?

Experiment number	Data on coffee taste			Experiment number	Data on coffee taste		
	good	normal	bad		good	normal	bad
1	2	7	1	10	9	0	1
2	7	3	0	11	1	8	1
3	1	8	1	12	8	1	1
4	2	6	2	13	4	5	1
5	0	7	3	14	1	8	1
6	9	1	0	15	1	9	0
7	4	6	0	16	8	2	0
8	6	4	0	17	9	1	0
9	9	1	0	18	9	1	0

(b) Assign the weights 0, 1 and 2 to good, normal and bad, respectively, and carry out ANOVA by the scoring method. Is there any difference in ANOVA results between (a) and (b)?

(c) Carry out ANOVA by the accumulating analysis method. Is there any difference in ANOVA results between the scoring method and accumulating analysis?

(d) When accumulating analysis is used, what is the optimum condition? In the optimum condition, what cumulative proportion is estimated to belong to class I? How about for class II?

7

Alternative Parameter Design and Other Considerations

7.1 PARAMETER DESIGN BY COMBINED ARRAY

In the previous parameter designs for continuous and discrete data, the control factors are assigned to an **inner array**, which is an orthogonal array. For each row in the inner array, the noise factors are assigned to an **outer array**, which is an orthogonal array or a factorial layout such as a one-way or two-way layout. The inner array and the outer array are often called the "**design array**" and the "**environmental array**", respectively, and the total arrangement is called "**cross-product array**", since it is a cross-product of the inner array and the outer array. In general, if there are n_1 runs in the inner array and n_2 runs in the outer array, and the runs are made independently, then we shall need $n_1 n_2$ runs for the total experiment. Thus, except in the case where n_1 and n_2 are both small, this would involve a large amount of experimental work.

In order to reduce the large number of experimental trials in Taguchi's cross-product array, the idea of combining the control and noise factors to make a single design matrix was suggested, this is called a "**combined array**". The combined array approach was first proposed by Welch et al. (1990), and related approaches were discussed by, for instance, Vining and Myers (1990), Box and Jones (1992), and Myers, Khuri and Vining (1992). The initial motivation for the combined array is the run-size saving. This approach is not as simple conceptually as the cross-product array approach, since it makes extensive use of regression and response surface analysis. However, it is a powerful tool for run-size savings, empirical model building and statistical optimization.

7.1.1 COMBINED ARRAY APPROACH FOR FIRST-ORDER MODEL

The combined array approach postulates a model of the type

$$y = f(x, z), \tag{7.1}$$

where x and z represent the setting in the control and noise variables, respectively. The noise variables are temporarily treated as fixed effects even though they are random effects in the process. We assume that the set of control variables x can be varied within some region of interest R_x, and the set of noise variables z can be varied within some region of interest R_z. This idea is illustrated geometrically in Figure 7.1 for single control variable x and a single noise variable z. We are looking for a choice of x that will give a section showing a reasonably high response by y over R_z, if τ is the target value for the response y.

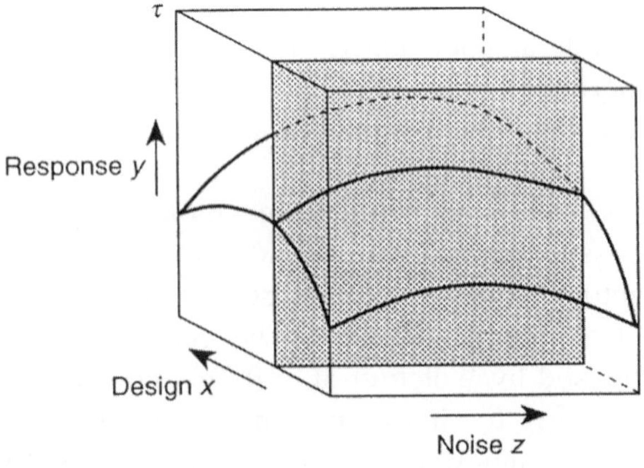

Figure 7.1 Geomteric representation of noise robustness.

By means of proper linear transformations on x and z, we usually define the R_x and R_z by

$$R_x = \{x : -1 \le x_i \le +1, i = 1, 2, \ldots, l\}$$

and

$$R_z = \{z : -1 \le z_i \le +1, i = 1, 2, \ldots, m\}.$$

For instance, the linear transformation

$$x_{i,trans} = (x_{i,orig} - a) / b_i,$$

where $x_{i,trans}$ and $x_{i,orig}$ are respectively the transformed and original values of x_i and

$$a_i = \{\max(x_i) + \min(x_i)\} / 2$$
$$b_i = \{\max(x_i) - \min(x_i)\} / 2$$

gives R_x as defined above.

Suppose a set of control variables is denoted by $\mathbf{x}' = (x_1, x_2, \ldots, x_l)$ and a set of noise variables by $\mathbf{z}' = (z_1, z_2, \ldots, z_m)$. Then the functional relationship f in (7.1) is well represented by the first-order polynomial regression model,

$$y(x, z) = \beta_0 + \mathbf{x}'\boldsymbol{\beta} + \mathbf{z}'\boldsymbol{\gamma} + \mathbf{z}'\mathbf{D}\mathbf{x} + \varepsilon, \tag{7.2}$$

where $\boldsymbol{\beta}$ is $l \times 1$, $\boldsymbol{\gamma}$ is $m \times 1$ and D is $m \times 1$, which are vectors or matrices of unknown regression parameters, and ε is a random error associated with the response y. For example, if $l = m = 2$, then the terms in the model (7.2) become $\mathbf{x}' = (x_1, x_2)$, $\boldsymbol{\beta} = (\beta_1, \beta_2)$, $\mathbf{z}' = (z_1, z_2)$, $\boldsymbol{\gamma} = (\gamma_1, \gamma_2)$ and

$$\mathbf{D} = \begin{bmatrix} d_{11} & d_{12} \\ d_{21} & d_{22} \end{bmatrix}.$$

The empirically fitted first-order model of (7.2) by the least-squares method may be written as

$$\hat{y}(\mathbf{x},\mathbf{z}) = b_0 + \mathbf{x}'\mathbf{b} + \mathbf{z}'\mathbf{r} + \mathbf{z}'\hat{\mathbf{D}}\mathbf{x}. \tag{7.3}$$

Let $\hat{\mathrm{m}}(\mathrm{x})$ be the estimated mean response at x averaged over the noise variables z, where z is usually uniformly distributed over R_z:

$$\begin{aligned}\hat{\mathrm{m}}(\mathrm{x}) &= k\int_{R_z} \hat{y}(x,z)dz \\ &= b_0 + \mathbf{x}'\mathbf{b}\end{aligned} \tag{7.4}$$

where $k^{-1} = \int_{R_z} dz$. Let $\hat{v}(x)$ be the mean square variation about the mean response

$$\begin{aligned}\hat{v}(\mathbf{x}) &= k\int [\hat{y}(\mathbf{x},\mathbf{z}) - \hat{m}(\mathbf{x})]^2\, d\mathbf{z} \\ &= \frac{1}{3}\left(\mathbf{r} + \hat{\mathbf{D}}\mathbf{x}\right)'\left(\mathbf{r} + \hat{\mathbf{D}}\mathbf{x}\right),\end{aligned} \tag{7.5}$$

which may be called the estimated variance response.

The following optimization measures can be proposed depending upon the type of quality characteristic.

1. Response-oriented measures:
   (a) Nominal-is-best case: $\min_{x\in R_x} [\hat{m}(x) - \tau]^2$ subject to $\hat{v}(x) \le c$, where c is some upper bound on the variance response.
   (b) Larger-the-better case:
   $$\max_{x\in R_x} \hat{m}(x) \text{ subject to } \hat{v}(x) \le c. \tag{7.6}$$
   (c) Smaller-the-better case: $\min_{x\in R_x} \hat{m}(x)$ subject to $\hat{v}(x) \le c$.
2. Variation-oriented measures:

(a) Nominal-is-best case: $\min_x \hat{v}(x)$ subject to $m_* \le \hat{m}(x) \le m^*$, where m_* and m^* are the minimum and maximum acceptable values of $\hat{m}(x)$, respectively.

(b) Larger-the-better case:

$$\min_{x \in R_x} \hat{v}(x) \text{ subject to } \hat{m}(x) \ge m_*. \tag{7.7}$$

(c) Smaller-the-better case: $\min_{x \in R_x} \hat{v}(x)$ subject to $\hat{m}(x) \le m^*$.

The methodology explained above will be demonstrated by means of an example in Section 7.2.

7.1.2 COMBINED ARRAY APPROACH FOR SECOND-ORDER MODEL

If each variable in x and z has more than two levels, we may use the second-order polynomial regression model, which may be expressed as

$$y(\mathbf{x},\mathbf{z}) = b_0 + \mathbf{x}'\boldsymbol{\beta} + \mathbf{x}'\mathbf{B}\mathbf{x} + \mathbf{z}'\mathbf{R}\mathbf{z} + \mathbf{z}'\boldsymbol{\gamma} + \mathbf{z}'\mathbf{D}\mathbf{x} + \varepsilon, \tag{7.8}$$

where B and R are appropriate matrices of unknown regression parameters. The empirically fitted second-order model of (7.8) may be written as

$$\hat{y}(\mathbf{x},\mathbf{z}) = b_0 + \mathbf{x}'\mathbf{b} + \mathbf{x}'\hat{\mathbf{B}}\mathbf{x} + \mathbf{z}'\hat{\mathbf{R}}\mathbf{z} + \mathbf{z}'\boldsymbol{\gamma} + \mathbf{z}'\hat{\mathbf{D}}\mathbf{x}. \tag{7.9}$$

Then the mean response can be obtained as

$$\begin{aligned} \hat{\mathrm{m}}(\mathbf{x}) &= k\int_{R_{\mathbf{z}}} \hat{y}(\mathbf{x},\mathbf{z})d\mathbf{z} \\ &= b_0 + \mathbf{x}'\mathbf{b} + \mathbf{x}\hat{\mathbf{B}}\mathbf{x} + \tfrac{1}{3}tr\hat{\mathbf{R}}, \end{aligned} \tag{7.10}$$

where $tr\hat{\mathbf{R}}$ is the trace of the matrix $\hat{\mathbf{R}}$. Also, we can find the variance response as

$$\begin{aligned}\hat{v}(\mathbf{x}) &= k\int_{R_z}\left[\hat{y}(\mathbf{x},\mathbf{z})-\hat{m}(\mathbf{x})\right]^2 d\mathbf{z}\\ &= (\mathbf{r}+\hat{\mathbf{D}}\mathbf{x})'(\mathbf{r}+\hat{\mathbf{D}}\mathbf{x})+\mathrm{A},\end{aligned} \tag{7.11}$$

where

$$\mathrm{A}=\frac{1}{45}\left[4\sum_{j=1}^{m} r_{jj}{}^{2}+5\sum_{j=1}^{m-1}\sum_{k=j+1}^{m} r_{jk}{}^{2}\right],$$

and r_{jk} is the jth row and kth column element of the matrix $\hat{\mathbf{R}}$. The detailed explanation for (7.10) and (7.11) is given in Box and Jones (1992).

The optimization measures in (7.6) and (7.7) can be also applied to the combined array approach for second-order models. An example for this case will be given in Section 7.3.

7.2 COMBINED ARRAY APPROACH FOR TWO-LEVEL FACTORS

If there are several two-level control and noise factors and an orthogonal array is used, we can fit the first-order response model through the combined array approach. Note that we cannot fit the second-order response model, since the second-order model needs at least three level for each variable. The way to use equations (7.1)-(7.6) will be explained by means of an example.

Suppose the quality characteristic of interest is the yield of a chemical reaction experiment, and there are five 2-level control factors A, B, C, D and F (Table 7.1) which are assigned to the OA L_{16} (see Table 7.2). Also suppose there is a noise factor N with two levels:

N_{-1} = environmental condition I

N_{+1} = environmental condition II.

Table 7.1 Factors and levels of a chemical reaction experiment.

Control factors	−1 level	+1 level
A: reaction time (min)	60	70
B: reaction temperature (°C)	130	150
C: cooling temperature (°C)	10	15
D: quantity of additive (%)	5	8
F: speed (r.p.m.)	1000	1200

Table 7.2 Assignment of factors on L_{16} and data.

Experiment number	Factor assignment and column number					Noise factor level		SN
	A 1	B 2	C 8	D 11	F 14	N_{-1}	N_{+1}	
1	−1	−1	−1	−1	−1	59	63	35.69
2	−1	−1	1	1	1	59	64	35.76
3	−1	−1	−1	−1	1	65	69	36.51
4	−1	−1	1	1	−1	51	56	34.54
5	−1	1	−1	1	1	69	75	37.12
6	−1	1	1	−1	−1	61	65	35.97
7	−1	1	−1	1	−1	67	74	36.93
8	−1	1	1	−1	1	55	59	35.10
9	1	−1	−1	1	−1	60	55	35.17
10	1	−1	1	−1	1	51	45	33.57
11	1	−1	−1	1	1	63	57	35.53
12	1	−1	1	−1	−1	47	44	33.15
13	1	1	−1	−1	1	64	61	35.91
14	1	1	1	1	−1	65	58	35.74
15	1	1	−1	−1	−1	68	62	36.23
16	1	1	1	1	1	56	51	34.54

Table 7.2 gives a set of yield data y, which is a cross-product array setting. To indicate levels, we use −1 and +1 instead of 0 and 1 in order to fit the first-order model easily. First the ordinary parameter design analysis will be provided for this set of data, and

then the combined array approach will be explained by using half of the data.

7.2.1 CROSS-PRODUCT ARRAY APPROACH

Since the yield is a larger-the-better characteristic, we can find the SN ratio for the ith row from

$$\mathrm{SN}_i = -10\log\left[\frac{1}{2}\left(\frac{1}{y_{i1}^2}+\frac{1}{y_{i2}^2}\right)\right].$$

To obtain the respective variations for SN ratios, Table 7.3 is provided, and the respective variations are computed as follows.

$$S_T = 35.69^2 + 35.76^2 + \ldots + 34.54^2 - 567.46^2/16 = 18.1148$$

$$S_A = (279.84 - 287.62)^2/16 = 3.7830$$

$$\vdots$$

$$S_F = (284.04 - 283.42)^2/16 = 0.0240$$

$$S_e = S_T - (S_A + S_B + S_C + S_D + S_F) = 2.8524.$$

We can therefore obtain the ANOVA table shown in Table 7.4 Note that the main effects of A (reaction time), B (reaction temperature), and C (cooling temperature) are very significant, and their optimum levels are $A_{-1}B_{+1}C_{-1}$.

7.2.2 COMBINED ARRAY APPROACH

In order to compare the cross-product array approach with the combined array approach, the data is selected on purpose from the combination of each level of factors in the cross-product array in Table 7.2. The noise variable N is arranged in column 4 of L_{16}, and the results of assignment and data are shown in Table 7.5. For instance, for the first experiment number, the yield is 59 since the levels of A, B, C, D and F are the same and the level of N is −1 only. Also as required in model (7.2), the interactions between the control factors and the noise factor are also assigned in the array.

Table 7.3 Supplementary table of SN ratios.

Factor level	Level total	Factor level	Level total
A_{-1}	287.62	D_{-1}	282.13
A_{+1}	279.84	D_{+1}	285.33
B_{-1}	279.92	F_{-1}	283.42
B_{+1}	287.54	F_{+1}	284.04
C_{-1}	289.09	Total	567.46
C_{+1}	278.37		

Table 7.4 ANOVA table for SN ratio of yield data.

Source of variation	Sum of squares	Degrees of freedom	Mean square	F_0
A	3.7830	1	3.7930	13.27**
B	3.6290	1	3.6290	12.73**
C	7.1824	1	7.1824	25.19**
D	0.6400	1	0.6400	2.24
F	0.0240	1	0.0240	0.08
E	2.8514	10	0.2851	
T	18.1148	15		

The combined array consists of five control variables x_1 (or A), x_2 (or B), x_3 (or C), x_4 (or D) and x_5 (or F), and one noise variable z (or N), and five interaction variables zx_1 (or $N \times A$), zx_2 (or N×B), zx_3, (or $N \times C$), zx_4 (or $N \times D$), and zx_5 (or $N \times F$), which are all assigned in L_{16}.

From the results of assignment in Table 7.5, we can obtain the estimates of regression coefficients by the least-squares method. The empirically fitted first-order model in (7.3) becomes

$$\begin{aligned}\hat{y}(\mathbf{x},\mathbf{z}) &= b_0 + \mathbf{x}'\mathbf{b} + \mathbf{z}'\mathbf{r} + \mathbf{z}'\hat{\mathbf{D}}\mathbf{x} \\ &= 60.00 - 3.25x_1 + 3.13x_2 - 4.25x_3 + 1.38x_4 - 0.13x_5 \\ &\quad - 1.00z - 2.25zx_1 - 0.63zx_2 - 2.25zx_3 - 0.88zx_4 + 0.13zx_5,\end{aligned} \tag{7.12}$$

with the coefficient of determination, $R^2 = 0.99$.

Table 7.5 Assignment of sources and data in the combined array of L_{16}.

Experiment number	Column number											Yield
	1	2	4	5	6	8	10	11	12	14	15	
1	−1	−1	−1	−1	−1	−1	−1	−1	−1	−1	−1	59
2	−1	−1	−1	−1	−1	1	1	1	1	1	1	59
3	−1	−1	1	1	1	−1	−1	−1	1	1	1	69
4	−1	−1	1	1	1	1	1	1	−1	−1	−1	56
5	−1	1	−1	−1	1	−1	1	1	−1	1	1	69
6	−1	1	−1	−1	1	1	−1	−1	1	−1	−1	61
7	−1	1	1	1	−1	−1	1	1	1	−1	−1	74
8	−1	1	1	1	−1	1	−1	−1	−1	1	1	59
9	1	−1	−1	1	−1	−1	−1	1	−1	−1	1	60
10	1	−1	−1	1	−1	1	1	−1	1	1	−1	51
11	1	−1	1	−1	1	−1	−1	1	1	1	−1	57
12	1	−1	1	−1	1	1	1	−1	−1	−1	1	44
13	1	1	−1	1	1	−1	1	−1	−1	1	−1	64
14	1	1	−1	1	1	1	−1	1	1	−1	1	65
15	1	1	1	−1	−1	−1	1	−1	1	−1	1	62
16	1	1	1	−1	−1	1	−1	1	−1	1	−1	51
Basic mark	a	b	c	ac	bc	d	bd	abd	cd	bcd	abcd	
Assignment	X_1	X_2	Z	ZX_1	ZX_2	X_3	ZX_5	X_4	ZX_1	X_5	ZX_4	

If we integrate the above equation with respect to z, since $R_z = \{z : -1 \le z \le +1\}$, we can obtain

$$\hat{m}(\mathbf{x}) = 0.5\int_{-1}^{1} \hat{y}(\mathbf{x},\mathbf{z})d\mathbf{z}$$

$$= 60.00 - 3.25x_1 + 3.13x_2 - 4.25x_3 + 1.38x_4 - 0.13x_5, \qquad (7.13)$$

$$\hat{v}(\mathbf{x}) = 0.5\int_{-1}^{1} [\hat{y}(\mathbf{x},\mathbf{z}) - \hat{m}(\mathbf{x})]^2 d\mathbf{z}$$

$$= (-1.00 - 2.25x_1 - 0.63x_2 - 2.25x_3 - 0.88x_4 + 0.13x_5)^2/3 \qquad (7.14)$$

as the mean and variance responses in (7.4) and (7.5).

If we assume as usual that the region of interest R_x is given by the inequality $-1 \le x_1, x_2, x_3, x_4, x_5 \le 1$, then the ranges for $\hat{m}(x)$ and $\hat{v}(x)$ over R_x can be obtained as

$$\min_{x\in R_x} \hat{m}(x) = 47.86 \le \hat{m}(x) \le 72.14 = \max_{x\in R_x} \hat{m}(x)$$
$$\min_{x\in R_x} \hat{v}(x) = 0.00 \le \hat{v}(x) \le 16.99 = \max_{x\in R_x} \hat{v}(x).$$

Since the yield in this experiment is a larger-the-better characteristic, we can use the response-oriented measure

$$\max_{x\in R_x} \hat{m}(x) \text{ subject to } \hat{v}(x) \le c \tag{7.15}$$

in (7.6), or the variation-oriented measure

$$\min_{x\in R_x} \hat{v}(x) \text{ subject to } \hat{m}(x) \ge m_* \tag{7.16}$$

in (7.7) as an optimization measure.

If c takes a value of 1.15 or greater than the maximum value of $\hat{m}(x)$, 72.14 is obtained at $x_1 = -1$, $x_2 = 1$, $x_3 = -1$, $x_4 = 1$, and $x_5 = -1$. Note that we obtained $A_{-1}B_{+1}C_{-1}$ for the optimum condition in the previous cross-product array approach. In fact, the optimum condition ($x_1 = -1$, $x_2 = 1$, $x_3 = -1$) is the same as $A_{-1}B_{+1}C_{-1}$. However, if c is less than 1.15, the maximum value of $\hat{m}(x)$ will be less than 72.14 and the optimum condition for at least one x_i will be different from the above optimum condition. For instance, if $c = 0$, the maximum value of $m(x)$ becomes 66.15 at approximately $x_1 = -1$, $x_2 = 0.8$, $x_3 = 0.1$, $x_4 = 0.5$ and $x_5 = -1$.

If we use the variation-oriented measure in (7.16) and if m_* takes value of 68, $\hat{v}(x)$ becomes 1.29, the minimum value, at $x_1 = -1$, $x_2 = 1$, $x_3 = -0.5$, $x_4 = -0.4$ and $x_5 = -1$ However, if we increase m_* to 69, the minimum value of $v(x)$ becomes 3.45 at $x_1 = -1$, $x_2 = 1$, $x_3 = -0.9$, $x_4 = -0.8$ and $x_5 = -1$.

In the combined array approach, we can observe that this approach uses only half of the data set of the cross-product array approach, but produces the same result for the optimum condition.

Furthermore, this approach provides more information on the variance caused by the noise variable.

7.3 COMBINED ARRAY APPROACH FOR THREE-LEVEL FACTORS

In the previous section, the combined array approach using two-level orthogonal arrays for two-level factors was examined by means of an example. In this section the combined array approach using three-level orthogonal arrays for three-level factors will be studied, also by means of an example. First, the ordinary parameter design analysis using the cross-product array approach will be provided in order to compare the two approaches.

7.3.1 CROSS-PRODUCT ARRAY APPROACH

Suppose that the objective is to find the optimum condition for increasing the strength of a plastic product. Suppose there are three control factors *A*, *B* and *C* (see Table 7.6), which are assigned to the OA $L_{18}(2^1 \times 3^7)$. The current operating condition for *A*, *B* and *C* is $A_0B_0C_0$. Suppose the environmental noise factor *N* with three levels.

N_{-1} : good condition

N_0 : normal condition

N_1 : bad condition.

Table 7.7 shows the strength data for the plastic experiment. Note that the three levels of the noise factor are run for every row in the control array, and the total number of runs of this cross-product array is $18 \times 3 = 54$. Three levels −1, 0 and 1 are used instead of 0, 1 and 2, since we want to use this set of data for empirical modeling between the strength response and the three factors later.

To obtain an ANOVA table, Table 7.8 is constructed, and the respective sums of squares are as follows. The ANOVA table is given in Table 7.9.

Table 7.6 Factors and levels of plastic experiment.

Control factors	−1 level	0 level	1 level
A: time (min)	120	125	130
B: temperature (°C)	60	70	80
C: stir speed (r.p.m.)	700	800	900

Table 7.7 Cross-product array design and strength data.

Experiment number	Control factor assignment and column number								Data (strength)			SN
	e 1	A 2	B 3	C 4	e 5	e 6	e 7	e 8	N_{-1}	N_0	N_1	
1	−1	−1	−1	−1	−1	−1	−1	−1	45	49	52	33.70
2	−1	−1	0	0	0	0	0	0	65	64	60	35.97
3	−1	−1	1	1	1	1	1	1	73	69	75	37.17
4	−1	0	−1	−1	0	0	1	1	63	60	69	36.08
5	−1	0	0	0	1	1	−1	−1	55	56	49	34.49
6	−1	0	1	1	−1	−1	0	0	68	72	72	36.97
7	−1	1	−1	0	−1	1	0	1	62	66	61	35.97
8	−1	1	0	1	0	−1	1	−1	55	49	56	34.49
9	−1	1	1	−1	1	0	−1	0	74	80	74	37.60
10	1	−1	−1	1	1	0	0	−1	69	55	66	35.91
11	1	−1	0	−1	−1	1	1	0	57	52	44	34.00
12	1	−1	1	0	0	−1	−1	1	78	76	68	37.34
13	1	0	−1	0	1	−1	1	0	50	52	46	33.83
14	1	0	0	1	−1	0	−1	1	51	45	46	33.47
15	1	0	1	−1	0	1	0	−1	66	75	69	36.87
16	1	1	−1	1	0	1	−1	0	56	51	59	34.81
17	1	1	0	−1	1	−1	0	1	50	45	48	33.54
18	1	1	1	0	−1	0	1	−1	73	67	55	36.07
	Sum								1110	1083	1070	638.28

Table 7.8 Sum of SN ratios at each level of each factor.

Level	Factor		
	A	B	C
-1	214.09	210.30	211.79
0	211.71	205.96	213.67
1	212.48	22.02	212.82
Total	638.28	638.28	638.28

Table 7.9 ANOVA table for plastic experiment.

Source of variation	Sum of squares	Degrees of freedom	Mean square	F_0
A	0.49	2	0.245	0.25
B	23.01	2	11.505	11.68**
C	0.29	2	0.145	0.14
E	10.83	11	0.985	
T	34.62	17		

** indicates 1% significance.

$$CT = \frac{638.28^2}{18} = 22633.41,$$

$$S_T = 33.70^2 + 35.97^2 + \ldots + 36.07^2 - 22633.41 = 34.62,$$

$$S_A = (214.09^2 + 211.71^2 + 212.48^2)/6 - 22633.41 = 0.49,$$

$$S_B = (210.30^2 + 205.96^2 + 222.02^2)/6 - 22633.41 = 23.01,$$

$$S_C = (211.79^2 + 213.67^2 + 212.82^2)/6 - 22633.41 = 0.29,$$

$$S_e = S_T - S_A - S_B - S_C = 10.83.$$

The ANOVA table shows that only B (temperature) is very significant for the strength data, and A (time) and C (stir speed) are not significant. From Table 7.8, B_1 is optimal. For the non-significant factors, suppose $A_{-1}C_0$ is optimal. The overall optimum condition is then $A_{-1}B_1C_0$, which means that A = 120min, B = 80°C and C = 800 rpm.

Table 7.10 Combines array design and strength data table for plastic experiment.

Experiment number	Variable assignment and column number				Strength data
	x_1 2	x_2 3	x_3 4	z 6	
1	-1	-1	-1	-1	45
2	-1	0	0	0	64
3	-1	1	1	1	75
4	0	-1	-1	0	60
5	0	0	0	1	49
6	0	1	1	-1	68
7	1	-1	0	1	61
8	1	0	1	-1	55
9	1	1	-1	0	80
10	-1	-1	1	0	55
11	-1	0	-1	1	44
12	-1	1	0	-1	78
13	0	-1	0	-1	50
14	0	0	1	0	45
15	0	1	-1	1	69
16	1	-1	1	1	59
17	1	0	-1	-1	50
18	1	1	0	0	67

7.3.2 COMBINED ARRAY APPROACH

In order to compare the cross-product array approach with the combined array approach, the data set in Table 7.7 will again be used here. The combined array experiment consists of three control variables x_1 (or A), x_2 (or B), and x_3 (or C) and one noise variable z (or N). However, the noise variable is assigned to a column in the inner array, L_{18} (2^1 x 3^7). As shown in Table 7.10, the noise variable is assigned to column 6, and the data comes from the combinations of variables. For instance, for the first experiment the strength should be 45 only since the level of noise variable is −1. Because there is no outer array in the combined array, we do not use the remaining data, 49 and 52, in the first experiment. The data summary is given in Table 7.10.

If we use the method of least squares to fit the second-order response model for the data set in Table 7.10, we can obtain the empirical model (7.9) as follows, which has coefficient of determination $R^2 = 0.90$.

$$\begin{aligned}\hat{y}(x_1, x_2, x_3, z) &= b_0 + \mathbf{x}'\mathbf{b} + \mathbf{x}'\hat{\mathbf{B}}\mathbf{x} + \mathbf{z}'\hat{\mathbf{R}}\mathbf{z} + \mathbf{z}'\boldsymbol{\gamma} + z'\hat{\mathbf{D}}\mathbf{x} \\ &= 53.15 + 1.24x_1 + 7.88x_2 - 0.87x_3 \\ &\quad + 6.28x_{12} + 11.04x_2^2 + 0.05x_3^2 - 7.93x_1x_2 \\ &\quad - 3.12x_1x_3 - 0.15x_2x_3 - 7.60z^2 - 1.10z \\ &\quad + 2.37zx_1 + 1.46zx_2 - 3.40zx_3 .\end{aligned} \tag{7.17}$$

Observe that

$$\begin{aligned}&\mathbf{b}' = (1.24,\ 7.88,\ -0.87), \\ &\hat{\mathbf{B}} = \begin{pmatrix} 6.28 & -7.93/2 & -3.12/2 \\ -7.93/2 & 11.04 & -0.15/2 \\ -3.12/2 & -0.15/2 & 0.05 \end{pmatrix}, \\ &\hat{R} = -7.60, \\ &\boldsymbol{\gamma} = -1.10, \\ &\hat{\mathbf{D}} = (2.37 \quad 1.46 \quad -3.40).\end{aligned}$$

Suppose the region of interest for the noise variable is (−1, + 1). Then the mean response equation (7.10) becomes

$$\begin{aligned}\hat{m}(x) &= \tfrac{1}{2}\int \hat{y}(x_1, x_2, x_3, z)\,dz \\ &= 48.09 + 1.24x_1 + 7.88x_2 - 0.87x_3 + 6.28x_1^2 + 11.04x_2^2 \\ &\quad + 0.05x_3^2 - 7.93x_1x_2 - 3.12x_1x_3 - 0.15x_2x_3 .\end{aligned} \tag{7.18}$$

The estimated variance response model (7.11) is obtained as

$$\begin{aligned}\hat{v}(\mathbf{x}) &= (\mathbf{r}+\hat{\mathbf{D}}\mathbf{x})'(\mathbf{r}+\hat{\mathbf{D}}\mathbf{x}) + \mathrm{A} \\ &= (-1.10+2.37x_1+1.46x_2-3.40x_3)^2/3+5.13\ , \qquad (7.19)\end{aligned}$$

since A = [4 × $(-7.60)^2$]/45 = 5.13.

Suppose the region of interest R_x is given by

$$R_x = \{(x_1, x_2) \mid -1 \le x_i \le +1, i = 1, 2\}\,.$$

Then we can obtain that

$$\min_{x\in R_x} \hat{m}(x) = 44.84 \le \hat{m}(x) \le \max_{x\in R_x} \hat{m}(x) = 82.13,$$
$$\min_{x\in R_x} \hat{v}(x) = 5.13 \le \hat{v}(x) \le \max_{x\in R_x} \hat{v}(x) = 28.26.$$

If the target value of plastic strength, τ, is greater than or equal to 82.13, then the first two response-oriented measures in (7.6) are in fact the same.

Suppose the plastic strength is a larger-the-better characteristic, and the upper bound on the variance response is c = 10. Then the optimization problem becomes:

$$\max_{x\in R_x} \hat{m}(x) \text{ subject to } \hat{v}(x) \le 10.$$

A numerical computation shows that the optimum is achieved at x_1 = 1 (120 min), x_2 = 1 (80°C), and x_3 = 0.53 (853 r.p.m.) with mean response 81.11. If the upper bound on the variance response is any value greater than or equal to 10.6, then the optimum is obtained at x_1 = −1 (120 min), x_2 = 1 (80°C), and x_3 = 1 (900 r.p.m.) with mean response 82.13.

Note that this combined array approach gives more information about optimality than the cross-product array approach, even though the number of experimental data is less in the combined array approach than that in the cross-product array approach.

7.4 ESTIMATION USING NONLINEAR REGRESSION

Multiple linear regression analysis (Section 2.4) is a powerful tool for fitting linear models in regression parameters. The limitations of this technique arise when the available data cannot adequately be represented by a linear model. **Nonlinear regression** models are therefore often employed in practice. Three examples of such models are

$$y = \exp(A + Bx + \varepsilon),$$
$$y = A[\exp(Bx) - \exp(Cx)]/(A - B) + \varepsilon,$$
$$y = ACx_1/(1 + Ax_1 + Bx_2) + \varepsilon,$$

where A, B and C are unknown parameters.

To solve these nonlinear problems, several methods have been proposed. Among them, the following methods are popularly used: the linearization method (or Taylor-series method or Gauss-Newton procedure); the steepest descent method, the Marquardt procedure; the quasi-Newton algorithm; and the simplex method. For more details, see Draper and Smith (1981) and Sen and Srivastava (1990).

For these nonlinear problems, Taguchi (1987) recommends a **multivariate successive approximation method**, which makes use of orthogonal arrays when there are several parameters. Taguchi's method is a valuable technique, especially when there are several parameters and a small number of observed data. The method uses a relatively small number of observed values to construct a satisfactory nonlinear model, which could even involve more parameters than the actual number of data values.

The basic steps of the method can be briefly outlined as follows. Initially a theoretical response model linking the response y and the independent variables $x_1, x_2, \ldots, x_k$ is assumed based on scientific or engineering knowledge. Some approximate ranges for the unknown parameters should be given. At successive stages, a few levels are chosen for each parameter from the ranges of the parameters, and the model's calculated response for various combinations of the levels of all the parameters is obtained. The rows of an orthogonal array will indicate the particular level combinations to be chosen. At every stage, a "temporary optimal" set of parameter levels is determined, and this produces the minimum sum of squares of the deviations of the theoretical response values from the actual response values. By choosing new levels near these "temporary optimal" ones for each parameter, the procedure is repeated until a convergence is achieved to a model which provides a reasonable fit to the observed data values. At this last stage, "temporary optimal" may be called "optimal". Usually no more than five to ten stages are needed to show some convergence in the sum of squares of deviations.

Insufficient convergence often occurs; this will indicate either that the ranges of the unknown parameters were incorrect, or that they were taken too wide, or perhaps that the assumed model should be changed. However, Taguchi's method is a useful technique for fitting complicated nonlinear regression models using only a limited number of actual observations.

The above steps can be demonstrated by the following example. A certain chemical reaction can be described by the nonlinear regression model

$$y = ACx_1/(1 + Ax_1 + Bx_2) + \varepsilon, \tag{7.20}$$

where y is the reaction rate, x_1 and x_2 are partial pressures of reactant and product respectively, A and B are absorption equilibrium constants for reactant and product respectively, and C is the effective reaction rate constant. The experimental data is given in Table 7.11. The unknown parameters are assumed, from past scientific knowledge, to fall in the ranges shown in Table 7.12.

First of all, considering A, B and C as factors, we create three equidistant levels for each factor as follows.

$$A_0 = 3.4 \quad B_0 = 12.5 \quad C_0 = 0.5$$
$$A_1 = 3.55 \quad B_1 = 12.75 \quad C_1 = 0.6$$
$$A_2 = 3.7 \quad B_2 = 13.0 \quad C_2 = 0.7.$$

Then we let *A*, *B* and *C* correspond to columns 1, 2 and 3 of the OA $L_9(3^4)$. Then for our nine experiments in the array, we will obtain nine equations as shown in Table 7.13.

If the equation in experiment 1 is a good fit to the actual data, the values of y thus calculated for every x_1 and x_2 should be close to the actual values of y in Table 7.11. To investigate "closeness", we calculate the sum of squares of the differences $(y-\hat{y})$ for experiment 1 as shown in Table 7.14. If the error sum of squares, *SSe*, for experiment 1 is small in comparison with the *SSe* for the other experiments, then the estimates of *A*, *B* and *C*, that is, $A_0B_0C_0$ (A = 3.4, B = 12.5, C = 0.5) are preferred over any other estimates. The *SSe* for all experiments are shown in Table 7.15.

Table 7.11 Reaction rate data (Box and Hunter, 1963).

Run	x_1	x_2	Y
1	1.0	1.0	0.126
2	2.0	1.0	0.219
3	1.0	2.0	0.076
4	2.0	2.0	0.126
5	0.1	0.0	0.186
6	3.0	0.0	0.606
7	0.3	0.0	0.318
8	3.0	0.8	0.298

Table 7.12 Ranges of unknown parameters.

Unknown parameters	Range
A	3.4–3.7
B	12.5–13.0
C	0.5–0.7

Table 7.13 Assignment by L_9 and regression equations.

Experiment number	Level combination	Equation for y
1	$A_0B_0C_0$	$(3.4)(0.5)x_1/(1+3.4x_1+12.5x_2)$
2	$A_0B_1C_1$	$(3.4)(0.6)x_1/(1+3.4x_1+12.75x_2)$
3	$A_0B_2C_2$	$(3.4)(0.7)x_1/(1+3.4x_1+13.0x_2)$
⋮	⋮	⋮
9	$A_2B_2C_1$	$(3.7)(0.6)x_1/(1+3.7x_1+13.0x_2)$

Table 7.14 Computation of sum of squares of the differences.

Run	x_1	x_2	y	$\hat{y}$	$y-\hat{y}$
1	1.0	1.0	0.126	$3.4\times0.5\times1.0/(1+3.4\times1.0+12.5\times1.0)=0.10059$	0.02541
⋮	⋮	⋮	⋮	⋮	⋮
8	3.0	0.8	0.298	$3.4\times0.5\times3.0/(1+3.4\times3.0+12.5\times0.8)=0.24057$	0.05743

Sum of squares of the differences (SSe) = $\sum_i (y_i-\hat{y}_i)^2$ = 0.037911

Table 7.15 Factor assignment in L_9 and SSe for first approximation.

Experiment number	Factor assignment and column number				SSe
	A 1	B 2	C 3	e 4	
1	0	0	0	0	0.0379110
2	0	1	1	1	0.0055997
3	0	2	2	2	0.0038186
4	1	0	1	2	0.0043903
5	1	1	2	0	0.0060844
6	1	2	0	1	0.0360040
7	2	0	2	1	0.0091201
8	2	1	0	2	0.0325770
9	2	2	1	0	0.0038295

If a formula fits well, error sum of squares should be small. We can now determine which of the levels for each factor is the best at this stage, by comparing the total for each level. For example,

$$SSe \text{ total for level } A_0 = 0.0379110 + 0.0055997 + 0.0038186 = 0.047329.$$

A similar calculation will give *SSe* the total levels A_1 and A_2, and all the results are shown in Table 7.16.

The question as to when a difference in the level total could be considered significant should be judged in comparison with a predetermined significant value *d*. Usually *d* will be 0.0001 or 0.00001, depending upon the magnitude of the total for the levels. In other words, if a difference is greater than *d*, the factor is significant, and successive approximation should continue until the difference in level totals becomes less than *d*.

Table 7.16 Total for the levels for first approximation.

Factor	Level 0	Level 1	Level 2	Significance ($d = 0.0001$)
A	0.047329	0.046479	0.045527	*
B	0.051421	0.044261	0.043652	*
C	0.106492	0.013820	0.019023	*

In Table 7.16 the significant value $d = 0.0001$ is used, and all the factors are judged significant. Note that the symbol "*" means significance; the symbol "–" will in later tables indicate non-significance. At the first stage, it is clear that the best factor levels (producing the minimum error sums of squares) are

$$A_2 = 3.7,\ B_2 = 13.0,\ C_1 = 0.6.$$

We can now take new levels of *A, B* and *C* close to their current optimum, $A_2B_2C_1$, and repeat the approximation procedure. The current optimum retains its value, and the distances between levels in the previous set of factors are halved for the new set. For example, for the second successive approximation, the following factor levels are used:

$$A_0 = 3.55 \quad A_1 = 3.625 \quad A_2 = 3.7$$
$$B_0 = 12.75 \quad B_1 = 12.75 \quad B_2 = 13.0$$
$$C_0 = 0.55 \quad C_1 = 0.6 \quad C_2 = 0.65.$$

We assign these new levels to orthogonal array L_9 and find the error variations for the nine new equations. As before, we investigate the significant differences for the error variations among the new A_0, A_1 and A_2. Table 7.17 gives the sums of the error variations between the three levels for the second approximation.

Table 7.17 Total for the levels for second approximation.

Factor	Level 0	Level 1	Level 2	Significance (d = 0.0001)
A	0.021489	0.020746	0.019903	*
B	0.021874	0.020119	0.020145	*
C	0.045308	0.012476	0.004356	*

In the second approximation, the error variation has become smaller than the first time, but the difference is still significant for each factor. We find that the optimum levels are $A_2B_1C_2$. Therefore, we take new levels for A, B and C as follows.

$$A_0 = 3.625 \quad A_1 = 3.6625 \quad A_2 = 3.7$$
$$B_0 = 12.8125 \quad B_1 = 12.875 \quad B_2 = 12.9375$$
$$C_0 = 0.6 \quad C_1 = 0.625 \quad C_2 = 0.65\,.$$

A simple computer program can handle the creation of new levels and as many successive approximation as required. In our example, the levels for the sixth approximation turn out to be

$$A_0 = 3.625 \quad A_1 = 3.6296875 \quad A_2 = 3.634375$$
$$B_0 = 12.90625 \quad B_1 = 12.921875 \quad B_2 = 12.9375$$
$$C_0 = 0.6375 \quad C_1 = 0.640625 \quad C_2 = 0.64375,$$

and the sums of error variations at the sixth approximation are as in Table 7.18.

Table 7.18 Total for the levels for sixth approximation.

Factor	Level 0	Level 1	Level 2	Significance (d = 0.0001)
A	0.003788	0.003824	0.003856	–
B	0.003858	0.003820	0.003790	–
C	0.003874	0.003790	0.003804	–

For our example, since no further iteration beyond the sixth approximation succeeded in giving a difference, say 0.0001 (the significance value), between the total of one level and the total of another level, we can therefore accept $A_1B_1C_1$ of the sixth approximation as the optimum levels of A, B and C. If the tolerance is the distance between adjacent two levels, then the estimates of A, B and C, with their tolerances, are:

$$A = 3.6296875 \pm 0.0046875,$$

$$B = 12.921875 \pm 0.015625,$$

$$C = 0.640625 \pm 0.005125.$$

Our formula therefore becomes

$$\hat{y} = \frac{3.6296875 \times 0.640625 x_1}{1 + 3.6296875 x_1 + 12.921875 x_2}. \tag{7.21}$$

The error sum of squares and the error variance of this formula are found to be:

$$SSe = 0.0012633,$$

$$V_e = \mathrm{SS}e/8 = 0.000157912.$$

Therefore, an approximate 95% confidence interval for $E(y)$ using (7.21) is,

$$\hat{y} \pm 2\sqrt{V_e} = \hat{y} \pm 2\sqrt{0.000157912}$$
$$= \hat{y} \pm 0.025133.$$

In the absence of a better alternative, experimental estimation of non-linear regression parameters using orthogonal arrays in the manner outlined above is an attractive method when there are only a limited number of actual observations.

7.5 SIMULTANEOUS OPTIMIZATION FOR MULTIPLE CHARACTERISTICS

For many experimental situations in practice, more than one response will be measured for the different combination of values which a set of design variables may take. When we have several responses, the optimum condition for one response is not very likely equal to the optimum condition for the other response. In this section, we will discuss how to find the overall optimum condition for several responses when we employ parameter designs using orthogonal arrays.

This **simultaneous optimization** can be best explained through an example. Suppose, in the plastic experiment in Tables 7.6 and 7.7, there are further responses of interest, say, wear, appearance and defectiveness, in addition to strength. For each experiment in Table 7.7, another set of wear data is observed at each level N_{-1}, N_0 and N_1. Also for each experiment, 100 plastic items are made and their appearance and defectiveness checked. The extra data are given in Table 7.19.

Wear is a smaller-the-better characteristic, and SN ratios are computed as follows. For instance, for the first experiment,

$$\mathrm{SN}_1 = -10\log\left(\frac{1}{3}\sum_{j=1}^{3} y_{ij}^2\right)$$
$$= -10\log[\tfrac{1}{3}(30^2 + 25^2 + 38^2)] = -39.96 \text{ dB}.$$

Table 7.19 Additional data for the plastic experiment.

Experiment number	Control factor and column number			Wear data			Appearance			Defectiveness (number of non-defectives out of 100 items)
	A 2	*B* 3	*C* 4	N_{-1}	N_{-0}	N_1	good	fair	poor	
1	-1	-1	-1	30	25	38	80	20	0	85
2	-1	0	0	15	11	10	85	15	0	90
3	-1	1	1	29	31	22	87	13	0	88
4	0	-1	-1	8	14	21	76	22	2	90
5	0	0	0	9	7	15	75	20	5	97
6	0	1	1	19	17	22	74	25	1	94
7	1	-1	0	9	12	5	90	9	1	90
8	1	0	1	14	20	17	96	4	0	95
9	1	1	-1	8	15	17	94	5	1	91
10	-1	-1	1	25	29	28	88	10	2	76
11	-1	0	-1	9	19	13	87	12	1	84
12	-1	1	0	12	15	14	85	15	0	78
13	0	-1	0	9	12	18	76	23	1	87
14	0	0	1	15	22	23	78	20	2	95
15	0	1	-1	12	13	8	79	19	2	90
16	1	-1	1	18	25	23	68	303	2	78
17	1	0	-1	11	19	23	70	28	2	83
18	1	1	0	11	7	10	77	20	3	82

For the appearance data, the weights 0, 1 and 2 are given for good, fair and poor appearance, respectively. For the first experiment, the SN ratio is computed from formula (6.7) as follows.

$$\mathrm{SN}_1 = -10\log\left(\frac{\sum_i w_i^2 n_i}{n}\right)$$

$$= -10\log\left(\frac{0^2 \times 80 + 1^2 \times 20 + 2^2 \times 0}{100}\right)$$

$$= -10\log(0.20) = 6.99\mathrm{dB}.$$

For non-defectiveness, the SN ratios are computed from (6.1). For the first experiment,

$$\mathrm{SN}_1 = -10\log\left(\frac{1}{p}-1\right)$$
$$= -10\log\left(\frac{1}{0.85}-1\right) = 7.533 \text{ dB}.$$

The SN ratios are tabulated in Table 7.20.

Table 7.20 SN ratios for wear, appearance and defectiveness data.

Experiment number	SN for wear data	SN for appearance	SN for defectiveness
1	-29.96	6.69	7.533
2	-21.73	8.24	9.542
3	-28.82	8.86	8.653
4	-23.69	5.23	9.542
5	-20.72	3.98	15.057
6	-25.77	6.38	11.950
7	-19.19	8.86	9.542
8	-24.70	13.98	12.788
9	-22.86	10.46	10.048
10	-28.75	7.46	5.006
11	-23.10	7.96	7.202
12	-22.74	8.24	5.497
13	-22.62	5.68	8.256
14	-26.16	6.53	12.788
15	-21.00	5.69	9.542
16	-26.93	4.20	5.497
17	-25.28	4.44	6.886
18	-19.54	4.95	6.585

7.5.1 ANALYSIS OF WEAR DATA

Sums of wear SN ratios at each level of *A*, *B* and *C* are given in Table 7.21. The respective variation of wear data for each factor is obtained as follows, and the ANOVA table is given in Table 7.22.

$$CT = \frac{(-433.36)^2}{18} = 10443.02,$$

$$S_T = (-29.96)^2 + (-21.73)^2 + \ldots + (-19.54)^2 - 10443.02 = 173.33,$$

$$S_A = \tfrac{1}{6}[(-155.10)^2 + (-139.96)^2 + (-138.50)^2] - 10443.02 = 28.16.$$

Similarly, we obtain S_B = 11.03 and S_C = 100.17, so that

$$S_e = S_T - S_A - S_B = 33.97.$$

From Tables 7.21 and 7.22, we find that the optimum condition for significant factors is A_1C_0.

Table 7.21 Sum of wear SN ratios at each level of each factor.

Level	Factor		
	A	B	C
-1	-155.10	-151.14	-145.89
0	-139.96	-141.69	-126.54
1	-138.50	-140.73	-161.13
Total	-433.56	-433.56	-433.56

Table 7.22 ANOVA table for wear data.

Source of variation	Sum of squares	Degrees of freedom	Mean square	F_0
A	28.16	2	14.08	4.56*
B	11.03	2	5.52	1.79
C	100.17	2	50.09	16.21**
E	33.97	11	3.09	
T	173.33	17		

** indicates 1% significance.
* indicates 5% significance.

7.5.2 ANALYSIS OF APPEARANCE DATA

Sums of appearance SN ratios at each level of *A, B* and *C* are given in Table 7.23. From this data we can obtain the ANOVA table as shown in Table 7.24. From Table 7.24, we judge that *A* is slightly significant, and A_{-1} is the optimum level. Recall that in the analysis of OAs, if the value of F_0 is greater than 2 but less than F(0.1), the factor may be judged to be slightly significant.

Table 7.23 Sum of appearance SN ratios.

Level	Factor		
	A	B	C
-1	47.74	38.42	40.77
0	31.50	44.13	39.96
1	46.89	43.58	45.40
Total	126.13	126.13	126.13

Table 7.24 ANOVA table for appearance data.

Source of variation	Sum of squares		Degrees of freedom	Mean square	F_0
A	27.85		2	13.93	2.52
B	3.31	○	2	1.65	
C	2.87	○	2	1.44	
e	76.85		11	6.99	
(*e*)	83.03		15	5.53	
T	110.88		17		

7.5.3 ANALYSIS OF DEFECTIVENESS DATA

Sums of defectiveness SN ratios at each level of *A, B* and *C* are given in Table 7.25. The ANOVA table for defectiveness SN ratio is constructed in Table 7.26. It shows that factor *C* is pooled into the error term, that *A* is significant at the 5% level, and *B* is significant at the 10% level. From Tables 7.25 and 7.26, we judge that the optimum condition for the significant factors A and B is A_0B_0.

Table 7.25 Sum of defectiveness SN ratios.

Level	Factor		
	A	B	C
-1	43.433	45.376	50.753
0	67.175	64.303	54.519
1	51.346	52.275	56.682
Total	161.354	161.354	161.354

Table 7.26 ANOVA table for defectiveness data.

Source of variation	Sum of squares	Degrees of freedom	Mean square	F_0
A	48.714	2	24.357	5.82*
B	30.584	2	15.292	3.65
C	3.001 ○	2	1.501	
e	51.402	11	4.673	
(e)	54.403	15	4.185	
T	133.701	17		

* indicates 5% significance.
○ indicates pooled variations.

7.5.4 SIMULTANEOUS OPTIMIZATION

We have examined four responses (strength, wear, appearance and defectiveness), and have found that each response has different significant factors and different optimal levels. How, then, can we find the overall simultaneous optimum levels for all the factors involved? We propose the following method for finding the simultaneous optimum condition through the above example.

Table 7.27 Overall summary table for optimal conditions.

Factor	Level	Strength B^{**}	Wear $A^{*}C^{**}$	Appearance A^{+}	Defectiveness $A^{*}B^{+}$	Overall optimum
A	-1		-155.10	47.74	43.433	
	0		-139.96	31.50	67.175	○
	1		138.50	46.89	51.346	
B	-1	210.30			45.376	
	0	205.96			64.303	
	1	222.02			52.275	○
C	-1		145.89			
	0		126.54			○
	1		161.13			

** indicates 1% significance.
* indicates 5% significance.
+ indicates 10% significance.

First of all, a summary table such as Table 7.27 needs to be constructed, in which only the level sums of SN ratios of significant factors appear. For factor C, only wear is significantly different, and the optimum condition is C_0. However, for factors A and B, more than one response is significantly different, and unfortunately the optimum levels are different. For factor A, A_1, A_{-1} and A_0 are optimal conditions for wear, appearance and defectiveness, respectively. However, since factor A is less meaningful in appearance than wear and defectiveness, we may ignore A_{-1} for the optimal condition. Note there is not much difference between A_0 and A_1 in wear, and A_0 is much better than A_1 in defectiveness. Therefore, A_0 can be judged to be optimal. For factor B, strength has more significance than defectiveness. Hence, we may select B_1 as the optimal condition.

As a conclusion, the overall simultaneous optimum condition can be determined to be $A_0B_1C_0$. For this condition, a confirmation test should be conducted to check whether it really is an acceptable condition.

Exercises

7.1 Prove equations (7.4), (7.5), (7.10) and (7.11).

7.2 Suppose we obtain the following response y for two control variables x_1 and x_2, and a noise variable z.

Run number	x_1	x_2	z	y
1	-1	-1	-1	80.6
2	-1	-1	1	74.9
3	-1	1	-1	83.1
4	-1	1	1	71.2
5	1	-1	-1	66.8
6	1	-1	1	74.2
7	1	1	-1	38.1
8	1	1	1	36.8
9	-1.41	0	0	80.9
10	1.41	0	0	42.4
11	0	-1.41	0	73.4
12	0	1.41	0	45.0
13	0	0	0	77.4
14	0	0	0	74.6

(a) Fit the second-order polynomial response model using the combined array approach (see (7.9)).

(b) Find the mean and variance response models (see (7.10) and (7.11)).

(c) Find the optimum condition of x_1, and x_2, which maximizes $\hat{m}(x)$.

7.3 Suppose we have the same problem as in Tables 7.1 and 7.2, but the yield data are different as follows.

(a) Construct a new ANOVA table for SN ratios of yield data to replace Table 7.4. What is the optimum level for significant factors?

(b) Suppose the noise variable z is assigned to column 4, and interactions are assigned to columns 5, 6, 10, 12 and 15 just as in Table 7.5. Select a yield response from the two responses for each experiment in the above yield data. Find $\hat{y}(\mathbf{x}, z)$, $\hat{m}(\mathbf{x})$ and $\hat{v}(x)$ using equations (7.12), (7.13) and (7.14).

Experiment number	N_{-1}	N_{+1}	Experiment number	N_{-1}	N_{+1}
1	73	79	9	80	85
2	80	87	10	81	86
3	77	83	11	83	91
4	74	79	12	70	76
5	86	84	13	81	75
6	71	68	14	82	76
7	82	80	15	74	70
8	79	75	16	84	81

(c) If there is no restriction on $\hat{v}(x)$, what is the optimum condition which maximizes $\hat{m}(\mathbf{x})$?

(d) What is the optimum condition which minimizes $\hat{v}(x)$ under the restriction that $\hat{m}(\mathbf{x}) \geq 83$?

7.4 In a chemical reaction experiment, there are five control factors (as in Table 7.1) and a noise factor. L_{16} is used and the control factors are assigned to columns 1, 2, 8, 11 and 14 just as in Table 7.2. Suppose we obtain responses as follows. Note that the yield data are the same as those in Table 7.2. For appearance and defectiveness, 200 items per experiment are inspected.

(a) Carry out an appropriate analysis for each response. What is the optimal condition for significant control factors?

(b) Construct an overall summary table like the one in Table 7.27. What is the overall optimum condition for the control factors?

Experiment number	Yield (%)		Wear data		Appearance			Defectiveness (number of non-defectives out of 200 items)
	N_{-1}	N_{+1}	N_{-1}	N_{+1}	good	fair	poor	
1	59	63	13	19	120	60	20	167
2	59	64	18	26	145	50	5	170
3	65	69	17	24	135	55	10	175
4	51	56	14	19	128	60	12	180
5	69	75	26	22	167	30	3	177
6	61	65	11	8	158	30	12	190
7	67	74	22	20	168	25	7	188
8	55	59	19	15	176	20	4	173
9	60	55	20	25	180	18	2	193
10	51	45	21	25	188	11	1	194
11	63	57	23	31	190	9	1	195
12	47	44	10	14	187	10	3	192
13	64	61	21	15	191	9	0	181
14	65	58	22	16	189	10	1	189
15	68	62	14	12	193	6	1	190
16	56	51	24	21	184	16	0	194

7.5 Suppose we have a nonlinear regression model

$$E(y) = A + (1 - A - B)/(1 + C x^{-D}) .$$

The ranges of interest for A, B, C and D are as follows.

A: 0.00–0.08, B: 0.00–0.04,

C: 0.2–1.0, D: 0.5–1.6.

Using the nonlinear regression technique of Section 7.4, estimate A, B, C and D. Use a significant difference of d = 0.0001. The observed data for x and y are as follows.

i	1	2	3	4	5	6	7
x_i	0.01	0.12	0.54	1.52	5.60	25.30	55.20
y_i	0.0072	0.0719	0.372	0.708	0.926	0.975	0.980

8

Parameter Design for Dynamic Characteristics

8.1 DYNAMIC CHARACTERISTICS

Suppose the objective in an experiment is to achieve an intention or a target value. To actualize the objective, a signal input should be dispatched. The output result to be actualized by an intention or a target value is called a **dynamic characteristic** (Figure 8.1). Usually the signal input is not transferred unaltered to the output result because of intervening noise.

Since the signal input, or signal factor, technically adjusts the dynamic characteristic as time changes, it is often called the adjusting factor. For example, in the case of a car, characteristics such as speed and steerability are typical dynamic characteristics. Suppose the driver's intention is to precisely drive or control a car by pressing the accelerator. Here the signal factor is the degree of pressure on the accelerator, and the resulting speed is the dynamic characteristic. However, there are various kinds of noise between the input and the output such as driver's lack of skill, the condition of the road, the car's mechanical condition and so on. Then the question is how a car should be designed so that it can be driven or controlled according to the driver's intention under noise conditions. How can an experiment be designed to test such characteristics? Answering these questions is one of the most important areas of application in the design of experiments.

So far, we have confined ourselves to treating quality characteristics which are not dynamic. These characteristics may be called static characteristics. In this chapter, we will deal only with dynamic characteristics.

Further examples of dynamic characteristics are the performance of skis, golf balls or tennis rackets, the athletic ability of humans, and the rolling ability of mills. In playing tennis, for example, the player shifts his position to the right or to the left depending on whether he wants to use a forehand or backhand stroke to hit the ball. Given an intention or target, a signal factor must be altered to carry it out. In playing tennis, the signal factor is the player's positioning which must be changed by an amount that matches the target. This is the reason for the term "dynamic".

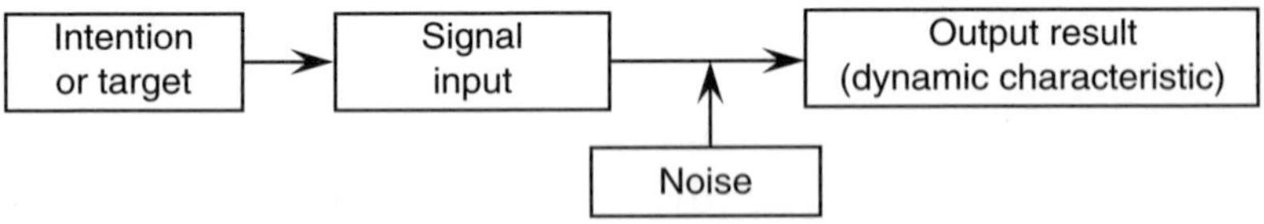

Figure 8.1 Dynamic characteristic resulting from signal and noise.

In parameter design for dynamic characteristics, product or process design should be conducted in such a way that an intention should result which is robust to noise factors. In other words, as the signal levels change, the power of the signal should be directly transmitted to the result which is robust to the power of noise. Therefore, the SN ratio for dynamic characteristic is defined as

$$SN = 10\log(\eta), \tag{8.1}$$

where η is the power of the signal divided by power of the noise.

Note that if the noise factor does not have much effect, and the signal input is almost entirely transmitted to the output result, then the SN value becomes large. The larger the SN ratio is, the better. In parameter design, we wish to find the values of design parameters which provide large SN ratios.

Classification of factors was studied in Chapter 3. The factors which are often used in practice are control factors, indicative factors, signal factors and noise factors. In parameter design for dynamic characteristics, the control and indicative factors are assigned to the inner array, and the signal and noise factors are assigned to the output array. The SN ratio is computed using equation (8.1), in which the power of the signal is the amount of variation occurring as the level of the signal factor changes. The power of the noise is the variation occurring as the level of the noise factor changes. Detailed explanation will be given by means of examples in the following sections.

8.2 FACTORIAL EXPERIMENTS

8.2.1 TWO-WAY LAYOUT

Suppose we have a signal factor M with l levels and a noise factor N with m levels, as shown in Table 8.1. Here y_{ij} represents the dynamic characteristics. Suppose the ideal relationship between y and M is

$$y = \beta_0 + \beta_1 M + \varepsilon, \qquad \text{var}(\varepsilon) = \sigma^2 . \tag{8.2}$$

Table 8.1 Two-way layout for a signal and a noise factor.

Level of noise factor	Level of signal factor				
	M_1	M_2	···	M_l	
N_1	y_{11}	y_{21}		y_{l1}	
N_2	y_{12}	y_{22}		y_{l2}	
⋮	⋮	⋮		⋮	
N_m	y_{1n}	y_{2n}		y_{ln}	
Sum	$T_{1.}$	$T_{2.}$	···	$T_{l.}$	T
Average	$\bar{y}_{1.}$	$\bar{y}_{2.}$	···	$\bar{y}_{l.}$	$\bar{\bar{y}}$

Then the total variation is

$$S_T = \sum_i \sum_j (y_{ij} - \bar{\bar{y}})^2 = \sum_i \sum_j y_{ij}^2 - \frac{T^2}{lm}, \qquad \phi = lm - 1. \quad (8.3)$$

Let S_β be the regression variation due to the signal factor M. It can be found from many regression textbooks that, if there is a data set $(x_i, y_j), i = 1, 2, \ldots, n,$

$$S_\beta = \frac{S_{(xy)}^2}{S_{(xx)}}, \qquad E(S_\beta) = \sigma^2 + S_{(xx)}\beta_1^2,$$
$$\hat{\beta}_1 = \frac{S_{(xy)}}{S_{(xx)}}, \qquad \hat{\beta}_0 = \bar{y} - \hat{\beta}_1 \bar{x}, \quad (8.4)$$

where $S_{(xx)} = \sum_i (x_i - \bar{x})^2, \quad S_{(xy)} = \sum_i (x_i - \bar{x})(y_i - \bar{y})$.

In Table 8.1, M is the independent variable x, and y is the dependent variable. Therefore, we obtain the following regression results:

$$S_\beta = \frac{\left[\sum_i \sum_j (M_i - \bar{M})(y_{ij} - \bar{\bar{y}})\right]^2}{m \sum_i (M_i - \bar{M})^2}, \qquad \phi = 1$$
$$= \frac{\left[\sum_i \sum_j M_i (y_{ij} - \bar{\bar{y}})\right]^2}{m \sum_i (M_i - \bar{M})^2}, \quad (8.5)$$

$$\hat{\beta}_1 = \frac{\sum_i \sum_j M_i (y_{ij} - \bar{\bar{y}})}{m \sum_i (M_i - \bar{M})^2}, \qquad \hat{\beta}_0 = \bar{\bar{y}} - \hat{\beta}_1 \overline{M}$$

$$S_e = S_T - S_\beta,$$

$$V_e = \hat{\sigma}^2 = \frac{S_e}{lm-2}. \tag{8.6}$$

Note that V_e is an unbiased estimator of σ^2. Also, from equation (8.4), we obtain

$$\hat{\beta}_1^2 = \frac{1}{r}(S_\beta - V_e), \tag{8.7}$$

where $r = m\sum_i (M_i - \bar{M})^2$. Here r is often called the effective divisor due to level changes of signal factor M.

By definition, the power of signal is represented by $\hat{\beta}_1^2$, and the power of noise by $\hat{\sigma}^2$. Therefore, the SN ratio for this experiment becomes, from (8.1),

$$\text{SN} = 10\log\left(\frac{\hat{\beta}_1^2}{\hat{\sigma}^2}\right)$$

$$= 10\log\left[\frac{\frac{1}{r}(S_\beta - V_e)}{V_e}\right]. \tag{8.8}$$

Very often in the simple regression model of (8.2), y should be zero when $M = 0$. In this case the model becomes

$$y = M\beta + \varepsilon. \tag{8.9}$$

In this simple situation, the quantities S_T, S_β, r, V_e and SN of (8.3)–(8.8) become:

$$S_t = \sum_i \sum_j y_{ij}^2, \qquad \phi = lm$$

$$r = m\sum_i (M_i)^2,$$

$$S_\beta = \frac{\left[\sum_i M_i T_{i.}\right]^2}{r}, \qquad \phi = 1$$

$$\hat{\beta} = \frac{\sum_i M_i T_{i.}}{r},$$

$$S_e = S_T - S_\beta,$$

$$V_e = \hat{\sigma}^2 = \frac{S_e}{lm-1}, \qquad \phi = lm-1$$

$$\text{SN} = 10\log\left[\frac{\frac{1}{r}(S_\beta - V_e)}{V_e}\right]. \qquad (8.10)$$

8.2.2 DYEING EXPERIMENT

In order to improve the color density of a textile product, a dyeing experiment is conducted in which the signal factor is the quantity of dye used and the color density is the dynamic characteristic. There is a control factor with two levels,

$$A_1 = \text{new dye}$$
$$A_2 = \text{current dye},$$

and a noise factor with three levels which reflects all kinds of error sources such as cleanliness of cloth, environmental conditions, dyeing temperature and so on.

$$N_1 = \text{good condition}$$
$$N_2 = \text{normal condition}$$
$$N_3 = \text{poor condition}.$$

The data set for color density is given in Table 8.2. Color density is analyzed for each level of A below. It is assumed that the simple regression model of (8.2) is appropriate.

Table 8.2 Color density of dyeing experiment.

Levels of control and noise factor		Levels of signal factor					
		M_1 (1%)	M_2 (2%)	M_3 (3%)	M_4 (4%)	M_5 (5%)	
A_1	N_1	6.8	8.6	10.4	12.2	13.8	
	N_2	6.7	8.4	10.2	12.0	13.5	$T = 152.9$
	N_3	6.7	8.3	10.1	11.9	13.3	$\bar{\bar{y}} = 10.193$
	Sum	20.2	25.3	30.7	36.1	40.6	
A_2	N_1	6.9	8.9	10.6	12.5	13.9	
	N_2	6.7	8.7	10.4	12.4	13.7	$T = 155.6$
	N_3	6.6	8.4	10.3	12.1	13.5	$\bar{\bar{y}} = 10.373$
	Sum	20.2	26.0	31.3	37.0	41.1	

Analysis of new dye (A_1)

$$S_T = \sum_i \sum_j y_{ij}^2 - \frac{T^2}{lm}$$

$$= 6.8^2 + 6.7^2 + \ldots + 13.3^2 - \frac{152.9^2}{5 \times 3} = 89.11,$$

$$S_\beta = \frac{\left[\sum_i \sum_j (M_i - \bar{M})(y_{ij} - \bar{\bar{y}})\right]^2}{m \sum_i (M_i - \bar{M})^2}$$

$$= \frac{[1 \times (6.8 - 10.193) + \ldots + 5 \times (13.3 - 10.193)]^2}{3\left[(1-3)^2 + \ldots + (5-3)^2\right]}$$

$$= \frac{51.615^2}{30} = 88.80,$$

$$S_e = S_T - S_\beta = 89.11 - 88.80 = 0.31,$$

$$V_e = \frac{S_e}{lm - 2} = \frac{0.31}{13} = 0.0238,$$

$$\mathrm{SN} = 10\log\left[\frac{\frac{1}{r}(S_\beta - V_e)}{V_e}\right]$$

$$= 10\log\left[\frac{\frac{1}{30}(88.80 - 0.0238)}{0.0238}\right] = 20.93\ \mathrm{dB}. \qquad (8.11)$$

Also note that, since

$$\hat{\beta}_1 = \frac{\sum_i \sum_j M_i (y_{ij} - \bar{\bar{y}})}{m\sum_i (M_i - \bar{M})^2} = \frac{51.165}{30} = 1.7205,$$

$$\hat{\beta}_0 = \bar{\bar{y}} - \hat{\beta}_1 \bar{M} = 10.193 - 1.7205 \times 3 = 5.0315,$$

the fitted regression equation is

$$\hat{y} = 5.0315 + 1.7205M \qquad (8.12)$$

with coefficient of determination

$$R^2 = \frac{S_\beta}{S_T} = \frac{88.80}{89.11} = 0.9965.$$

Analysis of current dye (A_2)

For the set of data for A_2, we can similarly obtain the following results:

$$S_T = 93.61, \qquad S_\beta = 92.98,$$

$$V_e = 0.0485, \qquad \mathrm{SN} = 28.06\ \mathrm{dB}, \qquad (8.13)$$

$$\hat{y} = 5.0915 + 1.7605M, \quad R^2 = 0.9933. \qquad (8.14)$$

Comparison of A_1 and A_2

Table 8.3 summarizes the above results. It shows that the new dye (A_1) has better R^2, V_e and SN than the current dye (A_2), which means that the color density can be better controlled by the quantity of new dye than that of current dye. Hence, if possible, the new dye should be used. However, it should be noted that the current dye has a slightly bigger slope than the new dye, which means that the current dye affects the color density slightly more than the new dye. But as far as controllability and variability of color density are concerned, we judge that the new dye is better than the current dye.

Table 8.3 Summary for analysis of two dyes.

Type of dye	R^2	V_e	SN (dB)	Fitted equation
New dye (A_1)	0.9965	0.0238	20.93	$\hat{y}$ =5.0315+ 1.7205M
Current dye (A_2)	0.9933	0.0485	18.06	$\hat{y}$ =5.0915+ 1.7605M

8.3 ORTHOGONAL ARRAY EXPERIMENT I

In the previous section, a two-way factorial parameter design for a dynamic characteristic was investigated. In this section we will examine a parameter design for another dynamic characteristic, which uses an orthogonal array.

8.3.1 DYNAMITE BLASTING EXPERIMENT

It is required that dynamite for blasting be delay-blasted in accordance with a specified time-lag which depends on the charge quantity. The dynamic characteristic is the time lag (measured in milliseconds), and the signal factor M is the charge quantity. The factor M was taken at the three levels $M_1 = 50\text{mg}$, $M_2 = 100\text{mg}$ and $M_3 = 150\text{mg}$. There are five control factors which may affect the time-lag as follows:

A : maker of dynamite, A_0 = company P, A_1 = company W
B : proportion of sodium nitrate, $B_0 = 20\%$, $B_1 = 25\%$
C : proportion of strontium nitrate, $C_0 = 10\%$, $C_1 = 12\%$
D : ground depth of dynamite, $D_0 = 50\text{cm}$, $D_1 = 100\text{cm}$
F : angle of dynamite pipe, $F_0 = 20°$, $F_1 = 30°$.

In all cases above, the first level is the current method and the second level is a new method. The four control factors were assigned to L_8; Table 8.4 shows the data on time to blasting obtained over two days (R_1 and R_2). We wish to find the optimum conditions of A, B, C, D and F under which the charge quantity M controls the time-lag with the minimum error. Assume there is no time-lag in the case of charge quantity 0.

Table 8.4 Parameter design and data for dynamic blasting experiment.

Experiment number	Factor/column					M_1 (50)		M_2 (100)		M_3 (150)		SN (dB)
	A	B	C	D	F	R_1	R_2	R_1	R_2	R_1	R_2	
	1	2	3	4	7							
1	0	0	0	0	0	14	13	27	28	40	42	-9.21
2	0	0	0	1	1	13	13	26	27	42	40	-10.40
3	0	1	1	0	1	15	16	30	32	46	47	-7.96
4	0	1	1	1	0	16	17	31	33	47	48	-8.83
5	1	0	1	0	1	14	13	29	28	40	42	-11.26
6	1	0	1	1	0	14	15	27	28	41	43	-10.07
7	1	1	0	0	0	17	18	33	32	48	46	-12.91
8	1	1	0	1	1	18	19	34	34	47	49	-14.95

8.3.2 ANALYSIS OF DATA

It is assumed that model (8.9), which has no intercept, is appropriate for this experiment. The formula in (8.10) will be used to obtain SN ratios. For the first experiment, we can compute the SN ratio as follows:

$$S_T = \sum_i y_i^2$$

$$= 14^2 + 13^2 + \ldots + 42^2 = 5242,$$

$$r = 2\sum_i M_j^2 = 2 \times (50^2 + 100^2 + 150^2) = 70000,$$

$$S_\beta = \frac{\left(\sum_i M_i T_i\right)^2}{r}$$

$$= \frac{(50 \times 27 + 100 \times 55 + 150 \times 82)^2}{70000}$$

$$= \frac{19150^2}{70000} = 5238.89,$$

$$S_e = S_T - S_\beta = 3.11,$$

$$V_e = \frac{3.11}{5} = 0.622,$$

$$\mathrm{SN} = 10\log\left[\frac{\frac{1}{r}(S_\beta - V_e)}{V_e}\right]$$

$$= 10\log\left[\frac{\frac{1}{70000}(5238.89 - 0.622)}{0.622}\right] = -9.21\mathrm{dB}.$$

Table 8.5 Sums (and means) of SN ratios at the various factor levels.

Level	Factor				
	A	*B*	*C*	*D*	*F*
0	-36.4 (-9.1)	-40.94 (-10.235)	-47.47 (-11.868)	-41.34 (-10.335)	-41.02 (-10.255)
1	-49.19 (-12.298)	-44.65 (-11.163)	-38.12 (-9.53)	-44.25 (-11.063)	-44.57 (-11.143)
Total	-85.59	-85.59	-85.59	-85.59	-85.59

Similar calculations yield the remaining SN ratios; their values are listed in the final column of Table 8.4. The sums and means of SN ratios at the various factor levels are summarized in Table 8.5.

The ANOVA table is shown in Table 8.6, which tells us that A is significant at the 5% level, C at the 10% level, and B and F cannot be ignored because of their high values of F_0 (recall that if the value of F_0 is greater than 2 in OA analysis, we may assume it is somewhat significant). Therefore, we may conclude that A, B, C and F are significant, and the optimum levels of the significant factors are $A_0B_0C_1F_0$.

The selection of levels for the insignificant factor D is usually based on cost differences, operability or other quality characteristics. Assume that D_0 is selected. The estimate of the SN ratio at the optimum condition $A_0B_0C_1D_0F_0$ is made by summing all the factorial effects shown in Table 8.5:

$$\begin{aligned}\hat{\mu}(A_0B_0C_1D_0F_0) &= \overline{A}_0+\overline{B}_0+\overline{C}_1+\overline{D}_0+\overline{F}_0-4\overline{T}\\ &= -9.1-10.235-9.53-10.335-10.255-4\times(-85.59)/8\\ &= -6.66\text{dB}.\end{aligned}$$

The SN ratio at the current condition (experiment number 1) was -9.21 dB (Table 8.4). Therefore, the gain that results from using the optimum condition is

$$-6.66-(-9.21)=2.55\text{dB}.$$

Note that SN ratio in (8.10),

$$\text{SN}=10\log(\eta),$$

$$\eta=\frac{\hat{\beta}_1^2}{\hat{\sigma}^2}=\frac{\frac{1}{r}(S_\beta-V_e)}{V_e}, \tag{8.15}$$

where η represents the strength of the straight-line relationship between the charge quantity and the time lag. Let η_0 be the strength of the straight-line relationship at the optimum condition, and η_c be that at the current condition. Then

$$\begin{aligned} 2.55 &= 10\log(\eta_0) - 10\log(\eta_c) \\ &= 10\log(\eta_0 / \eta_c). \end{aligned} \tag{8.16}$$

Hence, $\eta_0 / \eta_c = 1.8$. This means that, by changing the current condition to the optimum condition, the degree of linearity can be improved by 80%.

Table 8.6 ANOVA table for SN ratios for dynamic blasting experiment *A*.

Source of variation	Sum of squares	Degrees of freedom	Mean square	F_0
A	20.448	1	20.448	32.94*
B	1.721	1	1.721	2.77
C	10.928	1	10.928	17.60**
D	1.059	1	1.059	1.71
F	1.575	1	1.575	2.54
E	1.242	2	0.621	
T	36.973	7		

*indicates 5% significance.

**indicates 10% significance.

In order to predict the time lag at an arbitrary charge quantity at the optimum condition, we need to find an appropriate simple regression equation. First of all, the means of time lags at factor levels for each charge quantity should be computed as shown in Table 8.7.

For example in Table 8.7, at $M_1 = 50$, the mean at A_0 is calculated as

$$(14+13+13+13+15+16+16+17)/8 = 14.625.$$

Now, at the optimum condition the estimated time lags become, at $M_1 = 50$,

$$\begin{aligned} \hat{\mu} &= \overline{A}_0 + \overline{B}_0 + \overline{C}_1 + \overline{D}_0 + \overline{F}_0 - 4\overline{T} \\ &= 14.625 + 13.625 + 15 + 15 + 15.5 - 4 \times 15.3125 = 12.5; \end{aligned}$$

at $M_2 = 100,$

$$\hat{\mu} = 29.25 + 27.5 + 29.75 + 29.875 + 29.875 - 4 \times 29.9375$$
$$= 26.5;$$

and at $M_3 = 150,$

$$\hat{\mu} = 44 + 41.25 + 44.24 + 43.875 + 44.375 - 4 \times 44.25$$
$$= 40.75.$$

Table 8.7 Means of time lags at factor levels for each charge quantity.

Quantity of dye	Level	Factor					Total mean
		A	*B*	*C*	*D*	*F*	
M_1=50	0	14.625	13.625	15.625	15	15.5	15.3125
	1	16	17	15	15.625	15.125	
M_2=100	0	29.25	27.5	30.125	29.875	29.875	29.9375
	1	30.625	32.375	29.75	30	30	
M_3=150	0	44	41.25	44.25	43.875	44.375	44.25
	1	44.5	47.25	44.25	44.625	44.125	

Therefore, we have a data set for a simple regression line as follows:

Charge quantity (M)	M_1=50	M_2=100	M_3=150
Time lag (y)	12.5	26.5	40.75

If we fit a straight regression line, we obtain that

$$\hat{y} = 0.2682M$$

with the coefficient of determination, $R^2 = 0.9995$, which is almost a perfect straight-line. If we use this line, we can predict the time lag at an arbitrary charge quantity with very good precision.

8.4 ORTHOGONAL ARRAY EXPERIMENT II

In the previous section, a two-level OA was used in a parameter design for a dynamic characteristic. In this section, a three-level OA will be used for another parameter design for a dynamic characteristic.

8.4.1 OXIDATION OF SEMICONDUCTOR PROCESS

An experiment was performed on a thin oxidation semiconductor process in a vertical diffusion furnace. The function of the thin oxidation process is to grow a layer of silicon dioxide uniformly on the surface of a silicon wafer. An oxide is grown on a silicon wafer when it is exposed to high temperatures and an oxidizing atmosphere. The average thickness and the uniformity of the oxide grown on the silicon wafer are critical for robust device performance. This experiment was performed in a semiconductor company to find out which combinations of process parameters and their settings result in minimum thickness variations over a range of target thicknesses. We are therefore concerned with one dynamic characteristic, oxide thickness (Å).

Many control factors could have been considered. However, only four factors were chosen, based on their ranking in importance to the process. The four factors and their levels are as follows. The current setting is "normal" in each case.

Control factor	Level 0	Level 1	Level 2
Temperature(A)	-10%	nominal	+10%
Total gas flow(B)	-10%	nominal	+10%
N_2 dilution(C)	-15%	nominal	+15%
H_2/O_2 ration(D)	-5%	nominal	+5%

There is one signal factor: oxidation time (ms). The time was set at four levels: 15, 20, 25 and 30.

A number of noise factors can be considered: variations in environmental conditions such as ambient temperature, pressure and humidity, and also gas flows and pressure across the boat position and across the wafer within the furnace. These noises could be reflected in the measurements taken at various positions of the wafers in the boat. Here only the top (T) and bottom (B) positions are measured.

The L_9 OA was used to allocate the four control factors, as shown in Table 8.8. Using L_9 with four levels for the signal factor resulted in a total of 36 experimental runs. Each run had two factors, top and bottom, and each wafer measured using an ellipsometer. The oxidation times were chosen to cover a range of oxide thicknesses from $30\mathring{A}$ to $100\mathring{A}$.

Table 8.8 L_9 dynamic array for oxidation process experiment.

Experiment number	Factor/column				T_1 (15)		T_2 (20)		T_3 (25)		T_4 (30)		SN (dB)
	A 1	*B* 2	*C* 3	*D* 4	T	B	T	B	T	B	T	B	
1	0	0	0	0	35	36	44	43	54	56	67	68	2.88
2	0	1	1	1	40	42	60	62	85	82	99	96	3.12
3	0	2	2	2	50	53	65	66	80	83	98	95	5.50
4	1	0	1	2	30	32	50	52	76	74	90	91	4.84
5	1	1	2	0	35	34	48	46	57	58	65	64	1.07
6	1	2	0	1	45	43	60	61	76	74	90	92	8.69
7	2	0	2	1	55	56	66	64	76	78	85	84	2.94
8	2	1	0	2	60	62	70	67	80	78	91	93	-0.01
9	2	2	1	0	62	64	70	68	80	77	92	94	-2.98

8.4.2 ANALYSIS OF DATA

The ideal function for this process is a linear relation between time and thickness. We wish to find out the optimum conditions of the control variables in which high linearity with minimum variation can be obtained, as shown in Figure 8.2. Therefore, the simple linear regression model of (8.2) is assumed, and related variations

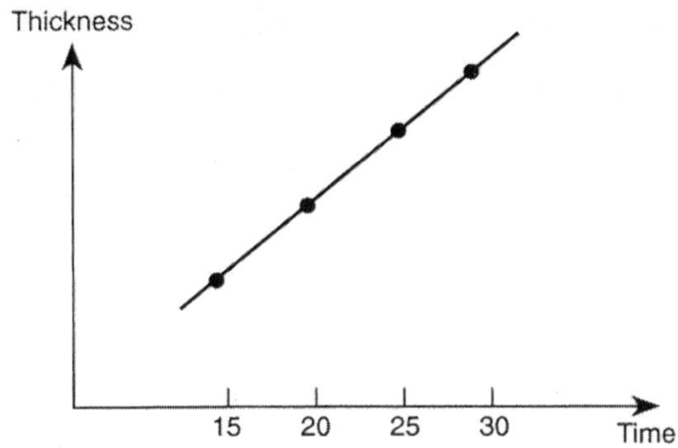

Figure 8.2 The ideal relation between time and thickness.

are computed by using equations (8.3)-(8.8). For example, for experiment 1, a two-way layout for the signal and noise factors can be constructed, and the SN ratio can be obtained accordingly.

Levels of noise factor	Levels of signal factor				
	15	20	25	30	
Top (T)	35	44	54	67	
Bottom (B)	36	43	56	68	
Sum	71	87	110	135	T = 403
Average	35.5	43.5	55	67.5	$\bar{\bar{y}}$ = 50.375

$$S_T = \sum_i \sum_j y_{ij}^2 - \frac{T^2}{lm}$$

$$= 35^2 + 36^2 + \ldots + 68^2 - 403^2/8 = 1169.875,$$

$$r = m\sum_i (M_i - \bar{M})^2 = 2\times[(15-22.5)^2 + \ldots + (30-22.5)^2] = 250,$$

$$\left[\sum_i \sum_j M(y_{ij} - \bar{\bar{y}})\right]^2 = \left(\sum_i M_i T_{i.} - lm\bar{\bar{y}}\bar{M}\right)^2 \quad (8.17)$$

$$= (15\times 71 + 20\times 87 + 25\times 110 + 30\times 135 - 4\times 2$$

$$\times 50.375 + 22.5)^2$$

$$= 537.5^2 = 288906.25,$$

$$S_\beta = 288906/250 = 1155.625,$$
$$S_e = S_T - S_\beta = 1169.875 - 1155.625 = 14.25,$$
$$V_e = S_e/(lm-2) = 14.25/6 = 2.375,$$

$$\text{SN} = 10\log\left[\frac{\frac{1}{r}(S_\beta - V_e)}{V_e}\right]$$
$$= 10\log\left[\frac{\frac{1}{25}(1155.625 - 2.375)}{2.375}\right] = 2.88\text{dB}.$$

We can obtain the SN ratios in a similar way for the remaining experiments; these SN ratios are listed in the final column of Table 8.8.

To analyze the SN ratios, Pareto ANOVA for three-level factors, explained in Chapter 5, will be used here. The Pareto ANOVA for this experiment is shown in Table 8.9, which is similar to Table 5.8. Note that Pareto ANOVA is a very useful tool for finding significant factors and their optimal levels.

The estimate of the SN ratio at the optimum condition is

$$\hat{\mu}(A_1B_2C_0D_1) = \overline{A}_1 + \overline{B}_2 + \overline{C}_0 + \overline{D}_1 - 3\overline{T}$$
$$= \frac{14.60}{3} + \frac{11.21}{3} + \frac{11.56}{3} + \frac{14.75}{3} - 3\times\frac{26.05}{9} = 8.69\text{dB}.$$

Note that this optimum condition happens to be the same as experiment 6 in Table 8.8, and the SN value happens to be the same, 8.69. This is pure coincidence, and such results do not often occur in practice. At the current condition the estimate for the SN ratio is

$$\hat{\mu}(A_1B_1C_1D_1) = \overline{A}_1 + \overline{B}_1 + \overline{C}_1 + \overline{D}_1 - 3\overline{T}$$
$$= \frac{14.60}{3} + \frac{4.18}{3} + \frac{4.98}{3} + \frac{14.75}{3} - 3\times\frac{26.05}{9} = 4.15\text{dB}.$$

Table 8.9 Pareto ANOVA for oxidation process experiment.

Factor and interaction		*A*	*B*	*C*	*D*	Total
Sum at factor level	0	11.5	10.66	11.56	0.97	T = 26.05
	1	14.60	4.18	4.98	14.75	
	2	-0.05	11.21	9.51	10.33	
Square of differences		357.635	91.714	68.020	297.034	
Contribution ratio (%)		43.9	11.3	8.3	36.5	100.0
Pareto diagram		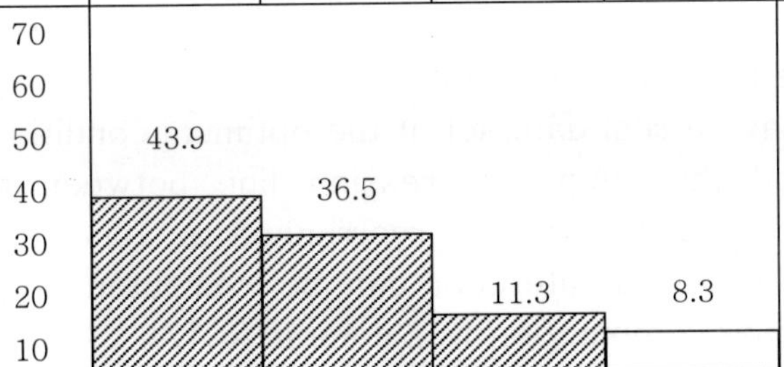				
Factor and interaction		*A*	*B*	*D*	*C*	
Cumulative contribution ratio (%)		43.9	80.4	91.7	100.0	
Optimum levels of significant factors		$A_1B_2D_1$				
Remarks on optimal conditions		The significant factors are chosen from the left-hand side in the above Pareto diagram which cumulatively contribute about 90%. For other non-significant factors, the optimal levels can be chosen from other economical and technical considerations. Suppose the optimal level of C is C_0.				
Overall optimal conditions for all factors		A_1: temperature, nominal level B_2: total gas flow, nominal +10% C_0: N_2 dilution, nominal −15% D_1: H_2/O_2 ratio, nominal				
Estimate of error		V_e = (sum of squares of non-significant factors)/ϕ $= (S_C/9)/2 = (68.0198/9)/2 = 3.779$				

Therefore, the gain that results from using the optimum condition is

$$8.69 - 4.15 = 4.54\text{dB}.$$

If we perform an analysis similar to (8.15) and (8.16) to check the improvement in linearity between and thickness, we find that

$$4.54 = 10\log(\eta_0/\eta_c),$$

so that $\eta_0/\eta_c = 2.85$. This means that, by changing from the current condition to the optimum condition, the degree of linearity can be improved by a factor of 2.85.

Since we have a real data set at the optimum condition, it is of interest to find the simple regression line between time and thickness. The data set can be rearranged as follows, and the estimates of slope and intercept can be computed accordingly. Note that a similar computation was done in (8.17).

Levels of noise factor	Levels of signal factor				
	15	20	25	30	
Top (T)	45	60	76	90	
Bottom (B)	43	61	74	92	
Sum	88	121	150	182	T = 541
Average	44	60.5	75	91	$\bar{\bar{y}}$ =67.625

$$S_T = 45^2 + 32^2 + \ldots + 92^2 - 541^2/8 = 2425.875,$$

$$r = m\sum_i (M_i - \bar{M})^2 = 250,$$

$$\sum_j \sum_i M_i(y_{ij} - \bar{\bar{y}}) = \sum_i M_i T_{i.} - lm\bar{\bar{y}}M$$

$$= 15\times 88 + 20\times 121 + 25\times 150 + 30\times 182$$

$$-4\times 2\times 67.625\times 22.5$$

$$= 777.5,$$

$$\hat{\beta}_1 = \left(\sum_i M_i T_{i.} - lm\bar{\bar{y}}\bar{M} \right) \Big/ r = 777.5/250 = 3.11,$$

$$\hat{\beta}_0 = \bar{\bar{y}} - \hat{\beta}_1 \bar{M} = 67.625 - 3.11 \times 22.5 = -2.35,$$

$$S_\beta = \left[\sum_i \sum_j M_i (y_{ij} - \bar{\bar{y}}) \right]^2 \Big/ r = (777.5)^2/250 = 2418.025,$$

$$R^2 = S_\beta / S_T = 2418.025/2425.875 = 0.9968,$$

$$S_e = S_T - S_\beta = 7.85,$$

$$V_e = S_e/(lm-2) = 7.85/6 = 1.3083.$$

Hence, the fitted straight line is

$$\hat{y} = -2.35 + 3.11M$$

with $R^2 = 0.9968$ and an error variance of 1.3083, which means that by using the fitted line we can predict the thickness at a given time with very good precision.

Exercises

8.1 Consider a coil product which is made by rolling in a rolling mill under two different conditions: immediately after removal from the furnace (R_1); and after being left to stand for a certain period of time after removal from the furnace (R_2). There are two types of materials, A_1 and A_2, to be compared. The rolled thickness (this is the dynamic characteristic) was measured while varying the reduction ratio of the rolling mill over the four levels M_1, M_2, M_3 and M_4 (see table below). Note that M is the signal factor, with levels taken at equal intervals, say, 1, 2, 3 and 4. It is assumed that the data obtained when the thickness was measured at two locations, center and edge, are given in the upper and lower rows of the table below. The working mean, 1000m, has been subtracted. In this case, R is in effect the two-level noise factor. Carry out an appropriate analysis, and select one material, A_1 or A_2. Which one is the better material for rollability? (Data source: Taguchi (1987).)

Material		Level of signal factor			
		M_1 (1)	M_2 (2)	M_3 (3)	M_4 (4)
A_1	R_1	102	86	58	34
		103	86	58	37
	R_2	110	88	64	40
		98	73	56	40
A_2	R_1	85	77	69	65
		85	77	72	64
	R_2	86	78	70	63
		86	79	70	62

8.2 A dyeing experiment was performed in which color density was measured as the dynamic characteristic. The color factors were:

A: kinds of dyes A_0 = new, A_1 = current
B: dyeing temperature B_0 = room temp., B_1 = high temp.
C: quantity of Glauber's salt C_0 = 50% of current, C_1 = current
D: pH D_0 = current -0.5, D_1 = current
F: liquid ratio F_0 = twice of current, F_1 = current.

There was a signal factor:

$$M\text{: quantity of dye}\quad \begin{aligned} M_1 &= \tfrac{1}{3}\% \text{ of fibre weight} \\ M_2 &= 1\% \text{ of fibre weight} \\ M_3 &= 3\% \text{ of fibre weight.} \end{aligned}$$

There was also a noise factor

$$N\text{: starching,}\quad N_1 = \text{not starched,}\quad N_2 = \text{starched.}$$

The five control factors and two interactions, $B \times C$ and $B \times F$ are assigned to an inner array L_8 table. The layout and results are shown below. (Date source: Wu and Moore (1985).)

Experiment number	B 1	F 2	$B\times F$ 3	D 4	A 5	C 6	$B\times C$ 7	M_1		M_2		M_3	
								K_1	K_2	K_1	K_2	K_1	K_2
1	0	0	0	0	0	0	0	5.2	5.9	12.3	12.4	22.6	22.5
								5.6	5.8	12.1	12.5	22.6	22.2
2	0	0	0	1	1	1	1	6.0	5.9	14.4	15.4	27.8	28.3
								5.7	6.0	14.8	14.4	26.9	28.8
3	0	1	1	0	0	1	1	5.1	5.5	10.5	10.9	17.8	18.4
								5.4	5.8	10.6	10.8	17.2	18.3
4	0	1	1	1	1	0	0	6.4	6.4	15.5	16.0	29.0	29.7
								6.7	5.8	14.9	15.8	28.3	31.1
5	1	0	1	0	1	0	1	5.1	5.4	12.1	11.6	25.2	25.5
								5.4	5.5	11.9	11.7	24.8	25.7
6	1	0	1	1	0	1	0	6.4	6.9	14.8	15.0	31.0	32.5
								6.6	6.6	15.0	15.0	31.3	31.4
7	1	1	0	0	1	1	0	5.5	5.9	11.8	13.6	23.2	24.0
								5.5	6.0	12.4	13.2	23.6	25.4
8	1	1	0	1	0	0	1	6.8	6.8	15.9	15.9	31.8	31.4
								6.6	6.5	16.3	16.7	31.7	32.4

Carry out the necessary analysis, and determine the optimum levels of the control factors. Find the fitted straight-line equation at the optimum conditions. Assume the straight line may have a non-zero intercept.

8.3 An experiment was carried out involving gold-to-gold thermosonic bonding in a semiconductor manufacturing process. There were six factors:

A: metallization device,	2 levels
B: ultrasonic power,	3 levels
C: bond force,	3 levels
D: stage temperature,	3 levels
F: bond time,	3 levels
G: bonding head velocity,	3 levels.

In addition, there was one signal factor, pad pitch (M), with three levels, $M_1 = 10$, $M_2 = 20$ and $M_3 = 30$. The two noise factors were tape-to-tape variation and positions on four sides of the packages. The dynamic characteristic was lead pull strength (g mm^{-2}).

The L_{18} OA was set up with the control factors A, B, C, D, F and G in columns 1, 2, 3, 4, 5 and 6, respectively. For each experiment 20 measurements per pitch were measured, and a total of $18 \times 3 \times 20$ pull strengths were obtained. The SN ratios shown below were obtained. The simple regression model with no intercept is assumed for the relationship between pad pitch and lead pull strength.

Experiment no.	SN	Experiment no.	SN	Experiment no.	SN
1	-26.49	2	-24.86	3	-23.29
4	-23.73	5	-22.69	6	-21.37
7	-26.26	8	-20.15	9	-23.03
10	-24.17	11	-25.58	12	-23.38
13	-24.31	14	-20.43	15	-20.10
16	-22.16	17	-21.36	18	-20.87

Find the optimum condition of the control factors which maximizes the SN ratio. Estimate the SN ratio at the optimum condition. Suppose the current condition is $A_0B_0C_0D_0F_0G_0$. What is the gain in SN ratio of the optimum condition over the current condition?

9

Tolerance Design

9.1 INTRODUCTION

As mentioned in Chapter 1, Taguchi's off-line quality control method has three steps in each of product and process design. These steps are system design, parameter design and tolerance design. System design is concerned with the basic methods for accomplishing a required output, and parameter design looks for the best settings to put the required output on target with low variation, with experimental design being extensively used in finding these best settings. If the reduction in quality variation achieved by the parameter design is not sufficient, then **tolerance design** is used. Tolerance design identifies the quality-sensitive components and applies tolerances to these components to meet the required level of variation on the output. In tolerance design, a trade-off is usually made between reduction in quality variation and increase in manufacturing cost. That is, we selectively specify higher-grade parts, materials or components to reduce tolerances in the order of their cost-effectiveness. Since there are many possible choices, experimental designs using tables of orthogonal arrays can also be favorably used in tolerance design.

We have devoted a considerable proportion of this book to the second stage, parameter design, and very little to system design. In fact, system design has nothing to do with statistic and this is why very little mention has been made of it. System design is the product of fertile and inventive minds that thrive on breaking rules and trying innovative approaches to solving challenging design problems. Most engineers tend to go straight from system design into tolerance design, probably because they do not know how to use experimental design and statistical optimization. In doing so, they miss out on the most important engineering activity.

After parameter design, tolerance design is necessary in most cases. Good engineering procedure requires that a sensitivity analysis be performed on a design to determine the relative contribution to the overall variation made by each component. Then, the cost of controlling the overall variation must be determined by taking control of these individual contributors to the variation. This is why a trade-off is made between reduction in the quality variation and increase in manufacturing cost.

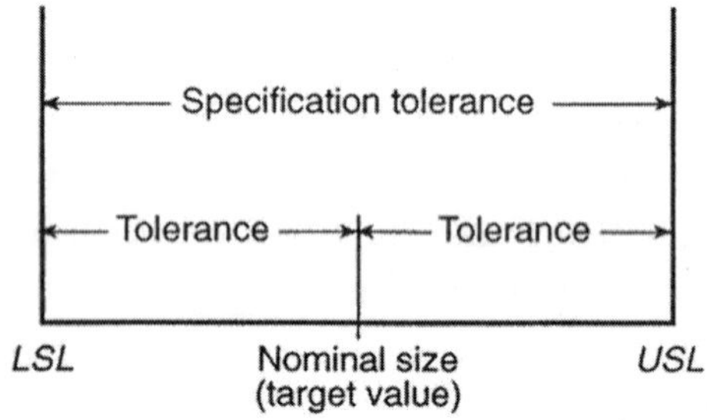

Figure 9.1 Specification and tolerance.

A technical **specification** consists of two parts, nominal size and **tolerance,** as shown in Figure 9.1. The nominal size is the standard target value for quality, and the tolerance is the difference between the nominal size and the specification limits. There are two specification limits: the **upper specification limit** (*USL*) and the **lower specification limit** (*LSL*).

Suppose the specification for a part dimension is 1.870 ± 0.005 cm. Then, the nominal size, *LSL* and *USL* are 1.870, 1.865 and 1.875 cm respectively. The tolerance is ± 0.005 cm, and the **specification tolerance** is 0.01 cm. Very often the specification tolerance is simply referred to as the "tolerance".

9.2 DETERMINATION OF TOLERANCES

For a manufacturer of components, parts or materials, the specification limits are usually set by the firm buying the products. For a manufacturer of end products, the product specifications are often set by the customer or the planning department of the manufacturer.

There are two specific sets of characteristics which determine the quality of a given product. These are the characteristics of component parts, called **lower-level characteristics**, and the characteristics of the product itself, called **higher-level characteristics**. It is important to determine the specification of lower-level quality characteristics (or cause parameters) properly, since these eventually affect the higher-level quality characteristics. Parameter design and tolerance design are included in the determination of these lower-level specifications. Parameter design is used for determining optimum component or parameter mid-values (optimum conditions) in such a way that quality variability is reduced. If, after these mid-values are determined, the level of variability is still not satisfactory, then components and materials must be upgraded in order to further reduce variability at a higher cost. Tolerance design is the determination of acceptable levels of variability and acceptable components or materials.

9.2.1 TOLERANCE LIMIT AND SAFETY DESIGN

There are two tolerance limits (specification limits): one is the **producer's tolerance limit**, and the other is the **customer's tolerance limit**, as shown in Figure 9.2. Let the nominal size be denoted by m, the producer's tolerance limits by $m \pm \Delta$, and the customer's tolerance limits by $m \pm \Delta_0$. It is always true that $\Delta \leq \Delta_0$, since the producer's tolerance limit is the producer's manufacturing specification and the customer's tolerance limit is the bound beyond which the functional characteristic of the product is not acceptable.

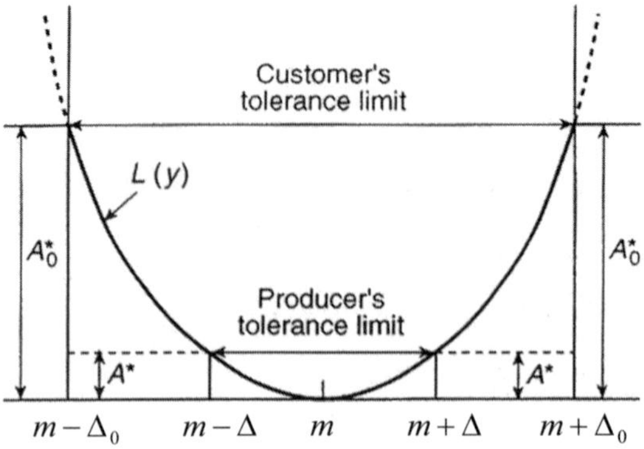

Figure 9.2 Tolerance limits for nominal-is-best characteristics.

Consider the following losses at the tolerance limits:

A^* = loss when the quality characteristic value is at the producer's tolerance limits;
A_0^* = loss when the quality characteristic value is at the customer's tolerance limits.

With regard to the safety design for the producers, the loss function can be effectively employed. The loss function for a nominal-is-best characteristic y may be written as

$$L(y) = \frac{A_0^*}{\Delta_0^2}(y-m)^2 . \tag{9.1}$$

Since the loss is A^* at the producer's tolerance limits,

$$A^* = \frac{A_0^*}{\Delta_0^2}(m \pm \Delta - m)^2$$

$$= \frac{A_0^*}{\Delta_0^2}\Delta^2 .$$

Therefore,

$$\Delta = \sqrt{\frac{A_0^*}{A^*}}\Delta_0 = \frac{\Delta_0}{\phi} , \tag{9.2}$$

where

$$\phi = \sqrt{\frac{A_0^*}{A^*}}$$

$$= \sqrt{\frac{\text{customer's loss when quality is beyond functional limit}}{\text{producer's loss when quality is beyond specification limit}}}, \qquad (9.3)$$

which is called the safety coefficient.

Example 9.1

Suppose the customer's tolerance limits for the output voltage of an electric circuit in a television set are 120 ± 24V, and the loss when it is outside these limits is A_0^* = \$45 (replacement cost plus customer's inconvenience cost).

If the circuit is not acceptable at factory, it can be adjusted by changing the resistance at a cost of A^* = \$5. Then, from (9.3), the safety coefficient is

$$\phi = \sqrt{\frac{A_0^*}{A^*}} = \sqrt{\frac{45}{5}} = 3,$$

and from (9.2), the producer's tolerance is

$$\Delta = \frac{\Delta_0}{\phi} = 24/3 = 8\,.$$

Therefore, the producer's specification becomes 120 ± 8V, which is considerably narrower than the customer's tolerance limits.

9.2.2 SMALLER-THE-BETTER AND LARGER-THE-BETTER CHARACTERISTICS

For a smaller-the-better characteristic, the formulas used for a nominal-is best characteristic can be used by setting $m = 0$. However, for a larger-the-better characteristic, the loss function takes a different form:

$$L(y) = A_0^* \Delta_0^2 \frac{1}{y^2}.$$

Hence, if the loss is A^* at the producer's tolerance limit, we find that

$$A^* = A_0^* \Delta_0^2 \frac{1}{\Delta^2}.$$

Therefore, the factory specification can be computed by

$$\Delta = \phi \Delta_0, \tag{9.4}$$

where $\phi = (A_0^* / A^*)^{1/2}$ is the safety coefficient.

Example 9.2

The strength of a pipe is a larger–the–better characteristic. If the strength is less than 5000 kgcm^{-2}, it is outside the function limit and the loss is A_0^* = \$150 per meter of pipe. If the pipe is scrapped in factory because of strength non–conformance, the loss is \$20 per meter. In this case, the safety coefficient is

$$\phi = (A_0^* / A^*)^{1/2} = \sqrt{150/20} = 2.74,$$

and the factory specification for pipe strength should be

$$\Delta = \phi \Delta_0 = 2.74 \times 5000 = 13700 \text{ kgcm}^{-2}.$$

In other words, the pipe should be manufactured in such a way that the strength should be greater than 13700 kgcm^{-2}.

9.2.3 LOWER-LEVEL AND HIGHER-LEVEL SPECIFICATIONS

To determine the tolerance of a lower–level characteristic (cause parameter), it is necessary first to write the loss function of the higher–level characteristic. Let the specification of the higher–level characteristic, y, be

$$L(y) = \frac{A_0^*}{\Delta_0^2}(y - m_0)^2, \tag{9.5}$$

where A_0^* is the loss when y takes a value at the specification limit. If the lower-level characteristic is x, and if the effect of x on y is b when x varies by a unit amount, then we can set up the regression relationship

$$y - m_0 = b(x - m),$$

where m is the mid-value of the lower-level characteristic, x.

The loss function (9.5) is then given by

$$L(y) = \frac{A_0^*}{\Delta_0^2}[b(x - m)]^2.$$

Suppose the loss due to the lower-level characteristic being out of specification is A_0^*; then we obtain the relationship

$$A^* = \frac{A_0^*}{\Delta_0^2}[b(x - m)]^2. \tag{9.6}$$

Therefore, the tolerance of the lower-level characteristic, denoted by $\Delta = (x - m)$, is found by solving equation (9.6):

$$\Delta = \sqrt{\frac{A^*}{A_0^*}}\frac{\Delta_0}{|b|}, \tag{9.7}$$

where

A_0^* = the loss caused by the higher-level characteristic being out of specification,

Δ_0 = tolerance of the higher-level characteristic,

A^* = the loss by the lower-level characteristic being out of specification,

b = the effect on the higher-level characteristic when the lower-level characteristic varies by a unit amount.

Example 9.3

A circuit has a nominal output value of 220 mV, with a tolerance of 5 mV. If the circuit is outside its specification limits, it is scrapped at a loss of $10. When a resistance in the circuit varies by 1 mΩ, the output voltage varies by 0.5 mV. The target value of this resistance is selected to give the target output voltage of 220 mV. The loss from rejecting the resistor is $2. Note that, in this problem, $\Delta_0 = 5$ mV, $A^* = \$2$, $A_0^* = \$10$ and $b = 0.5\ \text{mV}\,\Omega^{-1}$. Hence, the tolerance of the resistance is from (9.7).

$$\Delta = \sqrt{\frac{A^*}{A_0^*}}\frac{\Delta_0}{|b|}$$

$$= \sqrt{\frac{2}{10}}\frac{5}{0.5} = 4.47\text{m}\Omega.$$

9.2.4 DETERIORATING CHARACTERISTICS

If characteristics change over time, they are described as **deteriorating**. Parts may gradually wear out, or the resistance values of the resistors in an electric circuit may become larger and larger due to deterioration of the component parts.

Consider a deteriorating characteristic which varies linearly over time. Let the designed life of a product be T years and let b be the amount of deterioration per year. The amount of deterioration in T years is given by bT. When the characteristic of a product has an initial value equal to its target value m, and the amount of change is b per year, then the deviation from the target value in T years is bT. As it gradually drifts from the target value, the mean squared deviation is given by

$$MSD = \frac{1}{T}\int_0^T (bT)^2 dt. \tag{9.8}$$

Note that this mean squared deviation from the target value may be regarded as the variance.

Suppose a product has a design life of T years, and its coefficient of deterioration, denoted by b, is adjustable in manufacturing to be within specification limits at a cost of $\$A^*$. Also suppose the loss as a result of the product not conforming to the customer's specification, Δ_0, is $\$A_0^*$. Then the manufacturer's tolerance, denoted by Δ, for the period T, is obtained by solving the following equation for b:

$$A^* = \frac{A_0^*}{\Delta_0^2} \times \frac{b^2}{3} \times T^2 .$$

We can find the manufacturer's tolerance Δ from b

$$\Delta = \sqrt{\frac{3A^*}{A_0^*}} \frac{\Delta_0}{T}, \tag{9.9}$$

where

A^* = cost of adjusting the coefficient of deterioration in manufacturing,
Δ_0 = customer's tolerance in the market,
A_0^* = loss to society when the characteristic is out of tolerance,
T = designed life or calibration interval.

Example 9.4 (Taguchi 1986)

Suppose that the specified accuracy of a watch is ± 10 s per month. If the accuracy is within this specification and the user sets the watch once a month, it will always be accurate to within ± 10 s. If the watch does not meet this specification, the consumer will have to set it more often. Let the cost of adjusting the watch be \$12. Assume that the watch gains or loses time at a nearly constant rate of b seconds per month. Suppose that the cost of checking and adjusting the accuracy of the watch at the end of the manufacturing process is \$60. If the design life of the watch is 10 years (120 months), what should be the manufacturer's tolerance for the watch?

Note that in the above example, $A^* = \$60$, $\Delta_0 = 10\text{s}$, $A_0^* = \$12 \times 120 \text{ months} = \$1440$, and T = 1 month, since b is the tolerance per month. Substituting these values into (9.9), we find that the manufacturer's tolerance for the watch is

$$\Delta = \sqrt{\frac{3 \times 60}{1040} \cdot \frac{10}{1}} = 3.5 .$$

Therefore, the accuracy of the watch when shipped from the factory should be ± 3.5 s per month.

9.3 ORTHOGONAL POLYNOMIALS

9.3.1 ONE INDEPENDENT VARIABLE

The kth-order polynomial regression model for one independent variable is written as

$$y = \beta_0 + \beta_1 x + \beta_2 x^2 + \cdots + \beta_k x^k + \varepsilon , \tag{9.10}$$

where the error term ε is assumed to be identically and independently normally distributed with mean zero and variance σ^2. Estimates of the regression coefficients can be computed by using matrices as shown in (2.39) in Chapter 2, which is not easy to do by hand. However, if the levels of x are equally spaced, there is an easy way to obtain the estimates of the regression coefficients by using orthogonal polynomials. In this section, we will introduce orthogonal polynomials. We will show later that these polynomials can be used effectively in tolerance design.

Suppose there is a set of data for one-way layout where there are l levels of $x(x_1, x_2, \ldots, x_l)$ and there are m repeated observations at each level, as shown in Table 9.1. It is assumed that the levels are equally spaced, so that $x_2 = x_1 + c$, $x_3 = x_1 + 2c$, and so on.

For the qth-order term x^q in (9.10), if we use the qth-order orthogonal polynomial $p_q(x)$, we can express equation (9.10) as follows.

Table 9.1 Experimental data for one-way layout.

	Levels					
	x_1	x_2	x_3	$\cdots$	x_l	
Data	y_{11}	y_{21}	y_{31}	$\cdots$	y_{l1}	
	y_{12}	y_{22}	y_{32}	$\cdots$	y_{l2}	
	.	.	.	$\cdots$		
	.	.	.	$\cdots$		
	.	.	.			
	y_{1m}	y_{2m}	y_{3m}		y_{lm}	
Sum	$T_{1.}$	$T_{2.}$	$T_{3.}$	$\cdots$	$T_{l.}$	T

$$y = \beta_0 + \beta_1 p_1(x) + \beta_2 p_2(x) + \cdots + \beta_k p_k(x) + \varepsilon . \tag{9.11}$$

Note that k should be less than l. The orthogonal polynomials should satisfy the following conditions.

$$\begin{aligned} &\sum_{i=1}^{l} p_q(x_i) = 0, \quad q = 1,2,...,k \\ &\sum_{i=1}^{l} p_q(x_i) p_r(x_i) = 0, \quad q \neq r = 1,2,...,k \\ &\sum_{i=1}^{l} [p_q(x_i)]^2 \neq 0, \quad q = 1,2,...,k. \end{aligned} \tag{9.12}$$

Polynomials which satisfy the above conditions are called the **orthogonal polynomials**, and the following orthogonal polynomials are normally used in practice.

$$\begin{aligned} &p_1(x) = x - \bar{x}, \\ &p_2(x) = (x-\bar{x})^2 - \frac{c^2}{12}(l^2 - 1), \\ &p_3(x) = (x-\bar{x})^3 - \frac{c^2}{20}(3l^2 - 7)(x-\bar{x}), \\ &p_4(x) = (x-\bar{x})^4 - \frac{c^2}{14}(3l^2 - 13)(x-\bar{x})^2 + \frac{3c^4}{560}(l^2-1)(l^2-9), \\ &\cdots \end{aligned} \tag{9.13}$$

where $\bar{x}$ is the average of the l levels of x,

$$\bar{x} = x_1 + \frac{l-1}{2}c. \tag{9.14}$$

If we substitute (9.13) into (9.11) and the regression model is the second-order polynomial, we have

$$y = \beta_0 + \beta_1(x-\bar{x}) - \beta_2[(x-\bar{x})^2 - \frac{c^2}{12}(l^2-1)] + \varepsilon. \tag{9.15}$$

In this case, the normal equations may be written as follows:

$$\begin{bmatrix} lm & 0 & 0 \\ 0 & m\sum_i (x_i-\bar{x})^2 & 0 \\ 0 & 0 & m\sum_i [(x_i-\bar{x})^2 - \frac{c^2}{12}(l^2-1)]^2 \end{bmatrix} \begin{bmatrix} \beta_0 \\ \beta_1 \\ \beta_2 \end{bmatrix} = \begin{bmatrix} T \\ m\sum_i (x_i-\bar{x})T_{i.} \\ m\sum_i [(x_i-\bar{x})^2 - \frac{c^2}{12}(l^2-1)]T_i \end{bmatrix}$$

$$\mathbf{X'X\beta} = \mathbf{X'y}. \tag{9.16}$$

Then we can easily obtain the result that

$$\hat{\boldsymbol{\beta}} = \begin{bmatrix} \hat{\beta}_0 \\ \hat{\beta}_1 \\ \hat{\beta}_2 \end{bmatrix} = \mathbf{(XX)^{-1}X'y} = \begin{bmatrix} \dfrac{T}{lm} \\ \dfrac{\sum_i (x_i-\bar{x})T_{i.}}{m\sum_i (x_i-\bar{x})^2} \\ \dfrac{\sum_i [(x_i-\bar{x})^2 - \frac{c^2}{12}(l^2-1)]T_i}{m\sum_i [(x_i-\bar{x})^2 - \frac{c^2}{12}(l^2-1)]^2} \end{bmatrix}. \tag{9.17}$$

Let us consider the estimate $\hat{\beta}_i$ in (9.17) in detail:

$$\hat{\beta}_i = \sum_i (x_i - \bar{x})T_{i.} \Big/ m\sum_i (x_i - \bar{x})^2 .$$

For notational convenience, let

$$w_i = \lambda\left(\frac{x_i - \bar{x}}{c}\right),$$
$$\lambda^2 S = \sum_i w_i^2 = \lambda^2 \sum_i \left(\frac{x_i - \bar{x}}{c}\right)^2 . \tag{9.18}$$

Then we can write

$$\hat{\beta}_i = \frac{\sum_i w_i T_{i.}}{\lambda Smc},$$

where λ is the smallest positive value which makes w_i positive. For instance, if $l = 4$, then $\bar{x} = x_1 + 3c/2$ from (9.14) and

$$(x_1 - \bar{x})/c = -2/3, \quad (x_2 - \bar{x})/c = -1/2,$$
$$(x_3 - \bar{x})/c = 1/2, \quad (x_4 - \bar{x})/c = 3/2.$$

Hence, if we let $\lambda = 2$, then we can obtain $w_i = -3, -1, 1, 3$ which are all integers.

In general, we can find the least-squares estimates of β_k from the formula

$$\hat{\beta}_k = \frac{\sum_i w_i^{(k)} T_{i.}}{(\lambda S)_k mc^k}, \qquad k = 0,1,2,..,l-1. \tag{9.19}$$

Here, w_i are the kth-order coefficients of the orthogonal polynomial, which are listed in Table 9.2. Table 9.2 excludes the case of $k = 0$. For $k = 0$, we should use

$$w_i^{(0)} = 1,$$
$$\lambda_0 = 1, \tag{9.20}$$
$$(\lambda^2 S)_0 = (\lambda S)_0 = (S)_0 = l.$$

Table 9.2 Orthogonal polynomial coefficients.

Number of level (l)	2	3		4			5				6				
Order (k)	(1)	(1)	(2)	(1)	(2)	(3)	(1)	(2)	(3)	(4)	(1)	(2)	(3)	(4)	(5)
Coefficient															
w_1	-1	-1	1	-3	1	-1	-2	2	-1	1	-5	5	-5	1	-1
w_2	1	0	-2	-1	-1	3	-1	-1	2	-4	-3	-1	7	-3	5
w_3		1	1	1	-1	-3	0	-2	0	6	-1	-4	4	2	-10
w_4				3	1	1	1	-1	-2	-4	1	-4	-4	2	10
w_5							2	2	1	1	3	-1	-7	-3	-5
w_6											5	5	5	1	1
$(\lambda^2 S)_k$	2	2	6	20	4	20	10	14	10	70	70	84	180	28	252
$(\lambda S)_k$	1	2	2	5	4	6	10	14	12	24	35	56	108	48	120
$(S)_k$	$\frac{1}{2}$	2	$\frac{2}{3}$	2	1	$\frac{9}{5}$	10	14	$\frac{72}{5}$	$\frac{288}{35}$	$\frac{35}{2}$	$\frac{112}{3}$	$\frac{324}{5}$	$\frac{576}{7}$	$\frac{400}{7}$
λ_k	2	1	3	2	1	$\frac{10}{3}$	1	1	$\frac{5}{6}$	$\frac{35}{12}$	2	$\frac{3}{2}$	$\frac{5}{3}$	$\frac{7}{12}$	$\frac{21}{10}$

Also note that in Table 9.2, any two columns are orthogonal, and for any k, the following equalities hold for $k = 0,1,2 \cdots l-1$.

$$\sum_i w_i^{(k)} = 0,$$

$$(\lambda^2 S)_k = \sum_i \left[w_i^{(k)}\right]^2,$$

$$(\lambda S)_k = (\lambda^2 S)/\lambda_k,$$

$$(S)_k = (\lambda S)_k / \lambda_k.$$

The variation explained by the kth-order orthogonal polynomial can be obtained by

$$S_{(k)} = \frac{\left[\sum_i w_i^{(k)} T_i\right]^2}{(\lambda^2 S)_k m}, \qquad k = 0,1,2,\ldots,l-1, \tag{9.21}$$

which has one degree of freedom. It is known that the estimate $\hat{\beta}_k$ in (9.19) has the following properties:

$$E(\hat{\beta}_k) = \beta_k,$$

$$\mathrm{var}(\hat{\beta}_k) = \frac{\sigma^2}{(S)_k mc^{2k}}, \tag{9.22}$$

$$\mathrm{cov}(\hat{\beta}_k, \hat{\beta}_{k'}) = 0 \quad \textit{if } k \neq k'.$$

We can also state two known properties of the variation in (9.21). In the one-way layout with l levels, if the polynomial is fitted up to the (l - 1)th order, then

$$S_A = \sum_{k=1}^{l-1} S_{(k)}.$$

Furthermore,

$$E[S_{(k)}] = \sigma^2 + (S)_k mc^{2k} \beta_k^2. \tag{9.23}$$

9.3.2 MORE THAN ONE INDEPENDENT VARIABLE

When we have more than one independent variable, and each variable has equally spaced levels, we can again use the orthogonal polynomials. Suppose we have a set of two-way layout data, in which x_1 has l levels, x_2 has m levels, and there are r repeated observations at each level combination of x_1 and x_2. The data arrangement for this case can be found in Table 3.6 in Chapter 3. There are a total of lmr observations, and the second-order orthogonal polynomial regression model may be written as

$$\begin{aligned} y &= \beta_{00} + \beta_{10}(x_1 - \bar{x}_1) + \beta_{01}(x_2 - \bar{x}_2) \\ &\quad + \beta_{20}[(x_1 - \bar{x}_1)^2 - \frac{c_1^2}{12}(l^2 - 1)] + \beta_{02}[(x_2 - \bar{x}_2)^2 - \frac{c_2^2}{12}(m^2 - 1)] \\ &\quad + \beta_{11}(x_1 - \bar{x}_1)(x_2 - \bar{x}_2) + \varepsilon, \end{aligned} \tag{9.24}$$

where c_1 and c_2 are the intervals between two adjacent levels of x_1 and x_2, respectively. The regression coefficients can be estimated by

$$\begin{aligned} \hat{\beta}_{00} &= \frac{T}{lmr}, \\ \hat{\beta}_{k0} &= \frac{\sum_i w_i^{(k)} T_{i..}}{(\lambda S)_k mrc_1^k} \qquad k = 1, 2, \\ \hat{\beta}_{0k} &= \frac{\sum_j w_j^{(k)} T_{.j.}}{(\lambda S)_k lrc_2^k} \qquad k = 1, 2, \\ \hat{\beta}_{11} &= \frac{\sum_i \sum_j w_i^{(1)} w_j^{(1)} T_{ij.}}{(\lambda S)_1^l (\lambda S)_1^m mc_1 c_2}, \end{aligned} \tag{9.25}$$

where $(\lambda S)_1^l$ and $(\lambda S)_1^m$ are the corresponding quantities of $(\lambda S)_1$ for the levels l and m, respectively.

The variation due to each orthogonal term can be computed as follows:

$$S(\hat{\beta}_{00}) = CT = \frac{T^2}{lmr},$$

$$S(\hat{\beta}_{k0}) = \frac{\left[\sum_i w_i^{(k)} T_{i..}\right]^2}{(\lambda^2 S)_k \cdot mr},$$

$$S(\hat{\beta}_{0k}) = \frac{\left[\sum_j w_j^{(k)} T_{.j.}\right]^2}{(\lambda^2 S)_k lr}, \tag{9.26}$$

$$S(\hat{\beta}_{11}) = \frac{\left(\sum_i \sum_j w_i^{(1)} w_j^{(1)} T_{ij.}\right)^2}{(\lambda^2 S)_1^l (\lambda^2 S)_1^m r}.$$

Each of these has one degree of freedom. Note that $(\lambda^2 S)_1^l$ and $(\lambda^2 S)_1^m$ are the corresponding quantities of $(\lambda^2 S)_1$ for the levels of l and m, respectively.

Example 9.5

The percentage shrinking ratio of a plastic product may be influenced by process temperature (x_1) and concentration of process liquid (x_2). To check the influence quantitatively, the following two-way layout experiment was conducted.

$$A: A_1 = 30\,\text{min}, \quad A_2 = 40, \quad A_3 = 50$$
$$B: B_1 = 10\%, \quad B_2 = 15, \quad B_3 = 20, \quad B_4 = 25$$

	A_1	A_2	A_3	Total ($T_{\cdot j}$)
B_1	15	11	34	60
B_2	31	20	42	93
B_3	47	37	49	133
B_4	62	45	54	161
Total ($T_{i\cdot}$)	155	113	179	447

Note that since the number of repeated observation is r = 1, $T_{i\cdot}$ and $T_{\cdot j}$ are used instead of $T_{i\cdot\cdot}$ and $T_{\cdot j\cdot}$.

Table 9.3 ANOVA table for plastic product experiment.

Source of variation	Sum of squares	Degrees of freedom	Mean square	F_0
A	558	2	279.0	7.86*
B	1969	3	656.3	18.49**
E	213	6	35.5	
T	2740			

** indicates 1% significance.
* indicates 5% significance.

First of all, the analysis of variance for this two-factor factorial design without replication is shown in Table 9.3, which indicates that factors A and B are both significant:

$$S_T = \sum_i \sum_j y_{ij}^2 - CT$$

$$= 15^2 + 31^2 + \cdots + 54^2 - \frac{447^2}{12} = 2740,$$

$$S_A = \sum_i \frac{T_{i.}^2}{m} - CT = (155^2 + 113^2 + 179^2) - \frac{447^2}{12} = 558,$$

$$S_B = \sum_j \frac{T_{.j}^2}{l} - CT = (60^2 + 93^2 + 133^2 + 161^2) - \frac{447^2}{12} = 1969,$$

$$S_E = S_T - S_A - S_B = 213.$$

Since factors A and B are both significant, it is now of interest to find the regression equation. Note that x_1 (factor A) and x_2 (factor B) are both equally spaced, and we can fit the second-order orthogonal polynomial model in (9.24). Using the formulas in (9.25), Table 9.2, and l = 3, m = 4, r = 1, c_1 = 10, c_2 = 5, $T_{ij.}$ = y_{ij} , we obtain the following:

$$\hat{\beta}_{00} = \frac{T}{lmr} = \frac{447}{3\times4\times1} = 37.250,$$

$$\hat{\beta}_{10} = \frac{\sum_i w_i^{(1)} T_{i.}}{(\lambda S)_1 mrc_1^1} = \frac{(-1)T_{1.} + (0)T_{2.} + (1)T_{3.}}{2\times4\times1\times10} = \frac{-155+799}{80} = 0.300,$$

$$\hat{\beta}_{01} = \frac{\sum_j w_j^{(1)} T_{.j}}{(\lambda S)_1 lrc_2^1} = \frac{(-3)T_{.1} + (-1)T_{.2} + (1)T_{.3} + (3)T_{.4}}{10\times3\times1\times5}$$

$$= \frac{-180+93+133+483}{150} = 2.287,$$

$$\hat{\beta}_{20} = \frac{\sum_i w_i^{(2)} T_{i.}}{(\lambda S)_1 mrc_2^2} = \frac{T_{1.} + 2T_{2.} + T_{3.}}{2\times4\times1\times10^2} = \frac{155-226+179}{800} = 0.135,$$

$$\hat{\beta}_{02} = \frac{\sum_j w_j^{(2)} T_{.j}}{(\lambda S)_2 lrc_2^2} = \frac{T_{.1} - T_{.2} - T_{.3} + T_{.4}}{4\times3\times1\times5^2} = \frac{60-93+133+161}{300} = -0.017,$$

$$\hat{\beta}_{11} = \frac{\sum_i\sum_j w_i^{(1)} w_j^{(1)} y_{ij}}{(\lambda S)_1^l (\lambda S)_1^m mc_1 c_2} = \frac{1}{2\times10\times1\times10\times5}[(-1)(-3y_{11} - y_{12} + y_{13} + 3y_{14})$$

$$+ (0)(-3y_{21} - y_{22} + y_{23} + 3y_{24}) + (1)(-3y_{31} - y_{32} + y_{33} + 3y_{34})]$$

$$= \frac{1}{1000}(3y_{11} + y_{12} - y_{13} - 3y_{14} - 3y_{31} - y_{32} + y_{33} + 3y_{34})$$

$$= \frac{1}{1000}(3\times15 + 31 - 47 - 3\times62 - 3\times64 - 42 + 49 + 3\times54)$$

$$= -0.090.$$

Now note that $\overline{x}_1$ = 40, and $\overline{x}_2$ = 17.5. Therefore, the fitted second-order orthogonal polynomial model (9.24) becomes

$$\hat{y} = 37.250 + 0.300(x_1 - 40) + 2.287(x_2 - 17.5)$$
$$+ 0.135\left[(x_1 - 40)^2 - \frac{10^2}{12}(3^2 - 1)\right]$$
$$- 0.017\left[(x_2 - 17.5)^2 - \frac{5^2}{12}(4^2 - 1)\right] - 0.09000(x_1 - 40)(x_2 - 17.5).$$

In order to check the significance of each orthogonal term, let us find the variation of each term by using (9.26). The variation due to the first-order term of A is given by

$$S_{A_l} = S\left(\hat{\beta}_{10}\right) = \frac{\left[\sum_i w_i^{(1)} T_{i.}\right]^2}{(\lambda^2 S)_1 mr} = \frac{(-155 + 179)^2}{2 \times 4 \times 1} = 72.0\,.$$

That due to the first-order term of B is

$$S_{B_l} = S\left(\hat{\beta}_{01}\right) = \frac{\left[\sum_j w_j^{(1)} T_{.j}\right]^2}{(\lambda^2 S)_1 lr} = \frac{(-180 - 93 + 133 + 483)^2}{20 \times 3 \times 1} = 1960.8\,.$$

That due to the second-order term of A is

$$S_{A_q} = S\left(\hat{\beta}_{20}\right) = \frac{\left[\sum_i w_i^{(2)} T_{i.}\right]^2}{(\lambda^2 S)_2 mr} = \frac{(155 - 226 + 179)^2}{6 \times 4 \times 1} = 486.0\,.$$

That due to the second-order term of B is

$$S_{B_q} = S\left(\hat{\beta}_{02}\right) = \frac{\left[\sum_j w_j^{(2)} T_{.j}\right]^2}{(\lambda^2 S)_2 lr} = \frac{(60 - 93 - 133 + 161)^2}{4 \times 3 \times 1} = 2.1\,.$$

Finally, that due to the interaction term of A and B is

$$S_{A_l \times B_l} = S\left(\hat{\beta}_{11}\right) = \frac{\left(\sum_i \sum_j w_i^{(1)} w_j^{(1)} y_{ij}\right)^2}{(\lambda^2 S)_1^l (\lambda^2 S)_1^m r}$$

$$= \frac{1}{2 \times 20 \times 1}(3y_{11} + y_{12} - y_{13} - 3y_{14} - 3y_{31} - y_{32} + y_{33} + 3y_{34})^2$$

$$= \frac{1}{40}(45 + 31 - 47 - 186 - 102 - 42 + 49 + 162)^2 = 202.5.$$

By summarizing the above results, we can construct the ANOVA table for this experiment as shown in Table 9.4. Here, since B has four levels, S_B can be decomposed into S_{B_l}, S_{Bq} and S_{Bc}. Hence, we can obtain the variation due to the cubic term:

$$S_{B_c} = S_B - S_{B_l} - S_{B_q} = 1969 - 1960.8 - 2.1 = 6.1.$$

Also note that S_E can be obtained by writing

$$S_E = S_{A \times B} - S_{A_l \times B_l} = 213.0 - 202.5 = 10.5.$$

Table 9.4 ANOVA table with variations of individual terms.

Source of variation	Sum of squares	Degrees of freedom	Mean square	F_0
A	558.0	2	279.0	
A_l	72.0	1	72.0	132.9**
A_q	486.0	1	486.0	34.3**
B	1969.0	3	656.3	231.4**
B_l	1960.8	1	1960.8	312.5**
B_q	2.1 O	1	2.1	933.7**
B_c	6.1 O	1	6.1	1.0
A_lB_l	202.5	1	202.5	2.9
E	10.5	5	2.1	96.4**
(e)	(18.7)	(7)	(2.67)	
T	740.0 11			

** indicates 1% significance.

O indicates pooled variations.

In Table 9.4, since B_q and B_c are not significant, they were pooled into the error term. All the other terms are very significant. If we write the equation only with the significant terms, we have

$$\hat{y} = 37.250 + 0.300(x_1 - 40) + 2.287(x_2 - 17.5)$$
$$+ 0.135\left[(x_1 - 40)^2 - \frac{10^2}{12}(3^2 - 1)\right] - 0.090(x_1 - 40)(x_2 - 17.5).$$

9.4 TOLERANCE DESIGN BY FACTORIAL EXPERIMENTS

Tolerance design by factorial experiments is frequently met in practice. The approach will be explained by means of two-way layout electric circuit example. This is an extension of the example given in Taguchi (1986).

A 100 volt, f hertz power supply across a resistance, A, in series with an inductor, B, results in a current of y amperes.

$$y = \frac{100}{\sqrt{A^2 + (2\pi f B)^2}}. \tag{9.27}$$

The customer's tolerance for this circuit is $y = 10.0 \pm 4.0$ amperes. When the current is out of tolerance, the loss is \$150. The annual production of this circuit is 500000 units. The frequency f ranges from 50 to 55 Hz.

Currently, second-grade resistors, A, and inductors, B, are used, and their statistical properties are as follows:

Resistor, A (ohm)	Mean	$m_A = 9.92$
	Standard deviation	$\sigma_A = 0.30$
Inductor, B (henry)	Mean	$m_B = 0.04$
	Standard deviation	$\sigma_B = 0.0008$

The variation in the components due to several years of deterioration is included in the standard deviations σ_A and σ_B. The problem is to reduce the variability of the current by upgrading either the resistor or the inductance coil, or both. If first-grade components are used, the ranges of variation would be reduced by half. The cost increases would be \$0.30 for a resistor and \$2 for an inductor.

Taguchi (1986) recommended that the factor levels for a tolerance design case of this type should be set as follows: for two-level factors,

$$\text{first level} = \text{mean value} - \sigma$$
$$\text{second level} = \text{mean value} + \sigma;$$

For three-level factors

$$\text{first level} = \text{mean value} - \sqrt{\frac{3}{2}}\sigma$$
$$\text{second level} = \text{mean value}$$
$$\text{third level} = \text{mean value} + \sqrt{\frac{3}{2}}\sigma.$$

By means of these settings, two levels are located at the 15th and 85th percentile points in the cumulative distribution. Similarly three levels are located at the 10th, 50th and 90th percentile points, which seem to be very reasonable for level allocation.

The two-way layout with factors A and B of three levels each, and two repetitions (once at 50 Hz and twice at 55 Ht) at each factor level combination will be investigated. Therefore, the three levels of these components are set as follows:

$$\begin{aligned} A_1 &= 9.920 - \sqrt{\frac{3}{2}} \times 0.300 = 9.552, \\ A_2 &= 9.920, \\ A_3 &= 9.920 + \sqrt{\frac{3}{2}} \times 0.300 = 10.288, \end{aligned} \tag{9.28}$$

$$
\begin{aligned}
B_1 &= 0.00400 - \sqrt{\frac{3}{2}} \times 0.00080 = 0.000302, \\
B_2 &= 0.00400, \\
B_3 &= 0.00400 + \sqrt{\frac{3}{2}} \times 0.00080 = 0.00498.
\end{aligned}
\tag{9.29}
$$

For each combination of A and B levels, the current is computed twice by using f = 50 and 55, respectively. For example, the currents which result when the resistor/inductor combination A_1B_1 is used, are y_{111} and y_{112}.

$$y_{111} = \frac{100}{\sqrt{A^2 + (2\pi fB)^2}} = \frac{100}{\sqrt{9.552^2 + (314.16 \times 0.00302)^2}} = 10.428 \text{ amperes},$$

$$y_{112} = \frac{100}{\sqrt{A^2 + (2\pi fB)^2}} = \frac{100}{\sqrt{9.552^2 + (345.58 \times 0.00302)^2}} = 10.407 \text{ amperes}.$$

Similar computations are made for each configuration to obtain Table 9.5.

Table 9.5 Two-way layout experiment results.

	B_1	B_2	B_3	Total
A_1	10.418	10.380	10.331	62.200
	10.407	10.361	10.303	
A_2	10.035	10.001	9.958	59.935
	10.025	9.984	9.932	
A_3	9.679	9.648	9.610	57.827
	9.670	9.633	9.587	
Total	60.234	60.007	59.721	179.962

9.4.1 ANALYSIS OF VARIANCE

From Table 9.5, the following factorial effects are computed. Note that these are the constituents of the orthogonal polynomial model in (9.24): A_l, the linear term of A; A_q, the quadratic term of A; B_l, the linear term of B; B_q, the quadratic term of B; $A_1 \times B_1$, the interaction of the linear terms of A and B; and E, the error term, including all higher-order terms.

$$S_{Al} = \frac{\left[\sum_i w_i^{(1)} T_{i.}\right]^2}{(\lambda^2 S)_1 mr} = \frac{(-62.3 + 57.827)^2}{2 \times 3 \times 2} = 1.59359,$$

$$S_T = \sum_i \sum_j \sum_k y_{ijk}^2 - CT$$

$$= 10.418^2 + 10.407^2 + \cdots + 9.587^2 - \frac{179.962^2}{18} = 1.61808,$$

$$S_{Aq} = \frac{\left[\sum_i w_i^{(2)} T_{i.}\right]^2}{(\lambda^2 S)_2 mr} = \frac{(62.2 - 2 \times 59.953 + 57.827)^2}{6 \times 3 \times 2} = 0.00069,$$

$$S_{Bl} = \frac{\left[\sum_j w_j^{(1)} T_{.j}\right]^2}{(\lambda^2 S)_1 lr} = \frac{(-60.234 + 59.721)^2}{2 \times 3 \times 2} = 0.02193,$$

$$S_{Bq} = \frac{\left[\sum_j w_j^{(2)} T_{.j}\right]^2}{(\lambda^2 S)_2 lr} = \frac{(60.234 - 2 \times 60.007 + 59.721)^2}{6 \times 3 \times 2}$$

$$= 0.00010,$$

$$S_{Al \times Bl} = \frac{\left(\sum_i \sum_j w_i^{(1)} w_j^{(1)} T_{ij.}\right)^2}{(\lambda^2 S)_1^l (\lambda^2 S)_1^m r} = \frac{[T_{11.} - T_{13.} - T_{31.} + T_{33.}]^2}{2 \times 2 \times 2}$$

$$= \frac{(20.825 - 20.634 - 19.349 + 19.197)^2}{8} = 0.00019,$$

$$S_E = S_T - S_{Al} - S_{Aq} - S_{Bl} - S_{Bq} - S_{Al \times Bl} = 0.00158.$$

By summarizing the above results, we can construct the ANOVA table for this experiment as shown in Table 9.6. From the analysis of variance, we note that B_q and $A_l \times B_l$ are pooled with error to get (*e*). The pooled error sum of squares is 0.00187, with 14 degrees of freedom. The pooled error variance, 0.000134, is used to calculate the percentage contributions. For instance, the percentage of contribution of S_{Al} is

$$\rho_{Al} = \frac{S_{Al} - V_e}{S_T} \times 100 = \frac{1.59359 - 0.000134}{1.61808} \times 100 = 98.48\% .$$

The total variance when second-grade component parts are used is the mean squared deviation, which is

$$V_T = MSD = \frac{S_T}{17} = \frac{1.61808}{17} = 0.09518 .$$

Table 9.6 ANOVA table with variations of individual terms.

Source of variation	Sum of squares		Degrees of freedom	Mean square	F_0	ρ (%)
A_l	1.59359		1	1.59359	11892.5**	98.48
A_q	0.00069		1	0.00069	5.1*	0.03
B_l	0.02193		1	0.02193	163.7**	1.35
B_q	0.00010	O	1	0.00010		
$A_l \times B_l$	0.00019	O	1	0.00019		
E	0.00158		12	0.00132		
(*e*)	(0.00187)		14	(0.00134)		0.14
T	1.61808		17			100.00

** indicates 1% significance.

* indicates 5% significance.

O indicates pooled variations.

9.4.2 ANALYSIS OF LOSS FUNCTION

For tolerance design at the design stage where the grades of the individual component parts are determined, it is necessary to acquire information regarding the loss function for each component part. In this example, the loss function is

$$L(y) = \frac{A^*}{\Delta^2} \times MSD = \frac{150}{4.0^2} \times 0.09518 = \$0.8923\,.$$

The annual loss becomes \$0.8923 × 500000 = \$446150. When first-grade component parts were used, the varying range of variation would be reduced by half, i.e. the standard deviation would be reduced by half, and the level intervals of A and B would be reduced by half. Then, adapting (9.23) as required, it is understood that

$$E[S_{(1)}] = \sigma^2 + r(S)_1 c^2 \beta_1^2\,,$$
$$E[S_{(2)}] = \sigma^2 + r(S)_2 c^4 \beta_2^2\,.$$

Therefore, the variance would be reduced to $(1/2)^2 = 1/4$ for a linear term and to $(1/2)^4 = 1/16$ for a quadratic term.

In general we can use the following relationship between the variances before upgrading and after upgrading.

$$V_N = V_C\{\rho_{Al}(R_N/R_C)^2 + \rho_{Aq}(R_N/R_C)^4 + \rho_{Bl}(R'_N/R'_C)^2 + \cdots + \rho_e\}, \tag{9.30}$$

where

V_N = variance after upgrading (new),

V_C = variance before upgrading (current),

ρ_{Al} = percent contribution of linear term of A,

ρ_{Aq} = percent contribution of quadratic term of A,

ρ_{Bl} = percent contribution of linear term of B,

R_N = range of variation of new component A,

R_C = range of variation of current component A,

R'_N = range of variation of new component B,

R'_C = range of variation of current component B.

In the case of component A, the loss due to using second-grade resistors is

$$L(y) = \frac{A^*}{\Delta^2} \times V_T \times \frac{\rho_{Al} + \rho_{Aq}}{100} = \frac{150}{4.0^2} \times 0.09518 \times \frac{98.48 + 0.03}{100}$$
$$= 9.375 \times 0.09518 \times 0.9851 = \$0.8790.$$

The loss due to using first-grade resistors is

$$L(y) = 9.375 \times 0.09518 \times (0.9848 \times \tfrac{1}{4} + 0.0003 \times \tfrac{1}{16})$$
$$= \$0.2197.$$

The quality improvement by changing from second-grade resistors to first-grade resistors is

$$0.8790 - 0.2197 = \$0.6593.$$

Since a \$0.30 cost increase is incurred by upgrading, the net gain would be

$$0.6593 - 0.3 = \$0.3593,$$

and the annual improvement would be

$$0.3593 \times 500000 = \$179650.$$

In the case of component B, the loss due to using second-grade inductors is, since $\rho_{Bl} = 1.35\%$,

$$L(y) = 9.375 \times 0.09518 \times (\rho_{Bl} / 100) = \$0.0120,$$

and the loss due to using first-grade inductors is

$$L(y) = 9.375 \times 0.09518 \times (0.135 \times \tfrac{1}{4}) = \$0.0030.$$

Table 9.7 Summary results of tolerance design.

Part	Grade	Cost	Loss	Total loss	Selection
Resistor	Second	Base cost	0.8790	0.8790	
	First	+ 0.3	0.2197	0.5197	*
Inductor	Second	Base cost	0.0120	0.0120	*
	First	+ 2.0	0.0030	2.0030	

Therefore, Table 9.7 can be constructed to summarize tolerance design.

As the final results of tolerance design by means of this two-way layout, the first-grade resistor and second-grade inductor should be chosen. Although this choice increases the cost by \$0.30 for each resistor, the first-grade resistor yields improved quality of \$0.3593.

In this section two-way arrays were used for tolerance design, but usually there are more system parameters, and orthogonal arrays are used to facilitate such analysis. Using orthogonal arrays for tolerance design is the subject of the next section.

9.5 TOLERANCE DESIGN USING ORTHOGONAL ARRAYS

Tolerance design is usually done by using orthogonal arrays. In this section, a typical example will be illustrated in which tolerance design is done after parameter design. In parameter design, the optimum combination of levels of the control factors is chosen. When parameter design cannot sufficiently reduce the effect of internal and external noise, it becomes necessary to restrict the variation of the major noise factors to within narrower ranges, even though this raises the cost. This process is an important aspect of tolerance design. Since the effect of the error factors can often be substantially reduced by parameter design, it is important to carry out parameter design first and tolerance design second.

9.5.1 PARAMETER DESIGN

In the process-design stage of a new product, it is desired to determine the process operating conditions of four design variables which influence the tensile strength of the product. The levels of the variables are as shown in Table 9.8. There are four noise factors; each one corresponds to each design variable. Since each design variable does not keep its own nominal level and varies in practice, it also becomes a noise factor where the range of variation is noted as percentage of the level as in Table 9.9. Note that A is the design variable while A' is the noise factor.

Table 9.8 Design variables and experimental levels.

Design variable	Level		
	0	1	2
A: heating temp.	1200°C	1300°C	1400°C
B: heating time	2 hours	3 hours	4 hours
C: % of additive	3%	4%	5%
D: speed	700rpm	800rpm	900rpm

Table 9.9 Noise factors and levels.

Noise factor	Level (%)		
	0	1	2
A'	-1	0	+1
B'	-3	0	+3
C'	-8	0	+8
D'	-5	0	+5

For the parameter design, L_9 is used both for the inner array and for the outer array, as shown in Table 9.10. Since the tensile strength is a larger-the-better characteristic, the SN ratio for the ith experiment number is computed as follows:

$$SN_i = -10\log\left(\frac{1}{9}\sum_{j=1}^{9}\frac{1}{y_{ij}^2}\right).$$

For example, the SN ratio of the first experiment is

$$SN_1 = -10\log\left[\frac{1}{9}\left(\frac{1}{26^2}+\frac{1}{19^2}+\cdots+\frac{1}{28^2}\right)\right]$$
$$= -10\log(0.00135) = 28.75\,\text{dB}.$$

The other SN ratios can be similarly obtained, and their sums and means are summarized in Table 9.11.

Table 9.10 Tensile strength experiment and experimental data.

Factor	A	B	C	D	Data									SN
Column number					Experiment number									Noise factor
	1	2	3	4	1	2	3	4	5	6	7	8	9	
					0	0	0	1	1	1	2	2	2	A'
					0	1	2	0	1	2	0	1	2	B'
					0	1	2	1	2	0	2	0	1	C'
Experiment number					0	1	2	2	0	1	1	2	0	D'
1	0	0	0	0	26	19	20	40	34	37	31	37	28	28.75
2	0	1	1	1	28	20	18	39	41	30	38	42	49	29.17
3	0	2	2	2	31	34	38	50	40	28	27	26	35	30.21
4	1	0	1	2	45	68	71	47	50	67	33	49	61	33.99
5	1	1	2	0	50	47	29	43	49	61	44	38	62	32.79
6	1	2	0	1	43	48	60	39	42	29	34	58	61	32.47
7	2	0	2	1	12	18	20	41	19	20	21	30	32	35.97
8	2	1	0	2	23	41	29	40	33	28	27	34	38	39.80
9	2	2	1	0	18	33	29	34	28	19	16	15	30	26.63

Table 9.11 Summary of SN ratios at each factor level.

Factor		A	B	C	D	
Level sum	0	88.13	88.71	91.01	88.16	
	1	99.25	91.76	89.79	87.61	Total 269.78
	2	82.40	89.31	88.97	94.01	
Level average	0	29.38	29.57	30.34	29.39	
	1	33.08	30.59	29.93	29.20	Average 29.98
	2	27.47	29.77	29.66	31.34	

The variation of SN ratios for factor A can be calculated by

$$S_A = \tfrac{1}{3}(88.13^2 + 99.25^2 + 82.40^2) - \frac{269.78^2}{9} = 48.93.$$

The other variations can be obtained in a similar way, and the ANOVA table for this parameter design can be obtained as shown in Table 9.12. This shows that the contribution ratios of A and D

Table 9.12 ANOVA table for tensile strength experiment.

Source variation	Sum of squares		Degrees of freedom	Mean square	ρ (%)
A	48.93		2	24.47	79.87
B	1.74	O	2		
C	0.71	O	2		
D	8.35		2	4.18	11.93
(e)	(2.45)		(4)	(0.613)	8.20
T	59.73		8		100.00

O indicates pooled variation.

are greater than 10%, which means they are significant. The optimal levels of A and D are A_1D_2 (from Table 9.11). Factors B and C are not significant and are pooled into the error term. From other reasons such as economy and operability, the optimal levels of B and C are determined to be B_0C_0. Hence, the overall optimum condition is $A_1B_0C_0D_2$, for which the SN ratio is estimated to be

$$\begin{aligned}\hat{\mu} &= \bar{A}_1 + \bar{B}_0 + \bar{C}_0 + \bar{D}_2 - 3\bar{T} \\ &= 33.08 + 29.357 + 30.34 + 31.34 - 3 \times 29.98 \\ &= 34.39\,\text{dB}.\end{aligned} \tag{9.31}$$

9.5.2 TOLERANCE DESIGN

The optimal levels of A, B, C and D are A_1 = 1300, B_o = 2 hours, C_o = 3% and D_2 = 900 rpm. Accordingly, the levels of noise factors can be obtained from Table 9.9, as shown in Table 9.13. For example, the factor levels of A are obtained as follows:

$$A'_0 = 1300 - 1\% \times 1300 = 1287°\text{C}$$

$$A'_1 = 1300$$

$$A'_2 = 1300 + 1\% \times 1300 = 1313°\text{C}$$

An L_9 tolerance design using these four noise factors is conducted and the tensile strength data is obtained as in Table

9.14. Note that this tolerance design can be considered to be a confirmation test at the optimum condition of $A_1B_0C_0D_2$. The SN ratio for this data is

$$\begin{aligned} SN &= -10\log\left[\frac{1}{9}\left(\frac{1}{48^2}+\frac{1}{57^2}+\cdots+\frac{1}{38^2}\right)\right] \\ &= 33.49\,\text{dB}, \end{aligned}$$

which is close enough to 34.39 in (9.31). This means that the optimum condition, $A_1B_0C_0D_2$, is verified by this confirmation test.

Table 9.13 Levels of noise factors.

Noise factor	Level		
	0	1	2
A'	(1300−13)°C	1300°C	(1300+ 13)°C
B'	2 hours−3.6 min	2 hours	2 hours+ 3.6 min
C'	(3−0.24)%	3%	(3+ 0.24)%
D'	(900−45) rpm	900 rpm	(900+ 45) rpm

Table 9.14 Tolerance design and data.

Experiment number	Noise factor and column no.				Data
	A'	B'	C'	D'	
	1	2	3	4	
1	0	0	0	0	48
2	0	1	1	1	57
3	0	2	2	2	52
4	1	0	1	2	34
5	1	1	2	0	56
6	1	2	0	1	49
7	2	0	2	1	58
8	2	1	0	2	55
9	2	2	1	0	38

Mean = 49.67

Table 9.15 ANOVA table for noise factors.

Source of variation	Sum of squares	Degrees of freedom	Mean square	ρ (%)
A'	56.0	2	28.0	9.6
B'	180.7	2	90.35	31.0
C'	232.7	2	116.35	40.0
D'	112.6	2	56.30	19.4
T	582.0	8	72.75	100.0

Table 9.15 is the ANOVA table for the tolerance-design data in Table 9.14. Here, each contribution ratio is the ratio of each sum of squares to the total sum of squares, since there is no error term. In Table 9.14, the mean tensile strength, 49.67, is satisfactory. However, the variance V_T=72.75 in Table 9.15 is too big, and it is desired to reduce the total variance V_T by reducing the tolerance of A', B', C' and/or D'. At this stage, we need to carry out a cost/benefit analysis. In order to reduce the variance, we should reduce the tolerance (range of variation) of each factor, which usually demands some investment to take care of parts replacement or facility change. Suppose, in this experiment, noise factors B' and C' are chosen and their tolerances can be reduced to half and one-third, respectively, by some investment. Observe that factors B' and C' have the biggest contribution ratios in Table 9.15.

Then the expected total variance from the reduced tolerances of B' and C' will be, using the formula in (9.30),

$$V_{T'} = V_{T'} \times \left[\rho_{A'} + \rho_{B'} \left(\frac{1}{2}\right)^2 + \rho_{C'} \times \left(\frac{1}{3}\right)^2 + \rho_{D'} \right] \Big/ 100$$

$$= 72.75 \times \left[0.096 + 0.31 \times \left(\frac{1}{2}\right)^2 + 0.40 \times \left(\frac{1}{3}\right)^2 + 0.194 \right]$$

$$= 29.97.$$

If this level of variance is satisfactory and the money invested is tolerable, we have achieved the purpose of tolerance design.

Exercises

9.1 Suppose the upper specification limit of automobile exhaust gas NO_x is 0.48 mg km^{-1}. If this limit is not satisfied in the factory, the cost of rework is \$80. However, when the car is delivered to a customer and a claim occurs because of the exhaust gas system, the customer's loss is \$300. What should be the producer's tolerance limit for NO_x?

9.2 A circuit has a nominal output, M mV, with a tolerance of 4mV ($M \pm 4$ mV). If the circuit is outside its specification limits, it is scrapped at a loss of \$3. When a resistance in the circuit varies by 1 mΩ, the output voltage varies 0.4 mV. The target value of this resistance is selected to give the target output voltage of M mV. What should be the tolerance of the resistance? The loss from rejecting the resistor is \$0.50.

9.3 The yield of a tower in a chemical plant is influenced by the amount of catalyst used. A one-way experiment was conducted as follows to check the relationship between the yield and the amount of catalyst to be used.

	Amount of catalyst(A)			
	$x_1 = 1.0\%$	$x_2 = 1.5\%$	$x_3 = 2.0\%$	$x_4 = 2.5\%$
Yield data (%)	72.8	779.2	82.1	73.2
	70.2	78.4	80.3	74.6
	71.8	80.1	79.8	75.0

(a) Construct a one-way ANOVA table. Is the amount of catalyst significant?

(b) Fit the second-order polynomial regression model by using orthogonal polynomials. Carry out analysis of variance for this model. Are the terms A_l and A_q

significant? Test whether the third-order term A_c is necessary.

(c) Using the fitted equation in (b), find the optimal amount of catalyst which maximizes the yield.

9.4 If alternating current with a voltage of V volt and a frequency of f hertz is applied to a circuit with resistance of R ohms and self-inductance of L henrys, the resulting flow, in amperes, is given by the following formula:

$$y = \frac{V}{\sqrt{R^2 + (2\pi f L)^2}}.$$

Under the consumer's conditions of use, the tolerance range of the current value of this circuit is 10.0 ± 4.0 amperes. The loss incurred through after-sales service and the like when the specification is not met is $50. The monthly production volume is 100000 units. The current statistics for V, f, R and L are as follows:

Variable	Mean	Standard deviation
V	100	10
f	55	5
R	10	0.3
L	0.004	0.0008

(a) Take the variables V, f, R and L as factors A, B, C and D, respectively, and allocate them in the OA L_{18}. A, C and D have three levels, as follows.

$$\text{level } 0 = \text{mean} - \sqrt{\frac{3}{2}} \times \text{standard deviation}$$

$$\text{level } 1 = \text{mean}$$

$$\text{level } 2 = \text{mean} + \sqrt{\frac{3}{2}} \times \text{standard deviation}.$$

B has two levels, determined as follows.

level 0: mean − standard deviation

level 1: mean + standard deviation.

Find the current y for all 18 experiments, and carry out analysis of variance.

(b) At present, third-rate components of resistance and self-inductance in the circuit are used. If the components are upgraded to second-rate or first-rate components, standard deviations are reduced as follows. However, extra costs will be incurred.

	Change to second-rate component	Change to first-rate component
R	Standard deviation=0.15 Extra cost=$0.10	Standard deviation=0.5 Extra cost=$2.00
L	Standard deviation=0.4 Extra cost=$0.20	Standard deviation=0.1 Extra cost=$5.00

Carry out tolerance design to check whether R and/or L should be replaced by the second-rate or first-rate components from the viewpoint of loss function. In this analysis use orthogonal polynomials to find C_l, C_q, D_l and D_q, and carry out an appropriate analysis of variance.

10

Robust Response Surface Design and Analysis

10.1 RESPONSE SURFACE METHODOLOGY AND ITS ROLES IN QUALITY IMPROVEMENT

Response surface methodology (RSM) is a collection of statistical techniques that are useful for the modeling and analysis of problems in which one or more responses of interest are influenced by several variables, and the objective is to find the relationship between the response(s) and several variables, and to optimize the response(s). Suppose that a chemical engineer wishes to find the levels of temperature (x_1) and pressure (x_2) that maximize the yield (y) of a chemical process. The process yield is a function of the levels of x_1 and x_2, say,

$$y = f(x_1, x_2) + \varepsilon, \tag{10.1}$$

where ε represents the noise or error observed in the response variable y, and x_1 and x_2 are the independent variables (factors). If we denote the expected response by $E(y) = f(x_1, x_2) = \eta$, then the surface represented by

$$\eta = f(x_1, x_2) \tag{10.2}$$

is called a true response surface.

We may represent the response surface graphically as shown in Figure 10.1, where η is plotted versus the levels of x_1 and x_2.

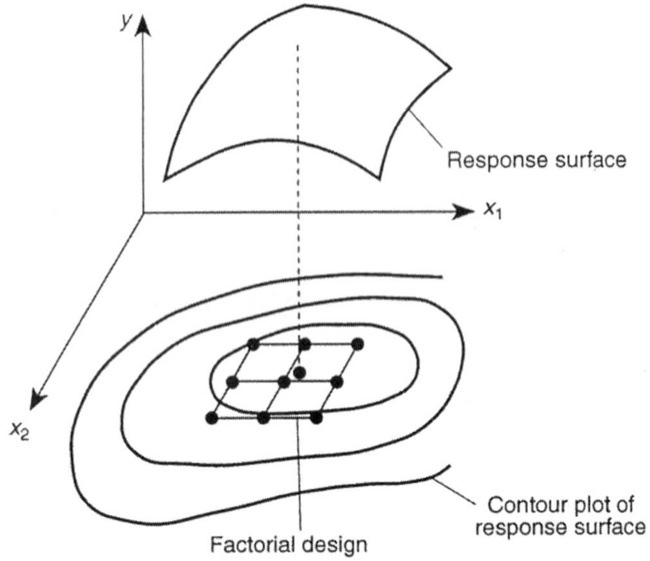

Figure 10.1 A three-dimensional response surface for $\eta = f(x_1, x_2)$.

To help visualize the shape of a response surface, we often plot its contours. In the contour plot, lines of constant response are drawn in the $x_1 - x_2$ plane. Each contour corresponds to a particular height of the response surface.

The form of the functional relationship f, between the response and the independent variables, is unknown. Usually, a low-order polynomial in some region of interest of the independent variables is employed. If the response is well modeled by a linear function of the independent variables, then the approximating function is the first-order model

$$y = \beta_0 + \beta_1 x_1 + \beta_2 x_2 + \cdots + \beta_k x_k + \varepsilon . \tag{10.3}$$

If there is curvature in the system, as shown in Figure 10.1, then the approximating function is the second-order model

$$y = \beta_0 + \sum_{i=1}^{k} \beta_i x_i + \sum_{i=1}^{k} \beta_{ii} x_i^2 + \sum_{i<j} \sum \beta_{ij} x_i x_j + \varepsilon . \tag{10.4}$$

If we have a set of n observations, (10.3) or (10.4) can be represented in matrix form as

$$y = X\beta + \varepsilon,$$

where $\mathbf{y}' = (y_1, y_2, \cdots, y_n)$, $\mathbf{X}$ is a known matrix of order $n \times p$ and rank p whose elements consist, in general, of powers and cross-products of the design settings for the independent variables, and $\boldsymbol{\varepsilon}$ is a random vector such that $E(\boldsymbol{\varepsilon}) = \mathbf{0}$, and $\mathrm{var}(\boldsymbol{\varepsilon}) = \mathbf{I}\sigma^2$. The least-squares estimator of $\boldsymbol{\beta}$ is given by

$$\hat{\boldsymbol{\beta}} = (\mathbf{X}'\mathbf{X})^{-1}\mathbf{X}'\mathbf{y},$$

where $E(\hat{\boldsymbol{\beta}} - \boldsymbol{\beta}) = \mathbf{0}$, and $\mathrm{var}(\hat{\boldsymbol{\beta}}) = (\mathbf{X}'\mathbf{X})^{-1}\sigma^2$.

The predicted response at a point $\mathbf{x}' = (x_1, x_2, \cdots, x_k)$ can be expressed as

$$\hat{y}(\mathbf{x}) = f'(\mathbf{x})\hat{\boldsymbol{\beta}}, \tag{10.5}$$

where $f'(\mathbf{x})$ is a vector of order $1 \times p$ of the same form as a row of $\mathbf{X}$. For example, the predicted response for the model (10.4) is

$$\hat{y} = \hat{\beta}_0 + \sum_{i=1}^{k} \hat{\beta}_i x_i + \sum_{i=1}^{k} \hat{\beta}_{ii} x_i^2 + \sum_{i<j}\sum \hat{\beta}_{ij} x_i x_j = f'(\mathbf{x})\hat{\boldsymbol{\beta}},$$

where $f'(\mathbf{x}) = (1, x_1, x_2, \cdots, x_k, x_1^2, x_2^2, \cdots, x_k^2, x_1 x_2, \cdots, x_{k-1}x_k)$. An important expression associated with $\hat{y}(\mathbf{x})$ is its prediction variance, given by

$$\mathrm{var}[\hat{y}(\mathbf{x})] = f'(\mathbf{x})(\mathbf{X}'\mathbf{X})^{-1} f(\mathbf{x})\sigma^2. \tag{10.6}$$

The regression parameters can be estimated most effectively if proper experimental designs are used to collect the data. Designs for fitting response surfaces are called **response surface designs**. The 3^2 factorial design, which is shown in Figure 10.1, is a typical response surface design.

The independent variables are usually coded to a (−1,1) interval, so that they are easy to handle. Suppose ξ is the natural variable, with ξ_{max} the maximum level of ξ, and ξ_{min} the minimum level of ξ. Let x be the coded variable. Then the usual coding scheme is

$$x = [(\text{real value of } \xi) - m\,]/c,$$

where $m = (\xi_{max} + \xi_{min})/2$, and $c = (\xi_{max} - \xi_{min})/2$. For example, if the range of time for experiments is (30, 40) minutes, then it is convenient to use the coded time variable

$$x = [\text{time} - 35]/5.$$

RSM is a very useful tool for quality and productivity improvement in industry. In industry, we often encounter quality and productivity problems in which we wish to discover some functional relationship between the response (quality or productivity) and several independent variables (factors that influence the response). Therefore, if we can find the functional relationship, we can easily approach the quality and productivity problems in practice.

RSM is a sequential procedure. Often, when we are at a point on the response surface that is remote from the optimum, there is little curvature in the system and the first-order model of (10.3) will be appropriate. By the method of steepest ascent — Myers (1971), Box and Draper (1987) and Khuri and Cornell (1987) — the experimenter efficiently approaches the general vicinity of the optimum. Once the region of the optimum has been found, a more elaborate model such as the second-order model of (10.4) may be employed, and an analysis may be performed to locate the optimum. In this chapter we will only deal with the second-order response surface model.

The eventual objectives of RSM may be summarized as follows.

1. Visualization of the relationship between one or more responses and a number of independent variables that influence the responses.

2. Finding and exploring the optimum operating conditions for the system, or determining a region of the factor space in which operating specifications are satisfied.
3. Finding appropriate robust designs that can be effectively used for the purposes of objectives 1 and 2, and finding robust conditions which produce product with robust quality characteristics.

10.2 ANALYSIS OF A SECOND-ORDER MODEL

When the experimenter is relatively close to the optimum, a model of degree 2 is usually required to approximate the response because of curvature in the true response surface. The fitted second-order model

$$\begin{aligned}\hat{y} &= \hat{\beta}_0 + \sum_{i=1}^{k} \hat{\beta}_i x_i + \sum_{i=1}^{k} \hat{\beta}_{ii} x_i^2 + \sum_{i<} \sum_{j} \hat{\beta}_{ij} x_i x_j \\ &= \hat{\beta}_0 + \mathbf{x}'\mathbf{b} + \mathbf{x}'\mathbf{B}\mathbf{x}, \end{aligned} \tag{10.7}$$

where

$$\mathbf{x} = \begin{pmatrix} x_1 \\ x_2 \\ \vdots \\ x_k \end{pmatrix}, \ \mathbf{b} = \begin{pmatrix} \hat{\beta}_1 \\ \hat{\beta}_2 \\ \vdots \\ \hat{\beta}_k \end{pmatrix} \text{ and } \mathbf{B} = \begin{pmatrix} \hat{\beta}_{11} & \hat{\beta}_{12}/2 & \cdots & \hat{\beta}_{1k}/2 \\ \hat{\beta}_{21}/2 & \hat{\beta}_{22} & \cdots & \hat{\beta}_{2k}/2 \\ \vdots & & \vdots & \vdots \\ \hat{\beta}_{1k}/2 & & & \hat{\beta}_{kk} \end{pmatrix},$$

is in most cases adequate for the relationship between a response y and a number of independent variables. In this section we will show how to use this fitted model to find the optimum set of operating conditions for the x_i and to characterize the nature of the response surface.

10.2.1 LOCATION OF THE OPTIMUM POINT

Suppose we want to find the levels of the x_i that optimize the predicted response in (10.7). This point, if it exists, will be that at which the partial derivatives are equal to zero:

$$\frac{\partial \hat{y}}{\partial \mathbf{x}} = \begin{pmatrix} \frac{\partial \hat{y}}{\partial x_1} \\ \frac{\partial \hat{y}}{\partial x_2} \\ \vdots \\ \frac{\partial \hat{y}}{\partial x_k} \end{pmatrix} = \frac{\partial}{\partial \mathbf{x}}(\hat{\beta}_0 + \mathbf{x}'\mathbf{b} + \mathbf{x}'\mathbf{B}\mathbf{x})$$

$$= \mathbf{b} + 2\mathbf{B}\mathbf{x} = 0. \tag{10.8}$$

This point, say, $\mathbf{x}_0' = (x_{10}, x_{20}, \cdots, x_{k0})$, is called the **stationary point.** The stationary point could represent a point of maximum response, a point of minimum response, or a saddle point. These three possibilities are shown in Figure 10.2, which demonstrate possible contour plots around the stationary point. The stationary point is the solution to (10.8):

$$\mathbf{x}_0 = -\frac{1}{2}\mathbf{B}^{-1}\mathbf{b}. \tag{10.9}$$

10.2.2 CANONICAL ANALYSIS

Once we have found the stationary point, it is usually necessary to determine whether this is a point of maximum or minimum response or a saddle point, and its relative sensitivity to the independent variables $x_1, x_2, \cdots, x_k$. If there are two or three independent variables, examining the contour plot of the fitted model could give enough information. If there are several variables, it is not easy to construct and interpret the contour plot, and we need a mathematical analysis. Even when there are only a few

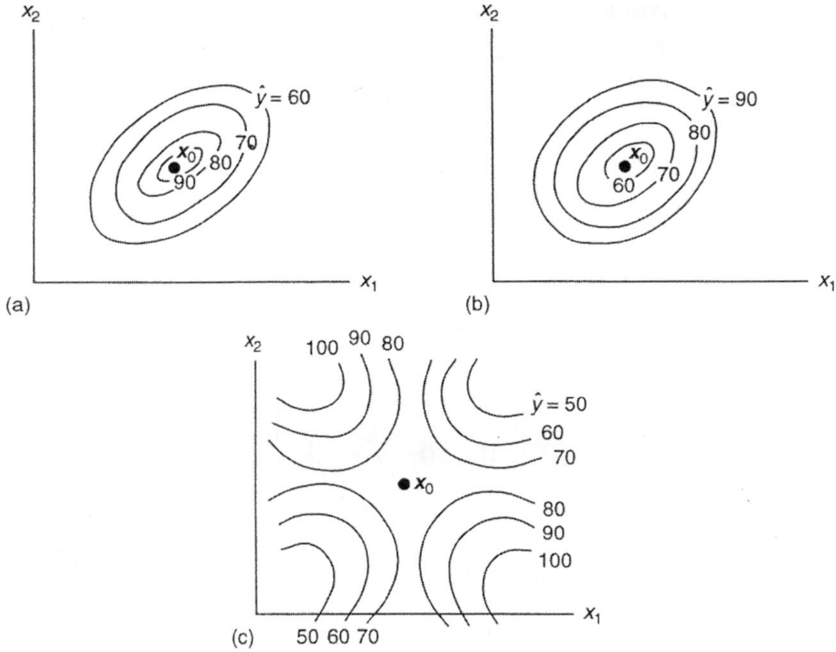

Figure 10.2 A stationary point $\mathbf{x_0}$: (a) maximum; (b) minimum; (c) saddle point.

variables, a more formal analysis can be useful. The rest of this section is devoted to one such method of analysis for response surfaces, canonical analysis.

It is helpful first to transform the model (10.7) into a new coordinate system, $\mathbf{z} = \mathbf{x} - \mathbf{x_0}$, with the origin at the stationary point $\mathbf{x_0}$ as shown in Figure 10.3:

$$\begin{aligned}\hat{y} &= \hat{\beta}_0 + \mathbf{x'b} + \mathbf{x'Bx} \\ &= \hat{\beta}_0 + (\mathbf{z'} + \mathbf{x'_0})\mathbf{b} + (\mathbf{z'} + \mathbf{x'_0})\mathbf{B}(\mathbf{z} + \mathbf{x_0}) \\ &= \hat{y}_0 + \mathbf{z'Bz},\end{aligned} \tag{10.10}$$

where y_0 is the predicted value of y at the stationary point, $\hat{y}_0 = \hat{\beta}_0 + \mathbf{x'_0b} + \mathbf{x'_0Bx_0}$. Let $\lambda_1, \lambda_2, \cdots, \lambda_k$ be the eigenvalues of the matrix $\mathbf{B}$, and $\mathbf{p_1}, \mathbf{p_2}, \cdots, \mathbf{p_k}$ be the eigenvectors of $\mathbf{B}$ associated with the eigenvalues. Then there exists an orthogonal transformation, $\mathbf{z} = \mathbf{P}\mathbf{w}$, which results in

$$\begin{aligned}\mathbf{z}'\mathbf{B}\mathbf{z} &= \mathbf{w}'\mathbf{P}'\mathbf{B}\mathbf{P}\mathbf{w} \\ &= \mathbf{w}'\mathbf{\Lambda}\mathbf{w},\end{aligned}$$

where $\mathbf{P} = (\mathbf{p}_1, \mathbf{p}_2, \cdots, \mathbf{p}_k)$, $\mathbf{p}_i'\mathbf{p}_i = 1$ for $i = 1, 2, \cdots, k$, and $\mathbf{\Lambda}$ is a diagonal matrix,

$$\mathbf{\Lambda} = \begin{pmatrix} \lambda_1 & 0 & \cdots & 0 \\ 0 & \lambda_2 & \cdots & 0 \\ \vdots & \vdots & \vdots & \vdots \\ 0 & 0 & \cdots & \lambda_k \end{pmatrix}.$$

Then equation (10.10) becomes

$$\begin{aligned}\hat{y} &= \hat{y}_0 + \mathbf{w}'\mathbf{\Lambda}\mathbf{w} \\ &= \hat{y}_0 + \lambda_1 w_1^2 + \lambda_2 w_2^2 + \cdots + \lambda_k w_k^2, \end{aligned} \tag{10.11}$$

which is called the **canonical form** of the second-order fitted response model. Figure 10.3 shows the case of two independent variables.

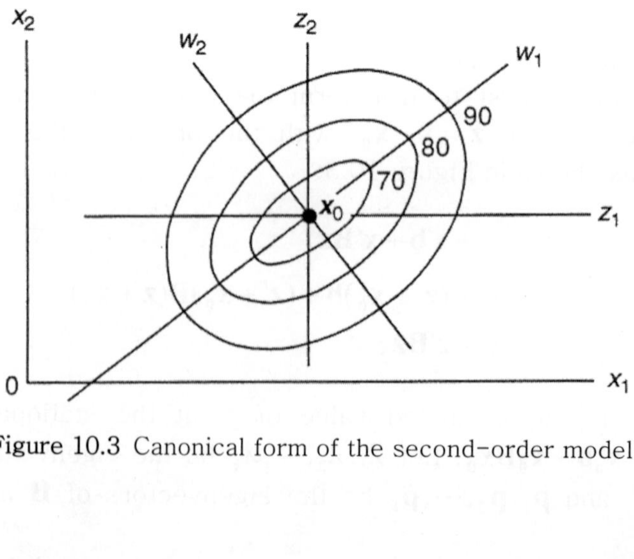

Figure 10.3 Canonical form of the second-order model.

The nature of the response surface can be determined from the stationary point and the sign and magnitude of the $\{\lambda_i\}$. Suppose that the stationary point is within the region of interest for fitting the second-order model. The region of interest is usually one of the following two in the coded independent variables.

$$R_1 = \{\mathbf{x} : -1 \le x_i^2 \le 1, i = 1, 2, \cdots, k\} \quad \text{square region of interest}$$
$$R_2 = \{\mathbf{x} : \sum_{i=1}^{k} x_i^2 \le 1\} \quad \text{circular region of interest.} \tag{10.12}$$

If the $\{\lambda_i\}$ are all positive, then $\mathbf{x_0}$ is a point of minimum response, since $y_0 \le y$ for all possible points of $\mathbf{x}$. If the $\{\lambda_i\}$ are all negative, then $\mathbf{x_0}$ is a point of maximum response; and if the $\{\lambda_i\}$ have different signs, $\mathbf{x_0}$ is a saddle point. Furthermore, the surface is steepest in the w_i direction for which λ_i is the greatest. For example, Figure 10.3 depicts a system for which $\mathbf{x_0}$ is a minimum, since λ_1 and λ_2 are both positive with $\lambda_1 \le \lambda_2$.

Example 10.1

A resin-manufacturing company wanted to improve the impact power of an ABS resin product. A quality improvement team was formed to handle the problem. The team found that the impact power was mainly determined in the extruding process, where the screw rotation rate in revolutions per minute (*RPM*) and the process temperature (*TEMP*) were the key factors. The operation conditions for these factors were *RPM*= 250 and *TEMP*= 240°C. From past experience, the team realized that higher *RPM* and *TEMP* would improve the impact power, and they decided to use an octagonal design in which the center point was (*RPM*= 270, *TEMP*= 250). Table 10.1 shows the octagonal design and experimental data. Since the resin liquidity is also an important characteristic, the liquidity was also observed. The order of experiments was randomly chosen.

In Table 10.1, note that the region of interest is 242 ≤ *RPM* ≤ 298, and 236 ≤ *TEMP* ≤ 264. The coding scheme for *RPM* and

Table 10.1 Octagonal design and experiment data.

Order of experiment	Experimental condition		Coded independent variables		Data	
	RPM	*TEMP*	X_1 (*RPM*)	X_2 (*TEMP*)	Impact power (kgcm^{-2})	Liquidity (g per 10min)
6	250	260	-0.714	0.714	20.0	52.5
3	250	240	-0.714	-0.714	19.8	50.8
2	290	260	0.714	0.714	20.8	52.8
11	290	240	0.714	-0.714	20.5	50.9
5	270	250	0	0	21.5	52.2
1	270	250	0	0	21.3	52.0
8	270	250	0	0	21.8	51.9
10	242	250	-1	0	19.6	51.5
4	298	250	1	0	21.0	51.3
9	270	236	0	-1	19.6	50.2
7	270	264	0	1	20.3	53.0

TEMP is

$$x_1 = \frac{RPM - 270}{28}, \quad x_2 = \frac{TEMP - 250}{14}. \tag{10.13}$$

The response surface analysis will be shown here for the impact power data only. Later on, the liquidity data analysis will be illustrated. The second-order model can be obtained by the regression approach.

$$\boldsymbol{\beta} = \begin{pmatrix} \hat{\beta}_0 \\ \hat{\beta}_1 \\ \hat{\beta}_2 \\ \hat{\beta}_{11} \\ \hat{\beta}_{22} \\ \hat{\beta}_{12} \end{pmatrix} = (\mathbf{X}'\mathbf{X})^{-1}\mathbf{X}'\mathbf{y} = \begin{pmatrix} 21.531 \\ 0.612 \\ 0.262 \\ -1.142 \\ -1.492 \\ 0.049 \end{pmatrix},$$

where

$$X = \begin{pmatrix} & x_1 & x_2 & x_1^2 & x_2^2 & x_1x_2 \\ 1 & -0.714 & 0.714 & 0.51 & 0.51 & -0.51 \\ 1 & -0.714 & -0.714 & 0.51 & 0.51 & 0.51 \\ 1 & 0.714 & 0.714 & 0.51 & 0.51 & 0.51 \\ 1 & 0.714 & -0.714 & 0.51 & 0.51 & -0.51 \\ 1 & 0 & 0 & 0 & 0 & 0 \\ 1 & 0 & 0 & 0 & 0 & 0 \\ 1 & 0 & 0 & 0 & 0 & 0 \\ 1 & -1 & 0 & 1 & 0 & 0 \\ 1 & 1 & 0 & 1 & 0 & 0 \\ 1 & 0 & -1 & 0 & 1 & 0 \\ 1 & 0 & 1 & 0 & 1 & 0 \end{pmatrix} \quad y = \begin{pmatrix} 20.0 \\ 19.8 \\ 20.8 \\ 20.5 \\ 21.5 \\ 21.3 \\ 21.8 \\ 19.6 \\ 21.0 \\ 19.6 \\ 20.3 \end{pmatrix}.$$

Therefore, the fitted second-order equation is

$$\hat{y} = 21.531 + 0.612x_1 + 0.262x_2 - 1.142x_1^2 - 1.492x_2^2 + 0.049x_1x_2 .$$

The contour plot of the above equation in the region of interest, $R = \{\mathbf{x} : -1 \le x_i \le 1, i = 1,2\}$, is shown in Figure 10.4. Observe from the graph that there is a point of maximum response somewhere in the middle. In order to locate the maximum point, we need canonical analysis.

From (10.9), the stationary point is

$$\mathbf{x_0} = -\frac{1}{2}\mathbf{B}^{-1}\mathbf{b}$$

$$= -\frac{1}{2}\begin{bmatrix} -1.142 & 0.0245 \\ 0.0245 & -1.492 \end{bmatrix}^{-1} \begin{bmatrix} 0.612 \\ 0.262 \end{bmatrix} = \begin{bmatrix} 0.270 \\ 0.092 \end{bmatrix}.$$

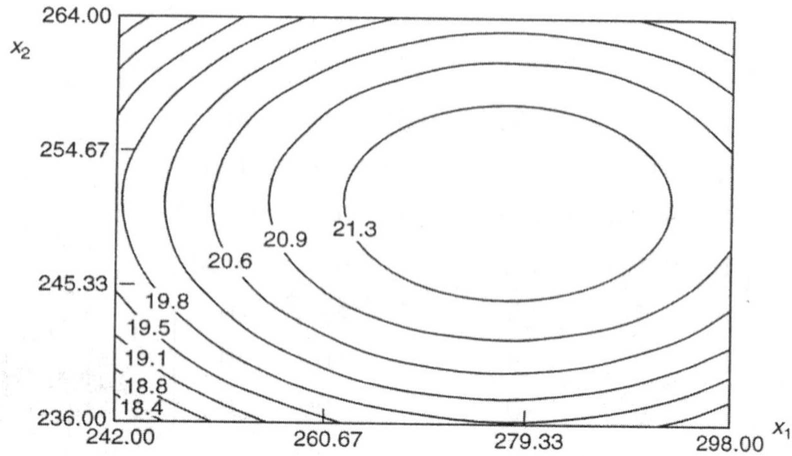

Figure 10.4 Contour plot for impact power.

The canonical form of (10.11) for this experiment is

$$\begin{aligned}\hat{y} &= \hat{y}_0 + \lambda_1 w_1^2 + \lambda_2 w_2^2 \\ &= 21.626 - 1.140 w_1^2 - 1.493 w_2^2 .\end{aligned}$$

Since both the eigenvalues are negative, we immediately observe that the stationary point is the maximum point. Also, from the coding scheme in (10.12), it follows that the maximum point in the real variables is (*RPM*, *TEMP*) = (278, 251).

10.3 RESPONSE SURFACE DESIGNS FOR FITTING SECOND-ORDER MODELS

Experimental designs for fitting a second-order response surface must involve at least three levels of each variable so that the coefficients in the model can be estimated. Probably the most often used response surface designs for fitting second-order models are 3^k factorial of 3^{k-p} fractional factorial designs, central composite designs and equiradial designs. These designs are briefly explained below.

10.3.1 3^k FACTORIAL AND 3^{k-p} FRACTIONAL FACTORIAL DESIGNS

The 3^k factorial is a factorial experiment with each of k factors at three levels. For small values of k, say 2 or 3, this design is well suited to fitting second-order models. However, when a large quantity of variables is under study, the number of observations required is excessive. In this case, the 3^{k-p} fractional factorial ($1/p$ fraction of 3^k factorial) is often used. In fact, three-level orthogonal arrays such as L_9 and L_{27} provide 3^{k-p} fractional factorial designs.

10.3.2 CENTRAL COMPOSITE DESIGNS

The central composite design consists of a 2^k factorial or fractional factorial (coded with the usual ± 1 notation) augmented by $2k$ axial points and n_c center points. The value of α is selected by the experimenter. Various criteria can be used in making the choice. The design matrices for $k=2$ and $k=3$ with two center points are as follows:

$$D = \begin{pmatrix} x_1 & x_2 \\ -1 & -1 \\ -1 & 1 \\ 1 & -1 \\ 1 & 1 \\ 0 & 0 \\ 0 & 0 \\ -\alpha & 0 \\ \alpha & 0 \\ 0 & -\alpha \\ 0 & \alpha \end{pmatrix} \begin{matrix} \\ \left.\begin{matrix} \\ \\ \\ \\ \end{matrix}\right\} 2^2 \text{ factorial points} \\ \left.\begin{matrix} \\ \\ \end{matrix}\right\} 2 \text{ center points} \\ \left.\begin{matrix} \\ \\ \\ \\ \end{matrix}\right\} 2\times 2 \text{ axial points} \end{matrix}$$

$$D = \begin{pmatrix} x_1 & x_2 & x_3 \\ -1 & -1 & -1 \\ -1 & -1 & 1 \\ -1 & 1 & -1 \\ -1 & 1 & 1 \\ 1 & -1 & -1 \\ 1 & -1 & 1 \\ 1 & 1 & -1 \\ 0 & 0 & 0 \\ 0 & 0 & 0 \\ -\alpha & 0 & 0 \\ \alpha & 0 & 0 \\ 0 & -\alpha & 0 \\ 0 & \alpha & 0 \\ 0 & 0 & -\alpha \\ 0 & 0 & \alpha \end{pmatrix} \begin{matrix} \\ \left.\begin{matrix} \\ \\ \\ \\ \\ \\ \\ \end{matrix}\right\} \\ \left.\begin{matrix} \\ \\ \end{matrix}\right\} 2 \text{ center points} \\ \left.\begin{matrix} \\ \\ \\ \\ \\ \\ \end{matrix}\right\} 2\times 3 \text{ axial points} \end{matrix}$$

The central composite design is probably the most widely used experimental design for fitting a second-order response surface. Central composite designs for $k = 2$ and $k = 3$ are shown in Figure 10.5.

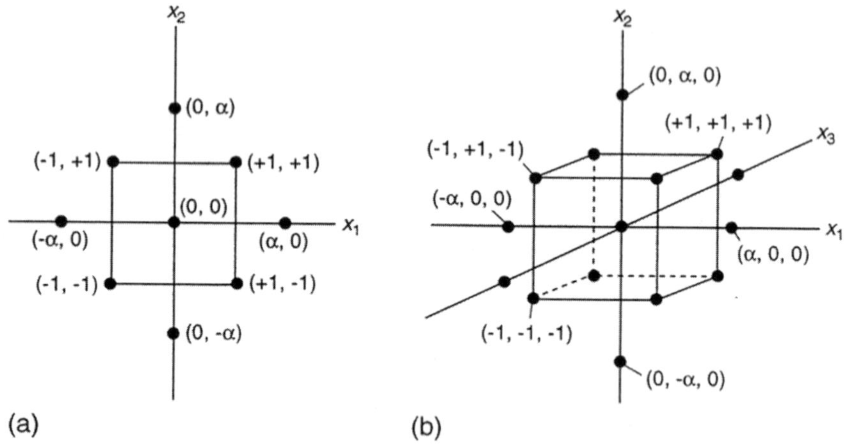

Figure 10.5 Central composite designs for (a) $k = 2$; (b) $k = 3$.

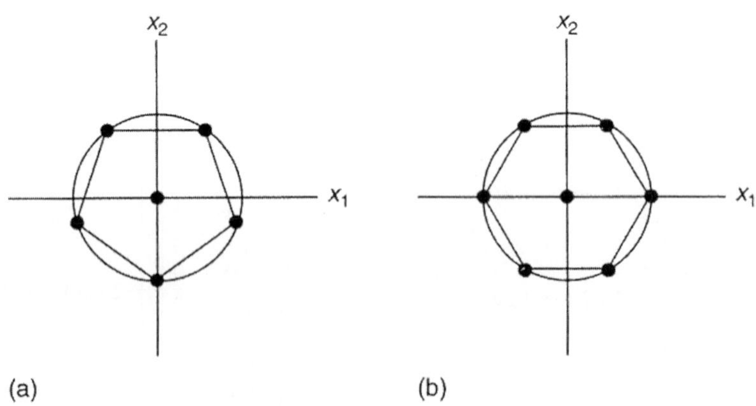

Figure 10.6 Equiradial designs for two variables: (a) pentagon; (b) hexagon.

10.3.3 EQUIRADIAL DESIGNS

The equiradial design consists of points that are equally spaced on a circle ($k = 2$) or a sphere ($k = 3$) and forms regular polygons of polyhedrons. The name "equiradial" comes from the fact that the design points are equidistant from the origin. Two equiradial designs for $k = 2$, hexagonal and pentagonal designs, are shown in Figure 10.6. The design used in Table 10.2 is an octagonal design. For $k = 3$, icosahedral and dodecahedral designs are equiradial. For these designs, refer to Myers (1971).

10.4 DESIRABLE PROPERTIES OF RESPONSE SURFACE DESIGNS

Fitting and analyzing response surfaces are greatly facilitated by the proper choice of an experimental design. In this section, we discuss some aspects of selecting appropriate designs for fitting response surfaces.

Some desirable properties of a response surface design, any or all of which might be important, depending on the experimental circumstances, are as follows. The design should:

1. Provide a reasonable distribution of data points (and hence information) throughout the region of interest;
2. Ensure that the fitted value at $\mathbf{x}, \hat{y}(\mathbf{x})$ is as close as possible to the true value at $\mathbf{x}, \eta(\mathbf{x})$;
3. Ensure also that the estimated slope at $\mathbf{x}, \partial\hat{y}(\mathbf{x})/\partial\mathbf{x}$ is as close as possible to the true slope at $\mathbf{x}, \partial\eta(\mathbf{x})/\partial\mathbf{x}$;
4. Be insensitive to wild observations, model misspecifications and non-normal errors;
5. Allow experiments to be performed in blocks;
6. Allow designs of higher order to be built up sequentially;
7. Provide an internal estimate of error;
8. Provide a minimum number of experimental runs;
9. Ensure simplicity of calculations of the model parameters;
10. Give good detectability of lack of fit.

These features are sometimes conflicting, so judgment is required to select the design that achieves the best compromise. For more information on the choice of a response surface design, refer to Box and Draper (1975; 1987).

The proper choice of design can have a profound effect on the success of any response surface exploration. A variety of criteria are used in RSM for design selection, among them D-optimality, A-optimality, E-optimality, G-optimality, rotatability, slope-rotatability and robustness.

10.4.1 ALPHABETIC OPTIMALITY

Design criteria such as D-, A-, E- and G-optimality are based on the theory of optimal design of experiments through the decision-theoretic approach. These criteria are often grouped together under the heading of "alphabetic optimality" since each is distinguished by a letter of the alphabet.

1. D-optimality: minimization of the determinant of $(\mathbf{X}'\mathbf{X})^{-1}$ or, equivalently, maximization of the determinant of $\mathbf{X}'\mathbf{X}$.
2. A-optimality: minimization of $\mathrm{tr}[(\mathbf{X}'\mathbf{X})^{-1}]$.
3. E-optimality: minimization of the largest eigenvalue of $(\mathbf{X}'\mathbf{X})^{-1}$.
4. G-optimality: minimization of the maximum prediction variance over R, the region of interest, or, equivalently, minimization of the maximum of $\mathrm{var}[\hat{y}(\mathbf{x})]$.

The books by Silvey (1980), Atkinson and Donev (1992), and Pukelsheim (1993) provide excellent surveys of the A-, D-, E- and G-optimality criteria.

10.4.2 ROTATABILITY

An experimental design is said to be **rotatable** if the variance of the predicted response $\hat{y}(\mathbf{x})$ in (10.6) is a function only of the distance of the point from the design center and is not a function of direction. Furthermore, a design with this property will leave the variance of $\hat{y}(\mathbf{x})$ unchanged when the design is rotated about the center $(0,0,\cdots,0)$. This criterion was introduced by Box and Hunter (1957). It is an important design property in the selection of a response surface design. Since the purpose of RSM is response optimization, and the location of the optimum is unknown before running the experiment, it is desirable to use a design that provides equal precision of estimation in all direction.

Suppose the information function $I(\mathbf{x})$ is defined to be the reciprocal of $\text{var}[\hat{y}(\mathbf{x})]$ in (10.6); that is,

$$I(\mathbf{x}) = 1/\text{var}[\hat{y}(\mathbf{x})].$$

Figure 10.7(a) shows the information surface and contours for a 3^2 design used to fit a second-order model. Notice that the 3^2 design is not rotatable. Figure 10.7(b) shows the information surface and contours for a central composite design. Notice that this design is rotatable.

10.4.3 SLOPE-ROTATABILITY

In many applications of RSM, adequate estimation of the partial derivatives of the mean response may be as important as estimation of the mean response itself. For example, it may be of interest to estimate the rates of reaction in a chemical experiment, or the rates of change in the yield of a crop. Furthermore, estimation of the optimum of a response surface depends on having good estimates of the partial derivatives of the mean response with respect to its input variables.

Hader and Park (1978) extended the notion of rotatability to cover the slope for the case of second-order models. They defined "slope-rotatability" as a design property under which the variance of $\partial\hat{y}(\mathbf{x})/\partial x_i, i = 1,2,\cdots,k$ is constant at all points that are equidistant from the design center. They also presented slope-rotatable central composite designs. Later, Park (1987) referred to this property as "slope-rotatability over axial directions". Park (1987) extended the concept of slope-rotatability over axial directions to "slope-rotatability over all directions".

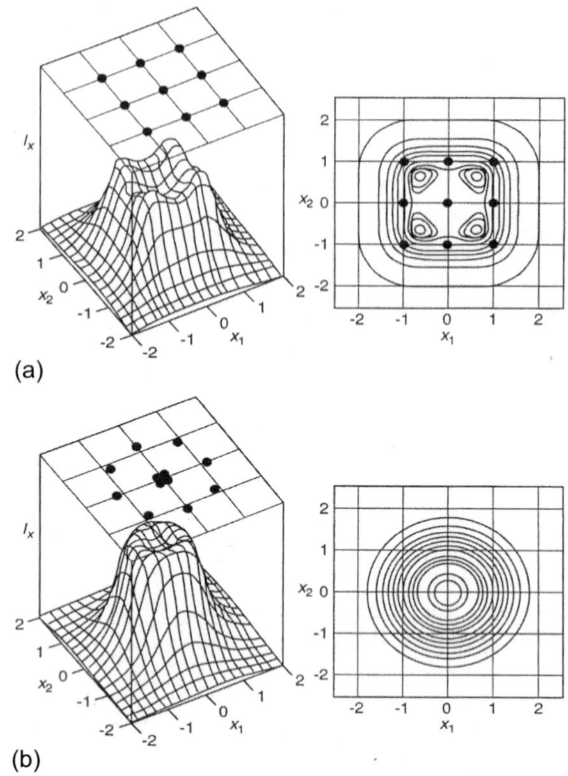

Figure 10.7 (a) Information surface and contours for 3^2 factorial used as a second-order design. (b) Information surface and contours for a second-order rotatable design consisting of eight points on a circle plus four center points.

Let the estimated slope vector be denoted by

$$\hat{g}(\mathbf{x}) = \begin{pmatrix} \dfrac{\partial \hat{y}}{\partial x_1} \\ \dfrac{\partial \hat{y}}{\partial x_2} \\ \vdots \\ \dfrac{\partial \hat{y}}{\partial x_k} \end{pmatrix} = \mathbf{D}\hat{\boldsymbol{\beta}}, \tag{10.14}$$

where $\mathbf{D}$ is a matrix arising from differentiating $\hat{y}(\mathbf{x}) = f'(\mathbf{x})\hat{\boldsymbol{\beta}}$ with respect to $x_1, x_2, \cdots, x_k$. For example, if $k=2$, since $f'(\mathbf{x}) = (1, x_1, x_2, x_1^2, x_2^2, x_1x_2)$,

$$\mathbf{D} = \begin{bmatrix} 0 & 1 & 0 & 2x_1 & 0 & x_2 \\ 0 & 0 & 1 & 0 & 2x_2 & x_1 \end{bmatrix}.$$

The estimated derivative at any point $\mathbf{x}$ in the direction specified by the $k \times 1$ vector of direction cosines,

$$\mathbf{v}' = (v_1, v_2, \cdots, v_k),$$

is $\mathbf{v}'\hat{g}(\mathbf{x})$, where $\sum_{i=1}^{k} v_i^2 = 1$. The variance of this slope is

$$\begin{aligned} V_v(\mathbf{x}) &= \text{var}[\mathbf{v}'\hat{g}(\mathbf{x})] \\ &= \mathbf{v}'\mathbf{D}\,\text{var}(\hat{\boldsymbol{\beta}})\mathbf{D}'\mathbf{v} \\ &= \sigma^2 \mathbf{v}'\mathbf{D}(\mathbf{X}'\mathbf{X})^{-1}\mathbf{D}'\mathbf{v}. \end{aligned}$$

If we are interested in all possible directions of $\mathbf{v}$, we want to consider the average of $V_v(\mathbf{x})$ over all possible directions, which will be referred to as the **average slope variance**. Park (1987) showed that the average slope variance is

$$\overline{V}(\mathbf{x}) = \frac{\sigma^2}{k} tr[\mathbf{D}(\mathbf{X}'\mathbf{X})^{-1}\mathbf{D}']. \tag{10.15}$$

Note that $\overline{V}(\mathbf{x})$ is a function of $\mathbf{x}$, the point at which the derivative is being estimated, and also a function of the design. By choice of design, it is possible to make this variance $\overline{V}(\mathbf{x})$ constant for all points equidistant from the design origin. This property is henceforth called slope-rotatability over all directions. Park (1987) also presented necessary and sufficient conditions for a design to be slope-rotatable over all directions. He concluded that a rotatable design in the Box-Hunter sense is also slope-rotatable over all directions. A rotatable design in the Box-Hunter sense will be called a $\hat{y}$-rotatable design.

As previously mentioned, all $\hat{y}$-rotatable designs are also slope-rotatable over all directions. The converse is not true, however. The following response surface designs are slope-rotatable designs over all directions, and they are also $\hat{y}$-rotatable.

1. Rotatable central composite designs, where $\alpha = n_f^{1/4}$ and n_f is the number of points in the factorial part.
2. Equiradial designs in two variables, where there are $n_1 (n_1 \geq 5)$ equally spaced points on a circle of radius r, augmented by $n_0 (n_0 \geq 1)$ center points.
3. Combinations of equiradial designs in two variables such as $s(s \geq 2)$ sets of equiradial points on concentric circles, the wth containing n_w points $(n_w \geq 5)$ and having radius r_w.
4. For three variables, 12 vertices of the icosahedron $(0, \pm a, \pm b)$, $(\pm b, 0, \pm a)$, $(\pm a, 0, \pm b)$, where a/b =1.618, plus some center points $(0,0,0)$.
5. For three variables, 20 vertices of the dodecahedron $(0, \pm c^{-1}, \pm c)$, $(\pm c, 0, \pm c^{-1})$, $(\pm c^{-1}, \pm c, 0)$, $(\pm 1, \pm 1, \pm 1)$, where c=1.618, plus some center points $(0,0,0)$.
6. Simplex-sum second-order rotatable designs proposed by Box and Behnken (1960).

The following response surface designs are slope-rotatable over all directions, however, but they are not generally $\hat{y}$-rotatable.

1. Central composite designs, where α is an arbitrary positive constant.
2. 3^k factorial designs and 3^{k-p} fractional factorial designs.
3. For two variables, a design with three equally spaced points on each of two circles plus an arbitrary number of center points; a design with four equally spaced points on each of two circles plus an arbitrary number of center points; and designs with three or four points on one circle and five or more points on a second circle.

4. For three variables, 12 vertices of the icosahedron in which a/b is an arbitrary positive constant, plus some center points; and 20 vertices of the dodecahedron in which c is an arbitrary constant, plus some center points.

From the preceding examples, it is clear that the designs most often used in response surface experiments such as the standard central composite designs, three-level factorials, certain Box-Behnken designs, equiradial designs, icosahedron and dodecahedron are all slope-rotatable designs over all directions. All $\hat{y}$-rotatable designs are simply a subset of slope-rotatable designs over all directions.

10.5 ROBUST RESPONSE SURFACE DESIGNS

Response surface analysis is usually carried out under specified conditions. For example, the error vector ε in a given model is assumed to have the normal distribution $N(\mathbf{0}, \mathbf{I}\sigma^2)$. A design is said to be **robust** if it does not cause the results of a particular response surface analysis to be severely affected by conditions that are not ideal. Some of these conditions include outliers, missing data points, non-normal errors, unequal response variances and model misspecifications.

The paper by Box and Draper (1975) was the first to address the choice of designs on the basis of minimizing the impact of outliers in the resulting data. Park (1976) considered design aspects of response surface experiments in which the emphasis is on robust estimation of the response variable, $\eta(\mathbf{x})$, to outliers. McKee and Kshirsagar (1982) studied the effects of missing observations on parameter estimates and their variances. Akhtar and Prescott (1986) examined the effects of missing observations on the D-optimality criterion for response surface designs. More recently, Yoo and Park (1991) discussed response surface designs that are robust with regard to outliers and model inadequacy. Defeo and Myers (1992) considered a new criterion for design robustness against misspecification of the fitted model. The work of Park (1976) on robustness of response surface designs is briefly reviewed below.

10.5.1 EFFECT OF OUTLIERS

Suppose that the ith observation has added to it an "aberration" d_i, making it an outlier. Generalizing, consider a discrepancy vector, $\mathbf{d}' = (d_1, d_2, \cdots, d_n)$, of non-random aberrations added to the $\mathbf{y}$ vector. It is easily shown, by assuming $\mathbf{T} = (\mathbf{X}'\mathbf{X})^{-1}\mathbf{X}'$, that

$$\begin{aligned} E(\hat{\boldsymbol{\beta}}) &= (\mathbf{X}'\mathbf{X})^{-1}\mathbf{X}'E(\mathbf{y}) \\ &= (\mathbf{X}'\mathbf{X})^{-1}\mathbf{X}'(\mathbf{X}\boldsymbol{\beta} + \mathbf{d}) \\ &= \boldsymbol{\beta} + \mathbf{Td}, \end{aligned}$$

and that, for any arbitrary point $\mathbf{x}$, the estimated response, $\hat{y}(\mathbf{x})$, is such that

$$\begin{aligned} E(\hat{y}(\mathbf{x})) &= E(\mathbf{x}'\hat{\boldsymbol{\beta}}) \\ &= \mathbf{x}'(\boldsymbol{\beta} + \mathbf{Td}) \\ &= \eta(\mathbf{x}) + \mathbf{x}'\mathbf{Td}. \end{aligned}$$

Therefore, the mean squared error, *MSE*, is

$$\begin{aligned} MSE[\hat{y}(\mathbf{x})] &= E[\hat{y}(\mathbf{x}) - \eta(\mathbf{x})]^2 \\ &= E[\hat{y}(\mathbf{x}) - E(\hat{y}(\mathbf{x}))]^2 + \{E(\hat{y}(\mathbf{x}) - \eta(\mathbf{x}))\}^2 \\ &= V(\mathbf{x}) + B(\mathbf{x}) \\ &= \mathbf{x}'\,\mathrm{var}(\hat{\boldsymbol{\beta}})\mathbf{x} + \mathbf{d}'\mathbf{T}'\mathbf{x}\mathbf{x}'\mathbf{Td}, \end{aligned}$$

where $V(\mathbf{x}) = \mathbf{x}'\,\mathrm{var}(\hat{\boldsymbol{\beta}})\mathbf{x}$ is the variance and $B(\mathbf{x}) = \mathbf{d}'\mathbf{T}'\mathbf{x}\mathbf{x}'\mathbf{Td}$ is the squared bias of $\hat{y}(\mathbf{x})$ due to the aberration vector $\mathbf{d}$.

Since we are interested in using $\hat{y}(\mathbf{x})$ over the entire region of interest, R, we will consider the integrated mean squared error, *IMSE*, under a weighting function $W(\mathbf{x})$. This is given by

$$\begin{aligned} IMSE &= \int_R MSE[\hat{y}(\mathbf{x})]dW(\mathbf{x}) \\ &= \int_R V(\mathbf{x})dW(\mathbf{x}) + \int_R B(\mathbf{x})dW(\mathbf{x}) \\ &= \sigma^2 tr[(\mathbf{X}'\mathbf{X})^{-1}M] + \mathbf{d}'\mathbf{T}'M\mathbf{T}\mathbf{d}, \end{aligned} \tag{10.16}$$

where $M = \int_R \mathbf{x}\mathbf{x}'dW(\mathbf{x})$ and tr indicates trace. The weight function $W(\mathbf{x})$ can be treated as a probability distribution function on R. It allows for differential importance of estimates of $\eta(\mathbf{x})$ at different points in the region and can be specialized to a discrete set of points if desired.

In equation (10.16), it is clear that *IMSE* is the sum of the integrated variance and the integrated squared bias $\mathbf{d}'\mathbf{T}'M\mathbf{T}\mathbf{d}$ with the latter term being the effect of the non-random discrepancies. The sensitivity of *IMSE* to a given discrepancy vector $\mathbf{d}$ depends on the experimental design through the matrix $\mathbf{T}'M\mathbf{T}$. Hence, robustness to wild observations can be achieved if $\mathbf{d}'\mathbf{T}'M\mathbf{T}\mathbf{d}$ is made as small as possible for any given vector $\mathbf{d}$.

10.5.2 MEASURE OF ROBUSTNESS

Let $\mathbf{S} = \mathbf{T}'M\mathbf{T}$. Clearly, $\mathbf{S}$ is positive semi-definite since $\mathbf{d}'\mathbf{S}\mathbf{d}$ is the integrated bias. Robustness of *IMSE* to aberrant observations is achieved by minimizing $\mathbf{d}'\mathbf{S}\mathbf{d}$. For single discrepancy of magnitude g in the ith observation,

$$\mathbf{d}'\mathbf{S}\mathbf{d} = s_{ii}g^2,$$

where s_{ii} is the ith diagonal element of $\mathbf{S}$. If we assume that the single discrepancy of magnitude g, $\mathbf{d}' = (0, 0, \cdots, 0, g, 0, \cdots, 0)$, has equal probability of occurring with any of the n observations, then the average value of $\mathbf{d}'\mathbf{S}\mathbf{d}$ is

$$\begin{aligned} avg(\mathbf{d}'\mathbf{S}\mathbf{d}) &= avg[tr(\mathbf{d}'\mathbf{S}\mathbf{d})] \\ &= avg[tr(\mathbf{S}\mathbf{d}'\mathbf{d})] \\ &= tr[\mathbf{S}avg(\mathbf{d}'\mathbf{d})] \\ &= tr[\mathbf{S}(g^2/n)\mathbf{I}] \\ &= (g^2/n)tr(\mathbf{S}). \end{aligned}$$

Thus, $tr(S)$ should be minimized to achieve robustness of *IMSE* to an aberrant observation. It is interesting to note that

$$tr(\mathbf{S}) = tr(\mathbf{T}'M\mathbf{T}) = tr[(\mathbf{X}'\mathbf{X})^{-1}M], \tag{10.17}$$

which is the integrated variance (divided by σ^2). Thus, it appears that to ensure insensitivity to an aberrant observation, the integrated variance should be made small.

If there are two or more aberrant observations, a similar conclusion can be deduced. Consider a general discrepancy vector, $\mathbf{d}' = (d_1, d_2, \cdots, d_n)$.

Then

$$\mathbf{d}'\mathbf{S}\mathbf{d} = \sum_{i=1}^{n}\sum_{j=1}^{n} s_{ij}d_i d_j .$$

A convenient measure of robustness might be

$$Q = \sum_{i=1}^{n}\sum_{j=1}^{n} s_{ij}^2 .$$

Since $\mathbf{S}$ is positive semi-definite, $s_{ij}^2 \le s_{ii}s_{jj}$, and therefore,

$$\begin{aligned} Q \le \sum_{i=1}^{n} s_{ii}^2 + 2\sum_{i=1}^{n}\sum_{j=1}^{n} s_{ii}s_{ii} &= (\sum_{i=1}^{n} s_{ii})^2 \\ &= [tr(\mathbf{S})]^2 . \end{aligned}$$

Thus, Q is made small by making $tr(S)$ small, that is, by making the integrated variance of $\hat{y}(\mathbf{x})$ small.

The approach taken in Park (1976) differs from that of Box and Draper (1975) primarily in that Park considered the integrated bias of $\hat{y}(\mathbf{x})$ over some region of interest, R, as the basic quantity to be made insensitive to non-random discrepancies. Box and Draper concentrated on the bias of $\hat{y}(\mathbf{x})$ at the design points only. They proposed that the sum of squares of diagonal elements of the matrix $\mathbf{X}(\mathbf{X}'\mathbf{X})^{-1}\mathbf{X}'$ should be made small. The basis of their criterion is that $\text{var}(\hat{y}(\mathbf{x}))$ at the design points can be made as uniform as possible if the sum of squares of diagonal elements of $\mathbf{X}(\mathbf{X}'\mathbf{X})^{-1}\mathbf{X}'$ is made small.

In general, the fitted equation $\hat{y}(\mathbf{x})$ in response surface experiments is intended to be used not only at the design points, but within some region of the $\mathbf{x}$ space of interest to the experimenter. If a criterion is to be developed based on insensitivity of $\hat{y}(\mathbf{x})$ to aberrant observations, it seems more satisfactory to evaluate its performance over the whole region than just over the design point. From this viewpoint, the proposed criterion proposed by Park (1976) possesses an advantage over that of Box and Draper (1975).

10.5.3 SOME ROBUST TWO-VARIABLE DESIGNS

Suppose that the second-order polynomial model is adequate over the region of interest, R, for a response surface experiment involving two variables. To illustrate the application of the $tr(S)$ criterion, we consider a useful class of two-variable designs with points equally spaced on one or two circles, and center points. This class includes the 3^2 factorial, the central composite, the equiradial designs such as the pentagon and hexagon, and many other common designs. We take the weight function, $W(\mathbf{x})$, to be constant and minimize $tr(S)$ either over a square or a unit circle region of interest:

$$R_1 = \{(x_1, x_2) \mid -1 \le x_1 \le 1, -1 \le x_2 \le 1\},$$

$$R_2 = \{(x_1, x_2) \mid x_1^2 + x_2^2 \le 1\},$$

$$\vdots$$

Recall that $M = \int_R \mathbf{x}\mathbf{x}' dW(\mathbf{x})$, and if we take $W(\mathbf{x})$ to be constant over the region, then

$$M = \begin{cases} (\frac{1}{4}) \int_{R_1} \mathbf{x}\mathbf{x}' dx_1 dx_2, & \text{if } R = R_1 \\ (\frac{1}{\pi}) \int_{R_2} \mathbf{x}\mathbf{x}' dx_1 dx_2, & \text{if } R = R_2 . \end{cases}$$

We assume that the experimental points are to be restricted to the region of interest, R.

For a fixed n, the design parameters which can be varied in the minimization are the number of points on each circle, the orientation angle of the points on each circle with respect to the positive direction of the x_1 axis, the radii of the circles and the number of center points, n_0. In Tables 10.2 and 10.3, the notation $(n_1, a_1)+(n_2, a_2)$ is used to indicate the numbers of points and orientation angles for the two circles, with n_1 and a_1 always referring to the larger circle. For example, from Table 10.2, with n=8, the optimum design has $n_1 = 4$ points on an outer circle of radius $r_1 = \sqrt{2}$. These points are in the corners of the square region $(a_1 = \pi/4)$. The design has $n_2 = 4$ points on a second circle of radius r_2=0.794. These points are on the x_1 axis and x_2 axis with distance 0.794 from the origin. The design does not have center points $(n_0 = 0)$. The design is illustrated in Figure 10.8.

Table 10.2 Designs with tr (S) minimized over square region R_1.

n	Design	n_0	R_1	r_2
6	(5, π/20)	1	1.012	
7	(4, π/4) + (2,0)	1	$\sqrt{2}$	1
8	(4, π/4) + (4,0)	0	$\sqrt{2}$	0.794
9	(4, π/4) + (4,0)	1	$\sqrt{2}$	0.906
10	(4, π/4) + (4,0)	2	$\sqrt{2}$	1
11	(4, π/4) + (4,0)	3	$\sqrt{2}$	1

Table 10.3 Designs with $tr(\text{S})$ minimized over circular region R_2.

N	Design	n_0	r_1
6	(5,0)	1	1
7	(5,0)	2	1
8	(6,0)	2	1
9	(6,0)	3	1
10	(7,0)	3	1
11	(8,0)	3	1

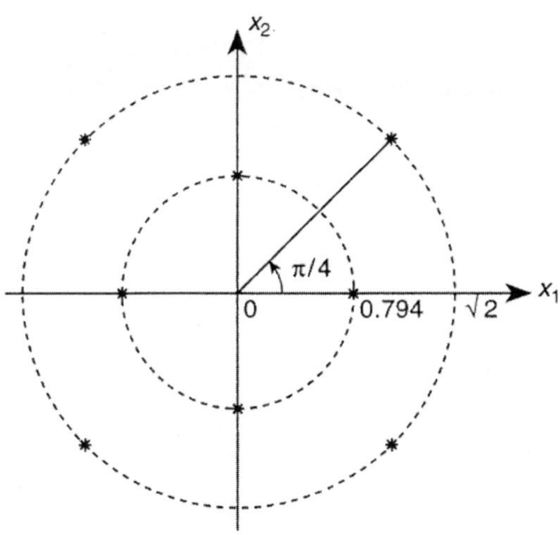

Figure 10.8 The design $(4, \pi/4) + (4,0)$ with $r_1 = \sqrt{2}$ and $r_2 = 0.794$.

Inspection of Tables 10.2 and 10.3 indicates that the experimental points tend to be located on the boundary of the experimental region, though center points seem to be necessary for most cases. By way of comparison with Box and Draper (1975), their two-variable central composite design had eight points equally spaced on a circle and had $n_0 = 2$. The $n = 10$ design of Table 10.3 has seven points on a circle and $n_0 = 3$. The n=11 design is a rotatable central composite with $n_0 = 3$. For the square region of interest, however, the designs of Table 10.2 for $n = 10$ and 11 are the traditional 3^2 factorials with added center points.

10.6 OPTIMIZATION OF MULTIRESPONSE EXPERIMENTS

In many experimental situations, it is quite common for several responses, rather than a single response, to be measured for each setting of a group of input variables. For example, a food technologist may be interested in determining optimum combinations of the various ingredients of a food product on the basis of acceptability, nutrition, economics and other considerations. Obviously, the analysis of a multiresponse experiment should be carried out using multivariate techniques. Univariate techniques applied to the individual responses are not appropriate because they do not take into consideration interrelationships that may exit among the responses. For example, optimal conditions for one response may be far from optimal or even physically impractical for the remaining responses.

The study of multiresponse experiments in RSM is considered to be fairly recent in comparison to the single response. Multiresponse surface techniques are therefore few in number, and unfortunately are neither frequently used nor even well-known to practitioners in general. This is partly due to the relative complexity of these techniques and to the lack of available software needed for their implementation.

Multiresponse optimization is the most visible and important aspect of multiresponse analysis. The object is to determine conditions on the input variables $x_1, x_2 \cdots, x_k$ that lead to optimal, or nearly optimal, values of the response variables, $y_1, y_2, \cdots y_m$. There are several approaches to multiresponse optimization; these are briefly sketched below. Recall that we dealt with a problem of simultaneous optimization for multi-characteristics in Section 7.5. There we used a parameter design for several responses, and a response surface analysis was not used. Here we will consider multiresponse optimization problem when we use response surface analysis.

10.6.1 GRAPHICAL SUPERIMPOSITION OF RESPONSE CONTOURS

This is a graphical technique which is easy to understand, and easy to use when there are only two or three input variables. First of all, for each response, a response contour is obtained by fitting, in general, the second-order response model. Then by superimposition of response contours, we arrive at optimal conditions. This approach will be illustrated in Example 10.2 for the data in Table 10.1. Even though this procedure is very useful in practice, it is difficult to apply when the number of input variables exceeds three.

10.6.2 USE OF DESIRABILITY

Derringer and Suich (1980) introduced the concept of **desirability**, whereby each response function is transformed into a desirability function. The choice of these transformations is subjective as it is governed by the experimenter's assessment of the importance of each response. A measure of the overall desirability of the response is obtained by using the geometric mean of the individual desirability functions. The geometric mean is then maximized over the region of interest.

Consider the case when the characteristic of interest is larger-the-better. Let $d_i(\mathbf{x})$ be the ith individual desirability function, which is usually defined by

$$d_i(\mathbf{x}) = \begin{cases} 0 & \hat{y}_i(\mathbf{x}) \le y_{i*} \\ \left[\dfrac{\hat{y}_i(\mathbf{x}) - y_{i*}}{y_i^* - y_{i*}} \right]^r & y_{i*} < \hat{y}_i(\mathbf{x}) < y_i^* \\ 1 & \hat{y}_i(\mathbf{x}) \ge y_i^*, \end{cases}$$

where y_{i*} is the minimum allowable value of $\hat{y}_i(\mathbf{x})$, y_i^* is the satisfactory value of $\hat{y}_i(\mathbf{x})$ for $i = 1, 2, \cdots, m$, and r is an arbitrary positive constant.

Then the overall desirability function $D(\mathbf{x})$ is

$$D(\mathbf{x}) = [d_1(\mathbf{x}) \times d_2(\mathbf{x}) \times \cdots \times d_m(\mathbf{x})]^{1/m}. \tag{10.18}$$

The problem of simultaneous optimization is to find the point of $\mathbf{x}$ which maximizes $D(\mathbf{x})$. Usually $r=1$ is used in practice. However, if we use different values of r, the individual desirability functions change, as shown in Figure 10.9.

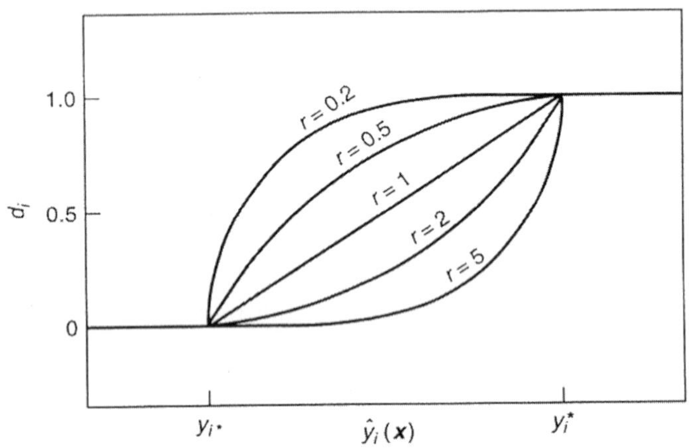

Figure 10.9 Shape of $d_i(\mathbf{x})$ as r changes.

If the characteristic is nominal-is-best, the individual desirability function is defined by

$$d_i(\mathbf{x}) = \begin{cases} \left[\dfrac{\hat{y}_i(\mathbf{x}) - y_{i*}}{t_i - y_{i*}} \right]^s & y_{i*} \le \hat{y}_i(\mathbf{x}) \le t_i \\ \left[\dfrac{\hat{y}_i(\mathbf{x}) - y_i^*}{t_i - y_i^*} \right]^c & t_i < \hat{y}_i(\mathbf{x}) \le y_i^* \\ 0 & \hat{y}_i(\mathbf{x}) < y_{i*} \text{ or } \hat{y}_i(\mathbf{x}) > y_i^*, \end{cases}$$

where t_i is the target value for the ith response, y_i^* is the maximum allowable value of $\hat{y}_i(\mathbf{x})$, and s and c are arbitrary positive constants. The overall desirability function is the same as in (10.18). This method will be explained in Example 10.2 for the example in Table 10.1.

10.6.3 PRIMARY AND SECONDARY FUNCTION APPROACH

Myers and Carter (1973) introduced an algorithm for determining conditions on $x_1, x_2, \cdots, x_k$ that maximize a primary response function subject to having an equality constraint on a secondary response function. This approach was extended by Biles (1975) to include several secondary response functions.

10.6.4 DISTANCE FUNCTION APPROACH

Another approach to multiresponse optimization was introduced by Khuri and Conlon (1981). They presented several distance functions (metrics) that measure the overall closeness of the response functions to achieving their respective optimal values at the same set of operating conditions (ideal optimum). Multi-response optimization is thus reduced to minimizing an appropriate distance function with respect to the input variables. This approach takes into consideration the variances and correlations among the responses.

Example 10.2

For the data in Table 10.1, we will show how simultaneously to optimize the responses by using graphical superimposition of response contours and also by using desirability functions. In Table 10.1 there were two responses; the first response, impact power, was analyzed in Section 10.2. The fitted second-order response equation was

$$\hat{y}_1 = 21.531 + 0.612x_1 + 0.262x_2 - 1.142x_1^2 - 1.492x_2^2 + 0.049x_1x_2,$$

and the response contour was shown in Figure 10.5. For the second response, liquidity, we can obtain the fitted second-order response equation

$$\hat{y}_2 = 52.030 + 0.021x_1 + 1.329x_2 - 0.500x_1^2 - 0.300x_2^2 + 0.098x_1x_2,$$

and the response contour shown in Figure 10.10.

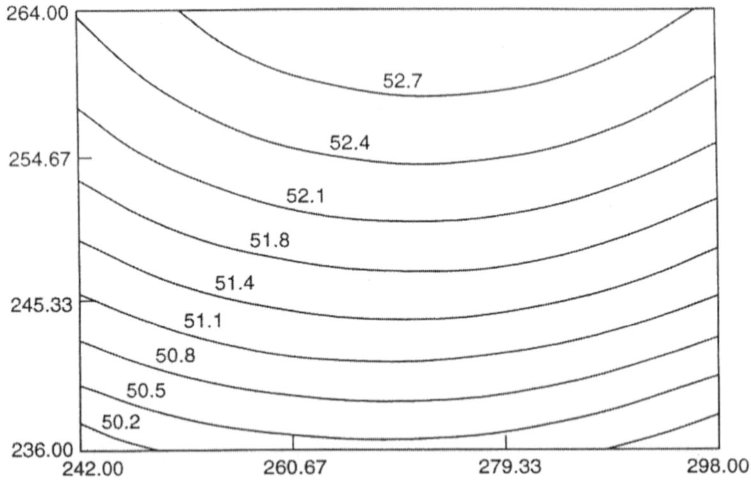

Figure 10.10 Contour plot for liquidity ($\hat{y}_2$).

The experimenter wishes to find conditions on (x_1, x_2) which give

$$\hat{y}_1 \geq 21.3, \quad \hat{y}_2 \geq 52.4. \tag{10.19}$$

From the contour plots in Figures 10.4 and 10.10, we can obtain the shaded area of (x_1, x_2) which satisfies (10.19) by graphical superimposition of the response contours as shown in Figure 10.11.

From Figure 10.11, we see that the optimal area of points is roughly

$$x_1 = -0.1 \sim 0.6 \ (RPM,\ 267.2 \sim 286.8),$$
$$x_2 = 0.4 \sim 0.6 \ (TEMP,\ 255.6 \sim 258.4),$$

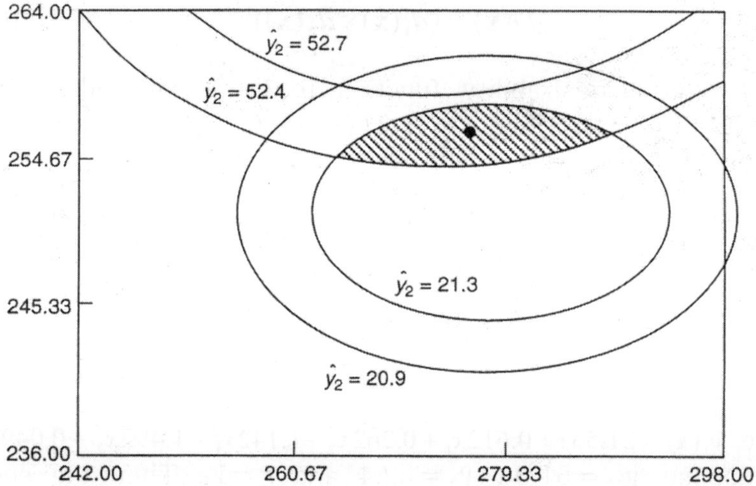

Figure 10.11 Graphical superimposition of two response contours.

The experimenter decides to choose $RPM = 277$ (the mid-point of 267.2 and 286.8) and $TEMP = 257°\text{C}$ (the mid-point of 256.6 and 258.4) as the optimal condition, and performs a confirmation test with five repetitions.

The experimental data obtained in the confirmation test are:

Impact power (y_1): 21.7, 21.5, 21.3, 21.6, 21.4,

Liquidity (y_2): 52.4, 51.7, 51.8, 52.6, 52.8,

which are quite satisfactory. Eventually, the condition $RPM = 277$ and $TEMP = 257°\text{C}$ is determined to be the optimal condition (the dot in the shaded area in Figure 10.11) for the ABS resin product manufacturing process.

If we use the desirability approach, we tackle the problem in Table 10.1 as follows. For each $\hat{y}_1(\mathbf{x})$ and $\hat{y}_2(\mathbf{x})$, we define the individual desirability functions, $d_1(\mathbf{x})$ and $d_2(\mathbf{x})$, where $0 \le d_i(\mathbf{x}) \le 1,\ i = 1, 2$. For each $d_i(\mathbf{x})$, the larger $d_i(\mathbf{x})$ becomes, the better it is. Then the geometric mean of equation (10.18),

$$D(\mathbf{x}) = (d_1(\mathbf{x}) \times d_2(\mathbf{x}))^{1/2} \tag{10.20}$$

is the overall desirability function to be maximized. Suppose $y_{1*} = 20.0$, $y_1^* = 21.3$ and $r = 1$. Then the first desirability function becomes

$$d_1(\mathbf{x}) = \begin{cases} 0 & \hat{y}_1(\mathbf{x}) \le 20.0 \\ \dfrac{\hat{y}_1(\mathbf{x}) - 20.0}{21.3 - 20.0} & 20.0 < \hat{y}_1(\mathbf{x}) < 21.3 \\ 1 & \hat{y}_1(\mathbf{x}) \ge 21.3, \end{cases}$$

where $\hat{y}_1(\mathbf{x}) = 21.531 + 0.612x_1 + 0.262x_2 - 1.142x_1^2 - 1.492x_2^2 + 0.049x_1x_2$.

Suppose $y_{2*} = 51.0$, $y_2^* = 52.4$ and $r = 1$. Then the second desirability function becomes

$$d_2(\mathbf{x}) = \begin{cases} 0 & \hat{y}_2(\mathbf{x}) \le 51.0 \\ \dfrac{\hat{y}_2(\mathbf{x}) - 51.0}{52.4 - 51.0} & 51.0 < \hat{y}_2(\mathbf{x}) < 52.4 \\ 1 & \hat{y}_2(\mathbf{x}) \ge 52.4, \end{cases}$$

where $\hat{y}_2(\mathbf{x}) = 52.030 + 0.021x_1 + 1.329x_2 - 0.500x_1^2 - 0.300x_2^2 + 0.098x_1x_2$. The point $\mathbf{x}$ which maximizes the overall desirability function $D(\mathbf{x})$ in (10.20) is the optimum point $\mathbf{x}$.

10.7 PARAMETER DESIGN IN RESPONSE SURFACE ANALYSIS

The response surface designs that we considered in Section 10.3 are mainly for control factors (design variables). If we have both control and noise factors, and we wish to find optimum conditions for the control factors from empirical model building, what kind of parameter design should be used? In this section we propose two approaches.

10.7.1 COMBINED ARRAY APPROACH USING THREE-LEVEL ORTHOGONAL ARRAYS

The control and noise factors are assigned to an inner three-level orthogonal array. An outer array is not used. Response surface models for the mean and variance are fitted and the optimum condition for the control factors can be found. This approach was explained in detail in Chapter 7 as an alternative parameter design.

10.7.2 CROSS-PRODUCT ARRAY APPROACH USING RESPONSE SURFACE DESIGNS

Both the inner array (design array) and the outer array (environmental array) are used. However, for the inner array, response surface designs such as 3^k factorial, 3^{k-p} fractional factorial (three-level orthogonal arrays), central composite designs and equiradial designs are used. For the outer array, one-way, two-way or two-level orthogonal arrays are used, depending on the number of noise factors. In this approach, since there are several responses for each experiment in the inner array, we can find the corresponding SN ratio for each experiment. Hence, we can analyze the SN ratios to find the optimum condition for the control factors. An example is shown below to illustrate this approach.

Example 10.3

A 3^2 factorial design was used to study the effect of welding temperature (°C) and flux quantity (cm min^{-1}) on the resulting weld strength (in units of 100 psi), which is a larger-the-better characteristic. The table below describes the conditions that were examined in the experiment and the associated measured weld strengths for three operators. Note that each unique combination of temperature and flux quantity was performed three times, once for each operator. The SN ratio was calculated for each unique combination.

Test number	Design factor		Weld strength			SN (dB)
	x_1 (temp.)	x_2 (flux)	Noise factor			
			Op.1	Op.2	Op.3	
1	300	10	31	28	29	29.32
2	300	15	40	38	36	31.57
3	300	20	28	32	29	29.40
4	400	10	35	38	34	31.02
5	400	15	39	42	44	32.36
6	400	20	34	33	31	30.26
7	500	10	27	25	22	27.75
8	500	15	32	33	30	29.99
9	500	20	30	27	26	28.79

Since the characteristic is larger-the-better, the computation of SN ratios for test number 1 is done as follows:

$$\begin{aligned} SN_1 &= -10\log\left(\frac{1}{n}\sum_j \frac{1}{y_{1j}^2}\right) \\ &= -10\log\left[\frac{1}{3}\left(\frac{1}{31^2}+\frac{1}{28^2}+\frac{1}{29^2}\right)\right] \\ &= -10\log[0.00117] = 29.32\text{dB}. \end{aligned}$$

The other SN ratios can be obtained in similar fashion. By using the following coded variables,

$$x_1 = \frac{temperature - 400}{100},$$

$$x_2 = \frac{flux - 15}{5},$$

we can obtain the fitted second-order response model,

$$\hat{y} = 32.4689 - 0.6267x_1 + 0.0600x_2 - 1.7433x_1^2 - 1.8833x_2^2 + 0.2400x_1x_2.$$

From equation (10.9), the stationary point is

$$\mathbf{x}_0 = \begin{bmatrix} -0.1794 \\ 0.0045 \end{bmatrix}.$$

And the canonical form in (10.11) becomes

$$\hat{y} = 32.5252 - 1.6744 w_1^2 - 1.9523 w_2^2 .$$

Since both eigenvalues, −1.6744 and −1.9523, are negative, the stationary point is the maximum point. The real values of temperature and flux quantity before coding are

$$temperature = 382.1°\text{C}, \qquad flux\ \ quantity = 15.0\text{cm}\ \ \text{min}^{-1}.$$

The contour plot of SN ratios by the fitted second-order response model is shown in Figure 10.12. To make sure that the optimum point is the real optimum condition which produces high weld strength with low variation, a confirmation test is required at the optimum condition.

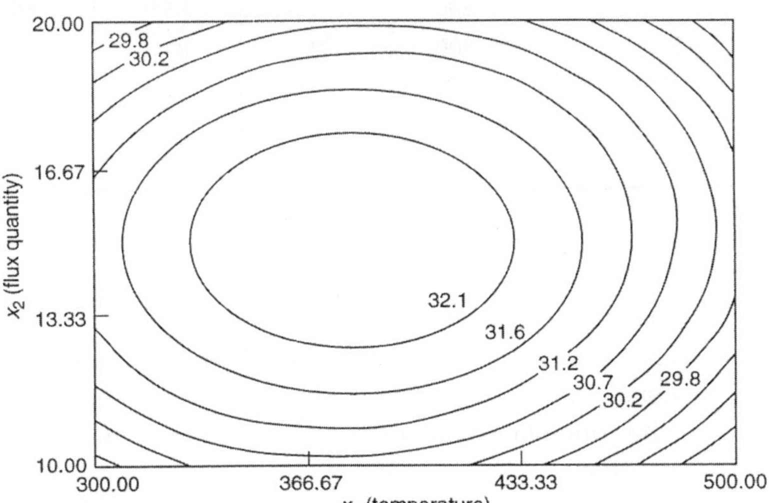

Figure 10.12 Contour plot for SN ratios of weld strength.

Exercises

10.1 A chemical plant produces oxygen by liquefying air and separating it into its component gases by fractional distillation. The purity of the oxygen is a function of the main condenser temperature and the pressure ratio between the upper and lower columns. Current operating conditions are –200°C for temperature and 1.5 for pressure ratio. A noise factor is the ambient condition, which causes some difference in purity between early morning and late evening. The following purity data was obtained from a 3^2 factorial experiment. Purity is a larger-the-better characteristic.

Experiment number	Temperature (x_1)	Ratio (x_2)	Purity (%)	
			Early morning	Late afternoon
1	-220	1.3	82.8	83.2
2	-220	1.5	85.5	84.7
3	-220	1.7	84.7	84.5
4	-200	1.3	82.5	82.2
5	-200	1.5	85.0	85.4
6	-200	1.7	83.5	83.9
7	-180	1.3	81.9	82.3
8	-180	1.5	84.3	83.8
9	-180	1.7	83.0	83.3

(a) Code the variables x_1 and x_2 so that the region of interest is $-1 \le x_i \le 1$, $i = 1,2$.

(b) Using all the data, fit the second-order model in (10.7) and analyze the fitted surface. What is the optimum condition of x_1 and x_2 which maximizes the purity? The optimum condition should be in the region of interest.

(c) Find the SN ratio for each experiment. Using these SN ratios, fit a second-order model and analyze the fitted surface. What is the optimum condition of x_1 and x_2 which maximizes the purity?

(d) The design used in this experiment is a 3^2 factorial design. Is it rotatable? Is it slope-rotatable?

(e) Find the variance of $\hat{y}(x_1, x_2)$, the fitted response by the second-order model. This variance is a function of x_1 and x_2. What is the point (x_1, x_2) which minimizes the variance?

10.2 Compare the following three designs from the standpoint of rotatability, slope-rotatability and robustness. The measure of robustness is given by (10.17) and the region of interest is square. What are the names of these designs?

Design 1		Design 2		Design 3	
x_1	x_2	x_1	x_2	x_1	x_2
-1	-1	-1	-1	1	0
-1	0	-1	1	0.5	0.866
-1	1	1	-1	-0.5	0.866
0	-1	1	1	-1	0
0	0	0	0	-0.5	-0.866
0	1	-2	0	0.5	-0.866
1	-1	2	0	0	0
1	0	0	-2	0	0
1	1	0	2	0	0

Suppose the variance of the estimated pure quadratic coefficient, $\hat{\beta}_{11}$ or $\hat{\beta}_{22}$ is of major interest. Which design gives the smallest variance?

10.3 In a certain forging process, two temperatures are identified as having a critical effect on the percentage of parts (y_1) with no faults (faults include listering, crushing, porosity, flakiness, and so on) and on the boiler efficiency (y_2). The two temperatures are metal temperature (x_1) and base temperature (x_2). The following experiment, using a central composite design, provided the data shown below. The coding scheme used is

$$x_1 = \frac{metal\ \ temp. - 250}{20},$$

$$x_2 = \frac{base\ \ temp. - 170}{10}.$$

Experiment number	x_1	x_2	y_1 (%)	y_2 (%)
1	-1	-1	56	89
2	-1	1	76	80
3	1	-1	70	79
4	1	1	80	73
5	0	0	69	75
6	0	0	70	73
7	1.414	0	79	70
8	-1.414	0	54	86
9	0	1.414	78	69
10	0	-1.414	61	85

(a) Fit second-order response models for each response. By using the graphical superimposition of response contours, find the optimum condition of x_1 and x_2 in the region of interest, $-1 \le x_i \le 1$, $i = 1$ and 2. Note that y_1 and y_2 are both larger-the-better characteristics.

(b) Set up some appropriate individual and overall desirability functions, and find the optimum condition.

11

Six Sigma for Management Innovation

11.1 WHAT IS SIX SIGMA?

Sigma(σ) is a letter in the Greek alphabet that has become the statistical symbol and metric of process variation. The sigma scale of measure is perfectly correlated to such characteristics as defects-per-unit, parts-per million defectives, and the probability of a failure. Six is the number of sigma measured in a process, when the variation around the target is such that only 3.4 outputs out of one million are defects under the assumption that the process average may drift over the long term by as much as 1.5 standard deviations.

Six Sigma may be defined in several ways. Tomkins (1997) defines Six Sigma to be "a program aimed at the near-elimination of defects from every product, process and transaction". Harry (1998) defines Six Sigma to be "a strategic initiative to boost profitability, increase market share and improve customer satisfaction through statistical tools that can lead to breakthrough quantum gains in quality".

Six Sigma was launched by Motorola in 1987. It was the result of a series of changes in the quality area starting in the late 1970s, with ambitious ten-fold improvement drives. The top-level management along with CEO Robert Galvin developed a concept called Six Sigma. After some internal pilot implementations, Galvin, in 1987, formulated the goal of "achieving 6σ capability by 1992" in a memo to all Motorola employees (Bhote, 1989). The results in terms of reduction in process variation were on-track and cost savings totaled US$13 billion and improvement in labor productivity achieved 204% increase over the period 1987-1997 (Losianowycz, 1999).

In the wake of successes at Motorola, some leading electronic companies such as IBM, DEC, and Texas Instruments launched Six Sigma initiatives in the early 1990s. However, it was not until 1995 when GE and Allied Sigma launched Six Sigma as strategic initiatives that a rapid dissemination took place in non-electronic industries all over the world (Hendricks and Kelbaugh, 1998). In early 1997, the Samsung and LG groups in Korea began to introduce Six Sigma within their companies. The results were amazingly good in those companies. For instance, Samsung SDI, which is the company who introduced Six Sigma first in the Samsung group, reported that the cost savings by Six Sigma projects totaled US$150 million (Samsung SDI, 2000). At the present time, the number of large companies applying Six Sigma in Korea is growing exponentially, with a strong vertical deployment in many small and medium size enterprises as well.

As of result of consulting experiences with Six Sigma in Korea, the authors (see, for instance, Park et al., 2000) believe that **Six Sigma is a "new strategic paradigm of management innovation for company survival in this 21st century, which implies three things: statistical measurement, management strategy and quality culture"**. It tells us how good our products, services and processes really are through statistical measurement of quality level. It is a new management strategy under leadership of the top-level management to create quality innovation and total customer satisfaction. It is also a quality culture. It provides a means of doing things right the first time and to work smarter by using data information. It also provides an atmosphere for solving many CTQ (critical-to-quality) problems through team efforts. CTQ could be a critical process/product result characteristic to quality, or a critical reason to quality characteristic. The former is termed as CTQ*y*, and the latter CTQ*x*.

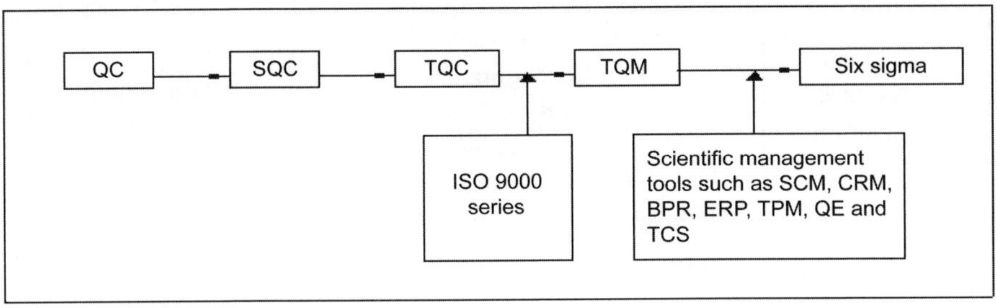

Figure 11.1 Development process of Six Sigma in quality management.

11.2 WHY IS SIX SIGMA FASCINATING?

Six Sigma has become very popular throughout the whole world. There are several reasons for this popularity. First, it is regarded as a fresh quality management strategy which can replace TQC, TQM and others. In a sense, we can view the development process of Six Sigma as shown in Figure 11.1, in which QC stands for quality control, SQC statistical quality control, TQC total quality control, TQM total quality management, ISO International Organization for Standardization, SCM supply chain management, CRM customer relations management, BPR business process reengineering, ERP enterprise resources planning, TPM total productive maintenance, QE quality engineering, and TCS total customer satisfaction, respectively. Many companies, which were not quite successful in implementing previous management strategies such as TQC and TQM, are eager to introduce Six Sigma.

Six Sigma is viewed as a systematic, scientific, statistical and smarter (4S) approach for management innovation which is quite suitable for use in a knowledge-based information society. The essence of Six Sigma is the integration of four elements (customer, process, manpower and strategy) to provide management innovation as shown in Figure 11.2.

Six Sigma provides a scientific and statistical basis for quality assessment for all processes through measurement of quality levels. The Six Sigma method allows us to draw comparisons among all processes, and tells how good a process is. Through this

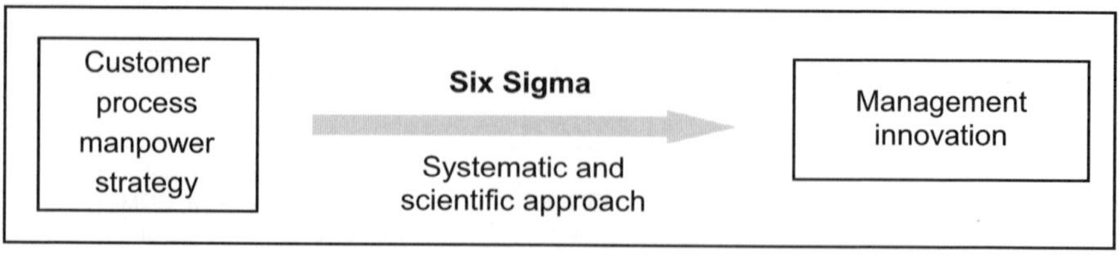

Figure 11.2 Essence of Six Sigma.

information, the top-level management learns what path to follow to achieve process innovation and customer satisfaction.

Second, Six Sigma provides efficient manpower cultivation and utilization. It employs a belt system in which the levels of mastery are classified as white belt, green belt, black belt, master black belt and champion. As a person in a company obtains certain training, he acquires a belt. Usually, a black belt is the leader of a project team and several green belts work together for the project team.

Third, there are many success stories of Six Sigma application in well-known world-class companies. As mentioned earlier, Six Sigma was pioneered by Motorola and launched as a strategic initiative in 1987. Since then, and particularly from 1995, an exponentially growing number of prestigious global firms have launched a Six Sigma program. It has been noted that many globally leading companies run Six Sigma programs (see Figure 3), and it has been well-known that besides Motorola, GE, Allied Signal, IBM, DEC, Texas Instruments, Sony, Kodak, Nokia, Samsung, LG and Philips Electronics among others have been quite successful in Six Sigma.

Lastly, Six Sigma provides flexibility in the new millennium of 3Cs, which are

- Change: Changing society.
- Customer: Power is shifted to customer and customer demand is high.
- Competition: Competition in quality and productivity.

The pace of change during the last decade has been unprecedented, and the speed of change in this new millennium is perhaps faster than ever before. Most notably, the power has shifted from producer to customer. The producer-oriented industrial society is over, and the customer-oriented information society has arrived. The customer has all the rights to order, select and buy goods and services. Especially, in the e-business, the customer has all-mighty power. Competition in quality and productivity has been ever increasing. Second-rate quality goods cannot survive anymore in the market.

11.3 KEY CONCEPTS OF MANAGEMENT

The core objective of Six Sigma is to improve the performance of processes. By improving processes, it attempts to achieve three things: the first is to reduce costs, the second is to improve customer satisfaction, and the third is to increase revenue, thereby, increasing profits.

11.3.1 PROCESS

A general definition of a process is an activity or series of activities transforming inputs to outputs in a repetitive flow as shown in Figure 11.3. For companies, the output is predominantly a product taking the form of hardware goods with their associated services. However, an R&D activity or a non-manufacturing service activity which does not have any form of hardware goods could also be a process.

Literally, the inputs can be anything from labor, materials, machines, decisions, information and measurements to temperature, humidity and weight. Inputs are either control factors which can be physically controlled, or noise factors which are considered to be uncontrollable, too costly to control, or not desirable to control.

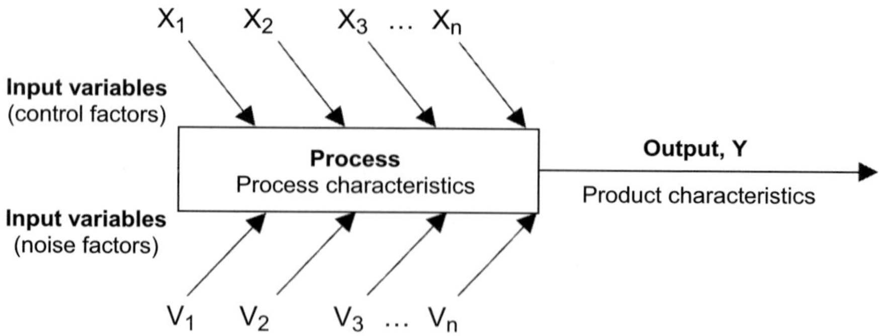

Figure 11.3 The process with inputs and outputs.

The model of Six Sigma in terms of processes and improvement is that y is a function of x and v:

$$y = f(x_1, x_2, \cdots, x_k; v_1, v_2, \cdots, v_m).$$

Here, y represents the result variable (characteristics of the process or product), x represents one or more control factors, and v represents one or more noise factors. The message in the process is to find the optimal levels of x variables which give desired values of y as well as being robust to the noise factors v. The word "robust" means that the y values are not changed much as the levels of noise factors are changed.

11.3.2 VARIATION

The data values for any process or product characteristic always vary. No two products or characteristics are exactly alike because any process contains many sources of variability. The differences among products may be large, or they may be immeasurably small, but they are always present. The variation, if the data values are measured, can be visualized and statistically analyzed by means of a distribution that best fits the observations. This distribution can be characterized by:

- Location (average value)
- Spread (span of values from smallest to largest)

- Shape (the pattern of variation – whether it is symmetrical, skewed, etc.)

Variation is indeed the number one enemy of quality control. It constitutes a major cause of defectives as well as excess costs in every company. Six Sigma, through its tracking of process performance and formalized improvement methodology, focuses on pragmatic solutions for reducing variation. Variation, which is the most important, relates to "how close are the measured values to the target value", cycle time to "how fast" and yield to "how much". Cycle time and yield are the two major elements of productivity.

There are many sources of variation for process and product characteristics. It is common to classify them into two types, common causes and special causes. Common causes refer to the sources of variation within a process that have a stable and repeatable distribution over time. This is called "in a state of statistical control". The random variation, which is inherent in the process, is not easily removable unless we change the very design of the process or product, and is a common cause found everywhere. Common causes behave like a stable system of chance causes. If only common causes of variation are present and do not change, the output of a process is predictable.

Special causes (often called assignable causes) refer to any factors causing variation that are usually not present in the process. That is, when they occur, they make a change in the process distribution. Unless all the special causes of variation are identified and acted upon, they will continue to affect the process output in unpredictable ways. If special causes are present, the process output is not stable over time.

11.3.3 CYCLE TIME, YIELD AND PRODUCTIVITY

Every process has a cycle time and yield. The cycle time of a process is the average time required for a single unit to complete the transformation of all input factors into an output. The yield of a process is the amount of output related to input time and pieces. A more efficient transformation of input factors into products will inevitably give a better yield.

Productivity for industrial activity is the degree of effective utilization of each element of production. Although the product is the main output of an enterprise, other tasks such as R&D activities, sale of products and other service activities are also closely linked to productivity. In economic terms, productivity refers to the extent to which a firm is able to optimize its management resources in order to achieve its goals.

11.3.4 CUSTOMER SATISFACTION

Customer satisfaction is one of the watchwords for company survival in this new 21st century. Customer satisfaction can be achieved when all the customer requirements are met. Six Sigma emphasizes that the customer requirements must be fulfilled by measuring and improving processes and products, and CTQ (critical-to-quality) characteristics are measured on a consistent basis to produce few defects in the eyes of the customer.

The identification of customer requirements is ingrained in Six Sigma and extended into the activity of translating requirements into important process and product characteristics. As customers rarely express their views on process and product characteristics directly, a method called QFD (quality function deployment) is applied for a systematic translation. Using QFD, it is possible to prioritize the importance of each characteristic based on input from the customer.

Having identified the CTQ requirements, the customer is usually asked to specify what the desired value for the characteristic is, i.e. target value and what a defect for the characteristic is, i.e. specification limits. This vital information is utilized in Six Sigma as a basis for measuring the performance of processes.

Six Sigma improvement projects are supposed to focus on improvement of customer satisfaction which eventually gives increased market share and revenue growth. As a result of

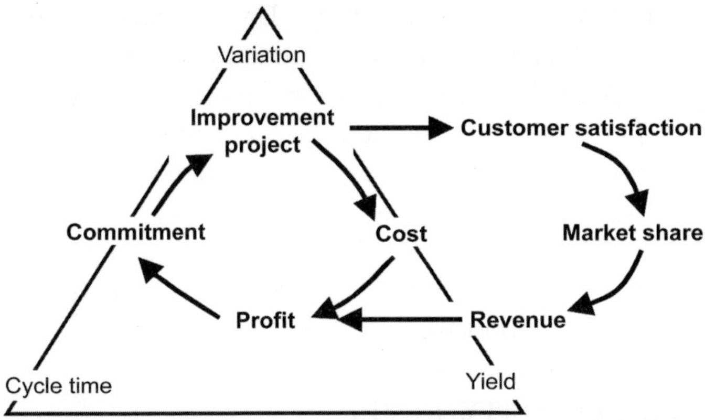

Figure 11.4 Six Sigma loop of improvement projects.

revenue growth and cost reduction, the profit increases and the commitment to the methodology and further improvement projects are generated throughout the company. This kind of loop is called "Six Sigma loop of improvement projects", and was suggested by Magnusson et al. (2001). This loop is shown in Figure 11.4.

11.4 MEASUREMENT OF PROCESS PERFORMANCE

Among the three dimensions (variation, cycle time and yield) of Six Sigma loop in Figure 11.4, variation is the preferred measurement for process performance in Six Sigma. Cycle time and yield could have been used, but they can be covered through variation. For example, if a cycle time has been specified for a process, the variation of the cycle time around its target value will indicate the performance of the process in terms of this characteristic. The same applies to yield. Besides variation, other measures are explained below.

11.4.1 DEFECT RATE, PPM AND DPMO

The defect rate, denoted by p, is the ratio of the number of defective items which are out of specification to the total number of items processed (or inspected). Defect rate or fraction of defective items has been used in industry for a long time. The number of defective items out of one million inspected items is called the ppm (parts-per-million) defect rate. Sometimes a ppm defect rate cannot be properly used, in particular, for the cases of service works. In this case, a DPMO (defects per million opportunities) is often used. DPMO is the number of defective opportunities which do not meet the required specification out of one million possible opportunities.

11.4.2 SIGMA QUALITY LEVEL

Specification limits are the tolerances or performance ranges that customers' demand of the products or processes they are purchasing. Figure 11.5 illustrates specification limits as the two major vertical lines in the figure. In the figure, LSL means the lower specification limit, USL means the upper specification limit and T means the target value. The sigma quality level (in short, sigma level) is the distance from the process mean (μ) to the closer specification limit.

In practice, we desire that the process mean be kept at the target value. However, the process mean during one time period is usually different from that of another time period for various reasons. This means that the process mean constantly shifts around the target value. To address typical maximum shifts of the process mean, Motorola added the shift value $\pm 1.5\sigma$ to the process mean. This shift of the mean is used when computing a process sigma level as shown in Figure 11.5. From this figure, we note that a 6σ quality level corresponds to a 3.4 ppm rate. Table 11.1 illustrates how sigma quality levels would equate to other defect rates and organizational performances. Table 11.2 shows the details of this relationship when the process mean is $\pm 1.5\sigma$ shifted.

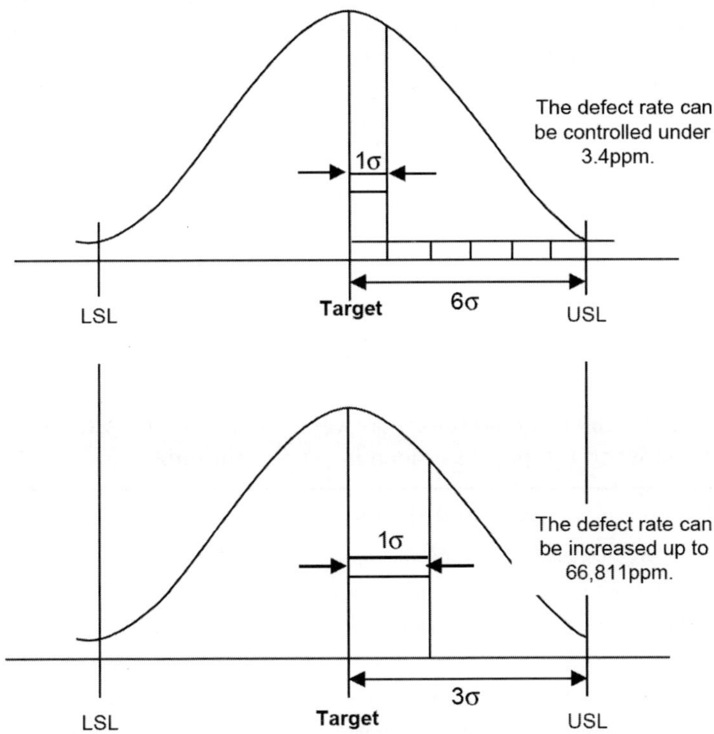

Figure 11.5 Sigma quality levels of 6σ and 3σ.

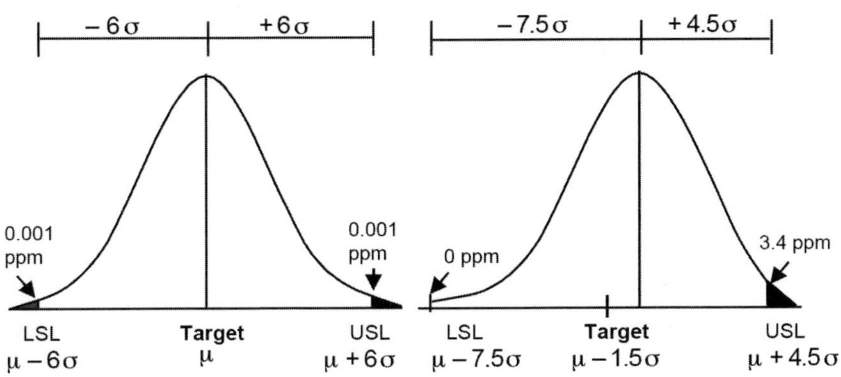

Figure 11.6 Effects of a 1.5σ shift of process mean when 6σ quality level is achieved.

Table 11.1 ppm changes when sigma quality level changes.

Sigma quality level	Process mean, fixed		Process mean, 1.5σ shifted	
	Non-defect rate(%)	Defect rate(ppm)	Non-defect rate(%)	Defect rate(ppm)
σ	68.26894	317,311	30.2328	697,672
2σ	95.44998	45,500	69.1230	308,770
3σ	99.73002	2,700	93.3189	66,811
4σ	99.99366	63.4	99.3790	6,210
5σ	99.999943	0.57	99.97674	233
6σ	99.9999998	0.002	99.99966	3.4

Table 11.2 Detailed conversion between ppm (or *DPMO*) and sigma quality level when the process mean is $\pm 1.5\sigma$ shifted.

Sigma level	0.00	0.01	0.02	0.03	0.04	0.05	0.06	0.07	0.08	0.09
2.0	308770.2	305249.8	301747.6	298263.7	294798.6	291352.3	287925.1	284517.3	281129.1	277760.7
2.1	274412.2	271084.0	267776.2	264489.0	261222.6	257977.2	254753.0	251550.2	248368.8	245209.2
2.2	242071.5	238955.7	235862.1	232790.8	229742.0	226715.8	223712.2	220731.6	217773.9	214839.2
2.3	211927.7	209039.6	206174.8	203333.5	200515.7	197721.6	194951.2	192204.6	189481.9	186783.0
2.4	184108.2	181457.4	178830.7	176228.0	173649.5	171095.2	168565.1	166059.2	163577.5	161120.1
2.5	158686.9	156278.0	153893.3	151532.9	149196.7	146884.7	144596.8	142333.2	140093.6	137878.1
2.6	135686.7	133519.3	131375.8	129256.3	127160.5	125088.6	123040.3	121015.7	119014.7	117037.0
2.7	115083.0	113152.2	111244.7	109360.2	107498.9	105660.5	103844.9	102052.1	100281.9	98534.3
2.8	96809.0	95106.1	93425.3	91766.6	90129.8	88514.8	86921.5	85349.7	83799.3	82270.1
2.9	80762.1	79275.0	77808.8	76363.2	74938.2	73533.6	72149.1	70784.8	69440.4	68115.7
3.0	66810.6	65525.0	64258.6	63011.3	61783.0	60573.4	59382.5	58210.0	57055.8	55919.6
3.1	54801.4	53700.9	52618.1	51552.6	50504.3	49473.1	48458.8	47461.2	46480.1	45515.3
3.2	44566.8	43634.2	42717.4	41816.3	40930.6	40060.2	39204.9	38364.5	37538.9	36727.8
3.3	35931.1	35148.6	34380.2	33625.7	32884.8	32157.4	31443.3	30742.5	30054.6	29379.5
3.4	28717.0	28067.1	27429.4	26803.8	26190.2	25588.4	24988.2	24419.5	23852.1	23295.8
3.5	22705.4	22215.9	21692.0	21178.5	20675.4	20182.4	19699.5	19226.4	18763.0	18309.1
3.6	17864.6	17429.3	17003.2	16586.0	16177.5	15777.7	15386.5	15003.5	14628.8	14262.2
3.7	13903.5	13552.7	13209.5	12873.8	12545.5	12224.5	11910.7	11603.9	11303.9	11010.7
3.8	10724.2	10444.1	10170.5	9903.1	9641.9	9386.7	9137.5	8894.1	8656.4	8424.2
3.9	8197.6	7976.3	7760.3	7549.4	7343.7	7142.8	6946.9	6755.7	6569.1	6387.2
4.0	6209.7	6036.6	5867.8	5703.1	5542.6	5386.2	5233.6	5084.9	4940.0	4798.8
4.1	4661.2	4527.1	4396.5	4269.3	4145.3	4024.6	3907.0	3792.6	3681.1	3572.6

Table 11.2 (Continued).

Sigma level	0.00	0.01	0.02	0.03	0.04	0.05	0.06	0.07	0.08	0.09
4.2	3467.0	3364.2	3264.1	3166.7	3072.0	2979.8	2890.1	2802.8	2717.9	2635.4
4.3	2555.1	2477.1	2401.2	2327.4	2255.7	2186.0	2118.2	2052.4	1988.4	1926.2
4.4	1865.8	1807.1	1750.2	1694.8	1641.1	1588.9	1538.2	1489.0	1441.2	1394.9
4.5	1349.9	1306.2	1263.9	1222.8	1182.9	1144.2	1106.7	1070.3	1035.0	1000.8
4.6	967.6	935.4	904.3	874.0	844.7	816.4	788.8	762.2	736.4	711.4
4.7	687.1	663.7	641.0	619.0	597.6	577.0	557.1	537.7	519.0	500.9
4.8	483.4	466.5	450.1	434.2	418.9	404.1	389.7	375.8	362.4	349.5
4.9	336.9	324.8	313.1	301.8	290.9	280.3	270.1	260.2	250.7	241.5
5.0	232.6	224.1	215.8	207.8	200.1	192.6	185.4	178.5	171.8	165.3
5.1	159.1	153.1	147.3	141.7	136.3	131.1	126.1	121.3	116.6	112.1
5.2	107.8	103.6	99.6	95.7	92.0	88.4	85.0	81.6	78.4	75.3
5.3	72.3	69.5	66.7	64.1	61.5	59.1	56.7	54.4	52.2	50.1
5.4	48.1	46.1	44.3	42.5	40.7	39.1	37.5	35.9	24.5	33.0
5.5	31.7	30.4	29.1	27.9	26.7	25.6	24.5	23.5	22.5	21.6
5.6	20.7	19.8	18.9	18.1	17.4	16.6	15.9	15.2	14.6	13.9
5.7	13.3	12.8	12.2	11.7	11.2	10.7	10.2	9.8	9.3	8.9
5.8	8.5	8.2	7.8	7.5	7.1	6.8	6.5	6.2	5.9	5.7
5.9	5.4	5.2	4.9	4.7	4.5	4.3	4.1	3.9	3.7	3.6
6.0	3.4	3.2	3.1	2.9	2.8	2.7	2.6	2.4	2.3	2.2
6.1	2.1	2.0	1.9	1.8	1.7	1.7	1.6	1.5	1.4	1.4
6.2	1.3	1.2	1.2	1.1	1.1	1.0	1.0	0.9	0.9	0.8
6.3	0.8	0.8	0.7	0.7	0.6	0.6	0.6	0.6	0.5	0.5
6.4	0.5	0.5	0.4	0.4	0.4	0.4	0.4	0.3	0.3	0.3
6.5	0.3	0.3	0.3	0.2	0.2	0.2	0.2	0.2	0.2	0.2
6.6	0.2	0.2	0.2	0.1	0.1	0.1	0.1	0.1	0.1	0.1
6.7	0.1	0.1	0.1	0.1	0.1	0.1	0.1	0.1	0.1	0.1
6.8	0.1	0.1	0.1	0.0	0.0	0.0	0.0	0.0	0.0	0.0

11.4.3 ROLLED THROUGHPUT YIELD (RTY)

Rolled throughput yield **(RTY)** is the final cumulative yield when there are several processes connected in series. **RTY** is the amount of non-defective products produced in the final process compared with the total input in the first process.

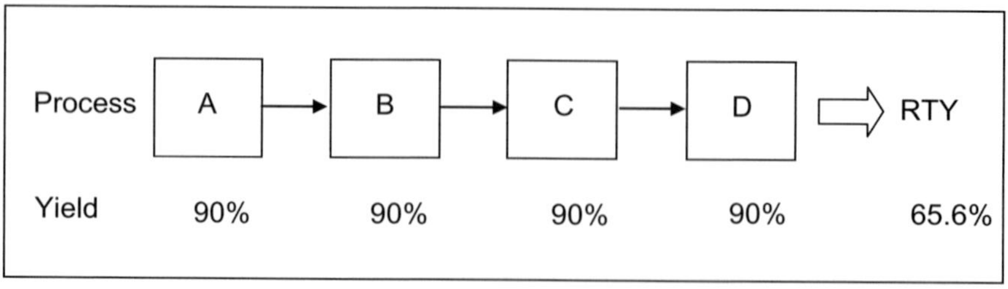

Figure 11.7 RTY and yield of each process.

For example, as shown in Figure 11.7, there are 4 processes (A, B, C and D) connected in consecutive series, and each process has 90% yield.

Then RTY of these processes is

$$\text{RTY} = 0.9 \times 0.9 \times 0.9 \times 0.9 = 0.656.$$

If there are k processes in series, and the ith process has its own yield y_i, then RTY of these k processes is

$$\text{RTY} = y_1 \times y_2 \times \cdots \times y_k.$$

11.5 SIX SIGMA FRAMEWORK

11.5.1 FIVE ELEMENTS OF THE SIX SIGMA FRAMEWORK

Management strategies, such as TQC, TQM, and Six Sigma, are distinguished from each other by their underlying rationale and framework. As far as the corporate framework of Six Sigma is concerned, it embodies the five elements of top-level management commitment, training schemes, project team activities, measurement system and stakeholder involvement as shown in Figure 11.8.

Stakeholders include employees, owners, suppliers and customers. At the core of the framework is a formalized improvement strategy with the following five steps; define, measure, analyze, improve and control (DMAIC), which will be explained in detail in Section 11.6. The improvement strategy is based on belt system, project team activities, measurement system and incentive system. Top-level management commitment and stakeholder involvement are all inclusive in the framework. Without these two, the improvement strategy functions poorly. All six elements support the improvement strategy and improvement project teams.

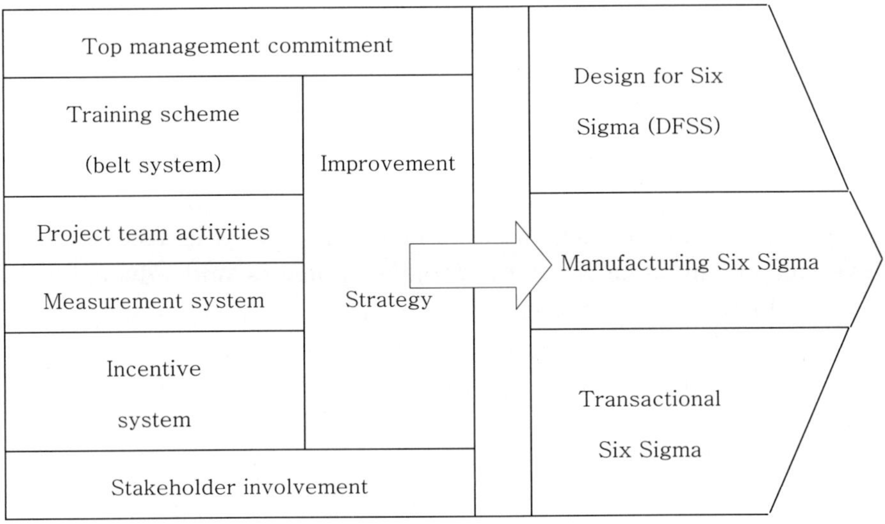

Figure 11.8 The corporate framework of Six Sigma.

Most big companies operate in three parts; R&D, manufacturing, and non-manufacturing service. Six Sigma can be introduced into each of these 3 parts separately. In fact, the color of Six Sigma could be different for each part. Six Sigma is often called "Design for Six Sigma (DFSS)" in R&D part, "Manufacturing Six Sigma" in manufacturing, and "Transactional Six Sigma (TSS)" in the non-manufacturing service part. All six elements in Figure 11.8 are necessary for each of the 3 different Six Sigma functions. However, the improvement methodology, DMAIC, could be modified in DFSS and TSS.

11.5.2 TOP-LEVEL MANAGEMENT COMMITMENT AND STAKEHOLDER INVOLVEMENT

(1) Top-level management commitment

Launching Six Sigma in a company is a strategic management decision that needs to be initiated by the top-level management. All the elements of the framework, as well as the formalized improvement strategy, need top-level management commitment for successful execution. Especially, without a strong commitment on the part of the top-level management, the training program and project team activities are seldom successful. Although not directly active in the day-to-day improvement projects, the role of top-level management as leaders, project sponsors and advocates is crucial. Pragmatic management is required, not just lip service, as the top-level management commits itself and the company to drive the initiative for several years and into every corner of the company.

There are numerous pragmatic ways for the CEO (chief executive officer) to manifest his commitment. First, in setting the vision and long-term or short-term goal for Six Sigma, the CEO should play a direct role. Second, the CEO should allocate appropriate resources in order to implement such Six Sigma programs as training schemes, project team activities, measurement system and incentive system. Third, the CEO should regularly check the progress of the Six Sigma program to determine whether there are any problems which might hinder its success. He should listen to Six Sigma reports and make comments on the progress of Six Sigma. Fourth, he should hold a Six Sigma presentation seminar regularly, say twice a year, in which the results of the project team are presented and good results rewarded financially by a proper incentive system. Finally, he should hold a Champion Day regularly, say once in every other month, in which Champions (upper managers) are educated by specially invited speakers and he should discuss the progress of Six Sigma with Champions.

Even before the first results start to come in, a high degree of personal faith and commitment from top-level management to the Six Sigma initiative are necessary. A good example is John F. Welch's elaboration on his five-year plan for Six Sigma. In his speech at the GE 1996 Annual Meeting in Charlottesville, he makes it clear that "... we have set for ourselves the goal of becoming, by the year 2000, a Six Sigma quality company which means a company that produces virtually defect-free products, services and transactions." His speech is a landmark one for Six Sigma, and it is cited in full in Appendix I.

It is also the responsibility of top-level management to set stretch goals for the Six Sigma initiative. Stretch goals are tough and demanding, but are usually achievable. Some companies set the stretch goal for process performance at 6 sigma or 3.4 DPMO for all critical-to-customer characteristics. However, the goals can also be set incrementally, by stating instead the annual improvement rate in process performance. The industry standard is to reduce DPMO by 50% annually.

(2) Stakeholder involvement

Stakeholder involvement means that the hearts and minds of employees, suppliers, customers, owners and even society should be involved in the improvement methodology of Six Sigma for a company. In order to meet the goal set for improvements in process performance and to complete improvement projects of a Six Sigma initiative, top-level management commitment is simply not enough. The company needs active support and direct involvement from stakeholders.

Employees in a company constitute the most important group of stakeholders. They carry out the majority of improvement projects and must be actively involved. The Six Sigma management is built to ensure this involvement through various practices, such as training courses, project team activities and evaluation of process performance.

Suppliers also need to be involved in a Six Sigma initiative. A Six Sigma company usually encourages its key suppliers to have their own Six Sigma programs. To support suppliers, it is common for Six Sigma companies to have suppliers sharing their performance data for the products purchased and to offer them participation at in-house training courses in Six Sigma. It is also common for Six Sigma companies to help small suppliers financially in pursuing Six Sigma programs by inviting them to share their experiences together in report sessions of project team activities. The reason for this type of involvement is to have the variation in the suppliers' products transferred to the company's processes so that most of the process improvement projects carried out on suppliers' processes would result in improvement of the performance.

Customers play key roles in a Six Sigma initiative. Customer satisfaction is one of the major objectives for a Six Sigma company. Customers should be involved in specific activities such as identifying the critical-to-customer (CTC) characteristics of the products and processes. CTC is a subset of CTQ from the viewpoint of the customers. Having identified the CTC requirements, the customers are also asked to specify the desired value of the characteristic, i.e. the target value and the definition of a defect for the characteristic, or the specification limits. This vital information is utilized in Six Sigma as a basis for measuring the performance of processes. In particular, the R&D part of a company should know the CTC requirements and should listen to the voice of customers (VOC) in order to reflect the VOC in developing new products.

11.5.3 TRAINING SCHEME, MEASUREMENT SYSTEM AND INCENTIVE SYSTEM

(1) Training scheme

In any Six Sigma program, a comprehensive knowledge of process performance, improvement methodology, statistical tools, process of project team activities, deployment of customer requirements and other facets is needed. This knowledge can be cascaded throughout the organization and become the shared knowledge of all employees only through a proper training scheme.

There are five different fairly standardized training courses in Six Sigma. To denote these courses, Six Sigma companies have adopted the belt rank system from martial arts which is shown in Figure 11.9. There are White Belt (WB), Green Belt (GB), Black Belt (BB), Master Black Belt (MBB) and Champion.

Course levels		Belts
Overall vision	☺	Champion
Most comprehensive	☺☺☺	Master Black Belt
Comprehensive	☺☺☺☺☺	Black Belt
Median	☺☺☺☺☺☺ ☺☺☺☺☺☺☺	Green Belt
Basic	☺☺☺☺☺☺☺☺☺ ☺☺☺☺☺☺☺☺☺☺	White Belt

Figure 11.9 Course levels and belts for Six Sigma training scheme.

The WB course gives a basic introduction to Six Sigma. Typically, it is a 2–3 day course and is offered to all employees. It covers a general introduction to Six Sigma, framework, structure of project teams and statistical thinking. The GB course is a median course in content and the participants also learn to apply the formalized improvement methodology in a real project. It is usually a 1–2 week course, and is offered to foremen and middle management. The BB course is comprehensive and advanced, and aims at creating full-time improvement project leaders. Black belts are the experts of Six Sigma, and they are the core group in leading the Six Sigma program. The duration of a BB course is around 4–6 months with about 20 days of study seminars. In-between the seminar blocks, the participants are required to carry out improvement projects with specified levels of DMAIC steps. The BB candidates are selected from the very best young leaders in the organization.

An MBB has BB qualifications and is selected from Black Belts who have much experience of project activities. An MBB course is most comprehensive as it requires the same BB training and additionally planning and leadership training. Champions are drivers, advocates and experienced sources of knowledge on Six Sigma. These people are selected among the most senior executives of the organization. A Champion course is usually a 3-4 day course, and it concentrates on how to guide the overall Six Sigma program, how to select good improvement projects and how to evaluate the results of improvement efforts.

The number of people who are trained at the different levels depends on the size of company and its resources. A common guideline is to have 1 BB for every 100 employees, around 20 GBs for every BB, and 20 BBs for every MBB. Therefore, if a company has 10,000 people, a good guideline is that there should be 5 MBBs, 100 BBs, 2,000 GBs and the remaining people are WBs.

Most Six Sigma companies, and also consulting organizations, which offer these training courses, issue a certificate to all participants successfully completing the courses. Just as the course contents differ among different Six Sigma companies, the certificates also differ in layout and content. After completing the courses, most companies require that GBs complete one improvement project and BBs 3-4 improvement projects annually. The consequence of not following these requirements would be withdrawal of the certificate.

(2) Measurement system

A Six Sigma company should provide a pragmatic system for measuring performance of processes using a sigma level, ppm or DPMO. The measurement system reveals poor process performance and provides early indications of problems to come. There are two types of characteristics — continuous and discrete. Both types can be included in the measurement system. Continuous characteristics may take any measured value on a continuous scale, which provides continuous data. In continuous data, normally the means and variances of the CTQ characteristics are measured for the processes and products. From the mean and variance, the sigma levels and process capability indices can be calculated.

Discrete characteristics are based on counts, such as yes/no, good/bad, which provide attribute data. A much larger number of observations is needed for a discrete characteristic compared to a continuous characteristic in measuring process performance by means of DPMO. A rule of thumb is to require at least 20 observations for assessing the performance of a continuous characteristic and at least 200 observations for a discrete characteristic.

The data for the characteristic selected for the Six Sigma measurement system is collected individually at predetermined time intervals such as hourly, daily, or weekly. Based on the data collected, the DPMO value for the individual characteristic is calculated. Although continuous data and discrete data need to be measured and analyzed differently, the results can be consolidated into one number for the process performance. The performance of the individual characteristic included in the measurement system can be tracked over time, as can the consolidated value for the company's goods, services, projects and processes. Most Six Sigma companies make use of spreadsheets and databases to collect, analyze and track results. Both standard software packages and tailor-made systems are used. The results, typically visualized in simple graphical illustrations such as a trend chart, are distributed within the company through intranet, newsletters, information stands and so on. Of particular importance is the consolidated DPMO value for the whole company. The measurement system brings process performance to the attention of the whole organization — simple to understand and easy to remember.

11.6 DMAIC PROCESS AND PROJECT TEAM ACTIVITIES

11.6.1 DMAIC PROCESS

The most important methodology in Six Sigma management is perhaps the formalized improvement methodology characterized by DMAIC (define - measure - analyze - improve - control) process. This DMAIC process works well as a breakthrough strategy. Six Sigma companies everywhere apply this methodology as it enables real improvements and real results. The methodology works equally well on variation, cycle time, yield, design and others. It is divided into five phases as shown in Figure 11.10. In each phase the major activities are as follows.

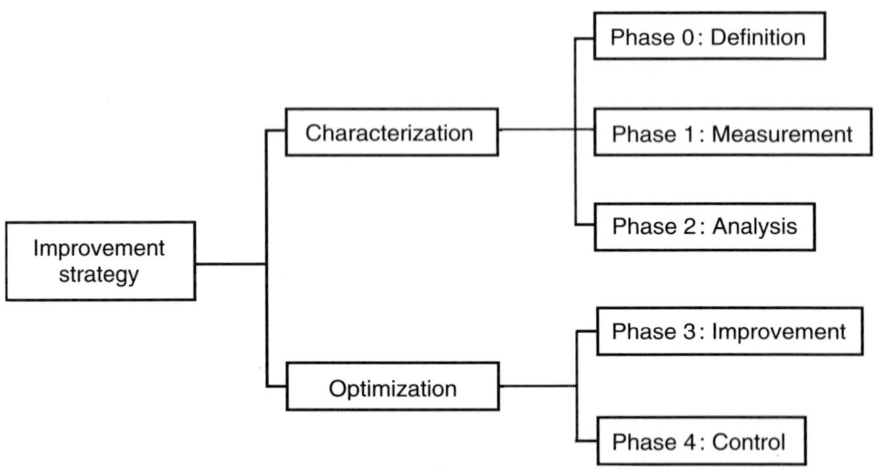

Figure 11.10 Improvement phases.

Phase 0: (Define). This phase is concerned with identification of the process or product that needs improvement. It is also concerned with benchmarking of key product or process characteristics of other world-class companies.

Phase 1: (Measure). This phase entails selecting product characteristics; i.e. dependent variables, mapping the respective processes, making the necessary measurement, recording the results and estimating the short- and long-term process capabilities. Quality function deployment (QFD) plays a major role in selecting critical product characteristics.

Phase 2: (Analyze). This phase is concerned with analyzing and benchmarking the key product/process performance metrics. Following this, a gap analysis is often undertaken to identify the common factors of successful performance; i.e. what factors explain best-in-class performance. In some cases, it is necessary to redefine the performance goal. In analyzing the product/process performance, various statistical and basic QC tools are used.

Phase 3: (Improve). This phase is related to selecting those product performance characteristics which must be improved to achieve the goal. Once this is done, the characteristics are diagnosed to reveal the major sources of variation. Next, the key process variables are identified usually by way of statistically designed experiments including Taguchi methods and other robust design of experiments (DOE). The improved conditions of key process variables are verified.

Phase 4: (Control). This last phase is initiated by ensuring that the new process conditions are documented and monitored via statistical process control (SPC) methods. After the "settling in" period, the process capability is reassessed. Depending upon the outcome of such a follow-on analysis, it may become necessary to revisit one or more of the preceding phases.

The flowchart for DMAIC quality improvement process is sketched in Figure 11.11.

11.6.2 SEVEN-STEP PROCEDURE FOR INTRODUCTION OF SIX SIGMA

For a company which wishes to introduce Six Sigma as the management strategy, the author would like to recommend the following seven-step procedure.

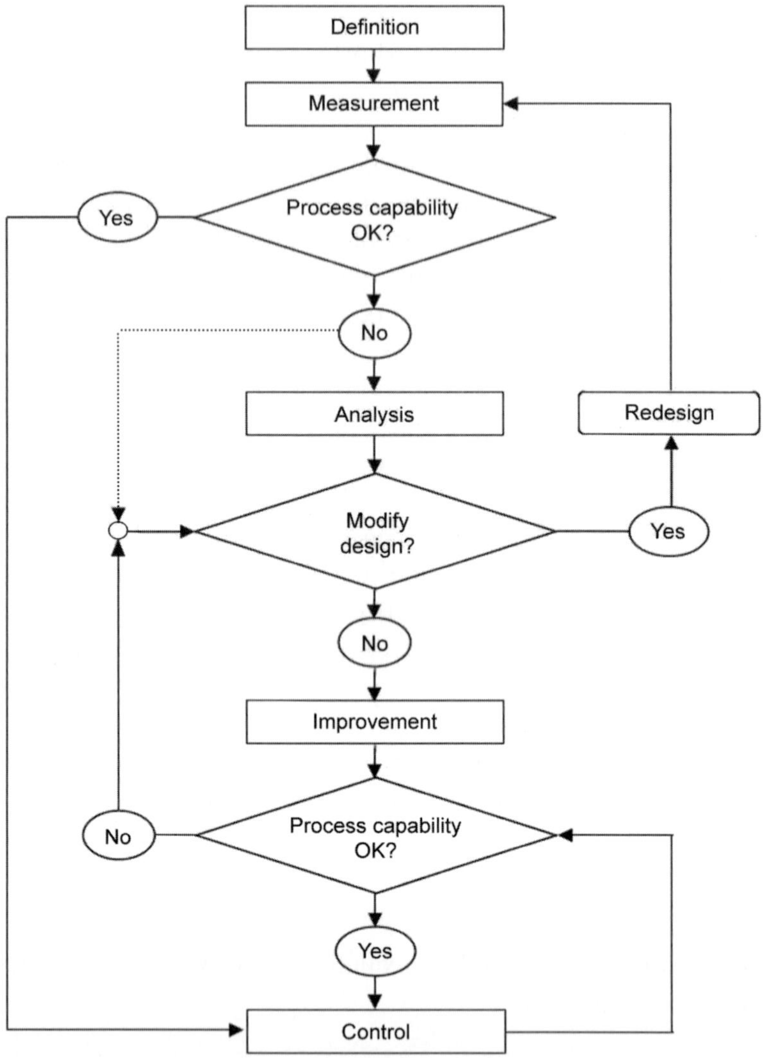

Figure 11.11 Flowchart of DMAIC process.

1) Organize a Six Sigma team and set up the long-term Six Sigma management vision for the company.
2) Start Six Sigma education for Champions first.
3) Choose the area for which a Six Sigma process is to be introduced first. For instance, in a manufacturing company,

choose the first area to introduce Six Sigma among R&D area, manufacturing area and service area.

4) Start the education for GBs and BBs.
5) Deploy CTQs for all areas concerned. Appoint a few or several BBs as full-time project team leaders and ask them to solve some important CTQ problems.
6) Strengthen the infrastructure for Six Sigma such as statistical process control (SPC), knowledge management (KM) and database management system.
7) Designate a "Six Sigma Day" each month, and have the top-level management check the progress of Six Sigma, and organize presentations or awards for accomplishments, if any.

First of all, a few or several members should be appointed as a Six Sigma team to handle all Six Sigma activities. Subsequently, the team should set up the long-term Six Sigma vision for the company under the supervision of top-level management. This is the first and the most important task for the team. It is said that this is the century of 3Cs, which are Change, Customer and Competition for quality. The Six Sigma vision should match these 3Cs well. Most importantly, all employees in the company must agree to and respect this vision.

Second, Six Sigma can begin with proper education for all levels of the company's employees. The education should begin with the top management and directors (Champions). If Champions do not understand the real meaning of Six Sigma, there is no way for Six Sigma to be disseminated within the company. Following the education of Champions, the training for GB, BB, and MBB (Master Black Belts) must be conducted in that sequence. However, the MBB education is done usually by professional organizations.

Third, Six Sigma can be divided into 3 parts according to its characteristics. They are Design for Six Sigma (DFSS) for the R&D area, Six Sigma for manufacturing processes, and Transactional Six Sigma (TSS). DFSS is often called R&D Six Sigma. It is not easy to introduce Six Sigma to all areas at the same time. In this case, the CEO should decide the order of introduction to those 3 areas. Usually it is easy to introduce Six Sigma to manufacturing processes first, followed by the service areas and the R&D areas. However, the order really depends on the circumstances of the company at the time.

Fourth, GB and BB educations are the most important ingredients for Six Sigma success. Fifth, deploy CTQs for all departments concerned. These CTQs can be deployed by policy management or by management by objectives. When the BBs are born, some important CTQ problems should be given to these BB to solve. In principle, the BB should be the project leaders and work as full-time workers for quality innovation.

Sixth, in order to firmly introduce Six Sigma, some basic infrastructure is necessary. The tools required include SPC, MRP (materials requirement planning), KM, and DBMS. In particular, efficient data acquisition, data storage, data analysis and information dissemination systems are necessary.

Lastly, a "Six Sigma Day" each month must be designated. On this day, the CEO must check the progress of Six Sigma personally. On this day, all types of presentations of Six Sigma results can be made, and rewards can be given to the persons who performed excellent jobs in support of the Six Sigma initiative.

(1) Problem-solving processes for project activities

The original Six Sigma process developed for problem solving at Motorola is MAIC which means measurement, analysis, improvement and control. Later, DMAIC instead of MAIC, was advocated at GE where D stands for definition. MAIC or DMAIC is mostly used as a unique problem-solving process in manufacturing areas. However, with DFSS, there are several proposed processes as follows.

1) DMADV (Define - Measure – Analyze - Design - Verify). MADV was suggested by Motorola for DFSS, and D was added to it for definition. DMADV is similar to DMAIC.
2) IDOV (Identify - Design - Optimize - Validate). This was suggested by GE and has been used most frequently in practice.

3) DIDES (Define - Initiate - Design - Execute - Sustain). This was suggested by Qualtec Consulting Company.

It seems that the above problem-solving processes for manufacturing and R&D are not quite suitable for service areas. The author believes that DMARIC (Define - Measure - Analyze - Redesign - Implement - Control) is an excellent problem-solving process of TSS for non-manufacturing service areas. Here, the "redesign" phase means that the system for service works should be redesigned in order to improve the service function.

(2) Difference between project teams and quality circles

In Six Sigma, the project teams leading by BBs are the backbone of group activities. However, in TQC or TQM, quality circles constitute the backbone of group activities. There are some basic differences between these two teams as shown in Table 1. In the old management strategies TQC and TQM, there are usually two types of team efforts, namely, the task-force-team and the quality circle team. The task-force-team mainly consists of engineers and scientists, and the quality circle team consists of the line operators. However, in Six Sigma, these two teams are merged into one, called the project team whose leader is usually a BB. For the theme selection and the problem-solving flow, the differences are also listed in Table 11.3.

Table 11.3 Differences between project team and quality circle.

Classification	Project team	Quality circle
Organization	Engineers(or scientists) + operators one BB + several GBs	Operators
Theme selection	top-down, company CTQs	Bottom-up, self-selection
Problem-solving, flow	DMAIC, DMADV, IDOV, DMARI	PDCA

Depending on the management policy, it is permissible for a company to have project teams and quality circle teams at the same time under the banner of Six Sigma. However, care should be exercised in controlling the two types of teams.

(3) How to select project themes?

As explained in Table 11.3, the project themes are selected essentially by a top-down approach, and company CTQs are nominated as themes most of the time. The deployment method in order to select project themes is shown in Figure 11.12.

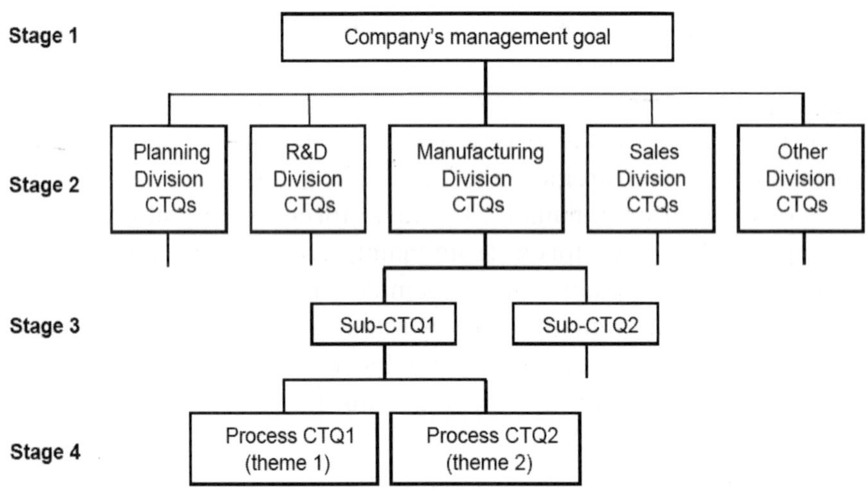

Figure 11.12 Deployment for selection of project themes.

For example, suppose that one of the company's management goals is to improve production capability without further investment. For this particular goal, each division must have its own CTQs. Suppose that the manufacturing division has such CTQs as machine down-time and rolled throughput yield (RTY). For instance, for the machine down-time, there may be more than two sub CTQs; heating machine down-time, cooling machine down-time, and pump down-time. For the sub CTQ of heating machine down–time, process CTQ1 (theme 1) could be "reduction of heating machine down–time from 10 hours/month to 5 hours/month", and process CTQ2 (theme 2) could be "10% improvement of heating process capability".

12

Further Issues for the Implementation of Six Sigma

12.1 DATA TECHNOLOGY

12.1.1 DEFINITION OF DATA TECHNOLOGY

The phenomenal growths in data storage, retrieval, communication and graphical representation capabilities have revolutionized the use of numbers and data aided by the enormous increase in the computing speed. Statisticians, industrial engineers and others in analytic disciplines are overwhelmed by the abundance of data and lack of tools for converting the massive amount of data and signals into useful information and decisions. This phenomenon is also true in the implementation of Six Sigma in practice. For these reasons, we propose to define a new technology called Data Technology as a new, integrated discipline of technology.

Key engines for moving the 21st century are said to be IT (information technology), BT (bio-technology), NT (nano-technology), ET (environment technology), ST (space technology), and CT (culture technology). What we have yet to recognize is a technology so deeply imbedded in every technology that deals with quantitative decision process — whether it is individual, organizational, societal, or scientific. This is what we will call Data Technology, or DT. The term DT was first coined by Park [2001a, 2001b], and he discussed the economics of DT and its applicability.

DT may be defined as a scientific methodology that deals with the following:

- systematic collection, storage and retrieval of required data,
- refinement and object-oriented analysis of data,
- conversion of data into information,
- statistical and computational modeling based on data,
- formation of inferences and relevant knowledge, and
- diagnosis of the present states and prediction of future events,

by use of statistical and probabilistic modeling techniques.

We believe that DT, as defined, would be an essential technology for promoting application of existing and new statistical methods imbedded in data for all aspects of modern society. They include organizational management, economic decisions at all levels and dimensions, creation and management of political and social systems, and forecasting for the enhancement of global competitiveness for business, industry and government. The importance of DT is expected to grow fast in this knowledge-driven information society.

Many believe that DT is a subset of information technology (IT). This is not true. DT is mainly concerned with data collection, statistical analysis of data, generation of information, and creation of necessary knowledge from information. However, IT is mainly concerned with data/information/image transmission and communication, and development of engineering devices and computers for information handling. Also IT is concerned with engineering tools for knowledge management. Generally speaking, DT forms the infrastructure of IT. Without DT, IT would have limitations in growth. DT is software-oriented, but IT is hardware-oriented and systems-oriented. Without IT, DT cannot be well visualized. We can say that IT is the vehicle for DT development.

12.1.2 KNOWLEDGE TRIANGLE

It is said that the 21st century is the knowledge-based information society. We can think about the knowledge triangle as shown in Figure 12.1 in which DT and IT play important roles.

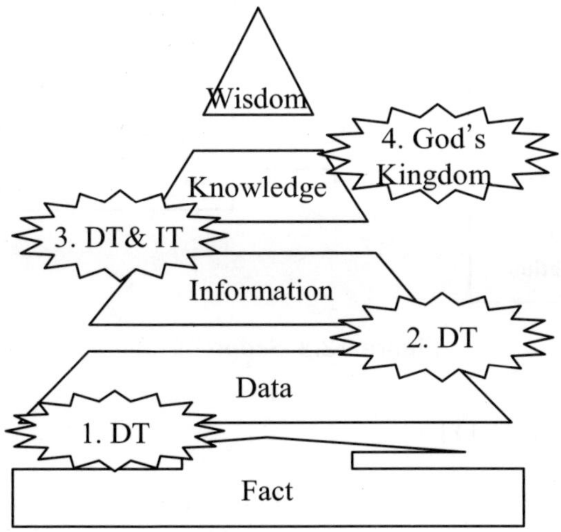

Figure 12.1 Knowledge triangle.

In each step, the following activities are usually implemented.

Table 12.1 Major activities in each step of knowledge triangle.

Step	Major activities
Step 1	Measurement, Data refinement, Sampling design, Design of experiments, Meta data management, Gage R&R test
Step 2	Data analysis and modeling, Data-mining, Data refinement for application, Diagnosis and control, Prediction modeling
Step 3	Output summary, Valuation, Remodeling, Information clustering
Step 4	Knowledge generation from information clustering

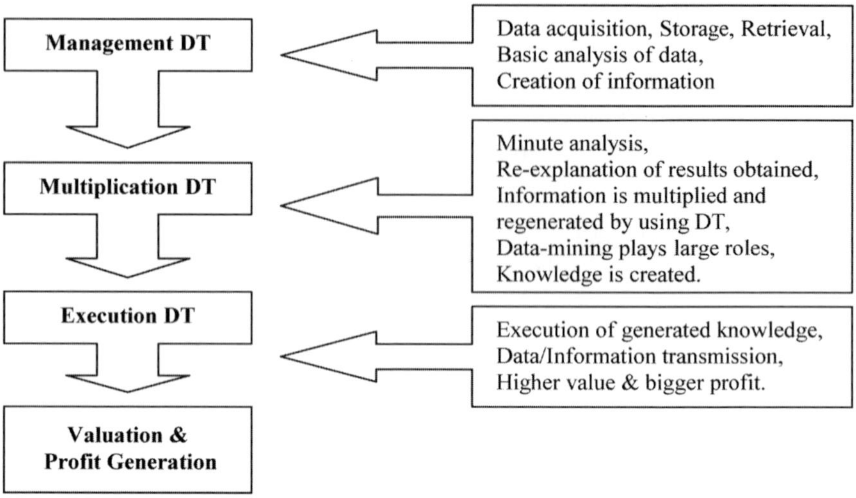

Figure 12.2 Scope of DT.

12.1.3 SCOPE OF DT

The scope of DT can be divided into three categories: management, multiplication and execution. Management DT comes first, and then multiplication DT comes next, and finally execution DT provides valuation and profit generation for the organization concerned. The scope can be shown sequentially as in Figure 12.2.

12.2 KNOWLEDGE-BASED DIGITAL SIX SIGMA

12.2.1 KDSS

Knowledge Management (KM) is a well-known terminology in management science. Hence the details of KM are not explained here. We believe that DT and KM are very important in this knowledge-based information society. If Six Sigma, DT and KM are combined, it could become a very powerful management strategy. We want to propose a so-called Knowledge-based Digital Six Sigma (KDSS) as the combination of Six Sigma, DT and KM.

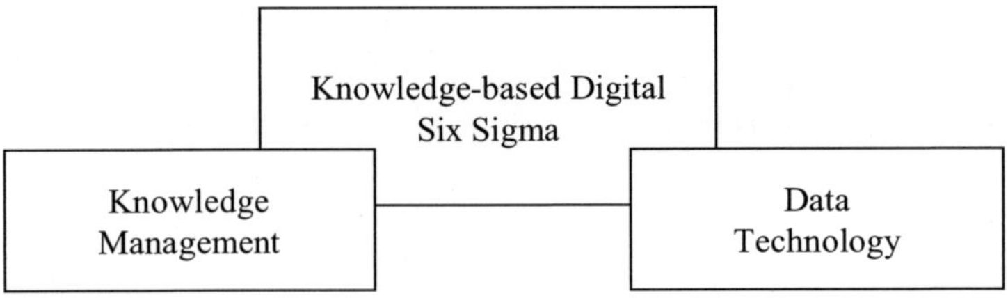

Figure 12.3 Knowledge-based Digital Six Sigma.

KDSS can be defined as "a company-wide management strategy whose goal is to achieve process quality innovation corresponding to 6σ level and customer satisfaction through such activities as systematic generation · storage · dissemination of knowledge by utilizing DT through statistical data bases, information creation and relevant tools". Since DT and KM can be useful only from appropriate utilization of computers and communication equipments of this digital society, we attached "digital".

As shown in Figure 12.3, there are some differences between Six Sigma and KM. However, there also exist some areas of intersections such as data acquisition and utilization, data analysis, generation of information, and so on.

KDSS is a combination of KM and DT and Six Sigma which can be developed as a new paradigm for management strategy in this digital society of the 21st century.

Some of the important methodologies in KDSS can be explained as follows.

(1) Process flow of improvement activities

In KM, it was proposed by Park (2002) that a good process flow of improvement activities is the CSUE cycle as shown in Figure 3. CSUE means Creating & Capturing, Storing & Sharing, Utilization and Evaluation. As explained previously, the well-known process flow of improvement activities in Six Sigma is MAIC. The Define (D) phase can be used in front of CSUE and MAIC. However, "Define" is omitted here.

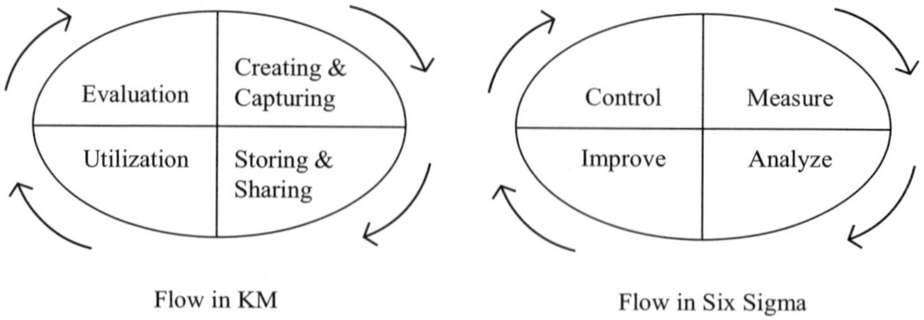

Figure 12.4 Process flow of improvement activities in KM and Six Sigma.

The CSUE and DMAIC cycles can be intermixed in order to create an efficient cycle in KDSS. One way is to use the DMAIC cycle in each step of CSUE, or to use the CSUE cycle in each step of the DMAIC cycle. We believe that CSUE and DMAIC are both complementary to each other.

(2) Project team activities

The project team activities by BBs and GBs for quality and productivity improvement are perhaps the most important activities in Six Sigma. If the concept of KM is added to these activities, more useful and profitable results could be made possible. We may call such activities KDSS project team activities. Through team efforts, we can create and capture information, store and share the information, and utilize it in the DMAIC process. Also by using the DMAIC process, we can create new information and follow the CSUE process.

(3) Education and training

Education and training is the most fundamental infrastructure in Six Sigma. A systematic training program for GB, BB, MBB and Champion levels is essential for the success of Six Sigma. Also in KM, without proper training, creation · storage · sharing · utilization would not be easy, and the process flow of knowledge would not be possible. It is often mentioned that the optimal education and training time in Six Sigma is about 5-7% of total working hours, and in KM it is about 6–8%. This means that more education and training time is necessary in KM than in Six Sigma. However, there is a lot of duplication in Six Sigma and KM, so the optimal education and training time in KDSS would be 8–10% of total working hours.

(4) Information management

Information in areas such as customer management, R&D, process management, quality inspection and reliability tests are essential elements in Six Sigma. In KM also, information management concerning storage, sharing and utilization of knowledge is the most important infrastructure. We believe that information management is essential in KDSS.

(5) Modeling and future prediction

Mathematical modeling based on past data is an important tool for prediction of a future outcome. DT is an essential tool for future prediction. Statistical methods such as regression modeling and time series modeling can be usefully utilized for future prediction. In Six Sigma, since future prediction is necessary, modeling and prediction can be very useful.

(6) Scientific tools

Basic QC and statistical tools such as 7 QC tools, process flow-charts, quality function deployment, hypothesis testing, regression and design of experiments can be used in KDSS. Also some advanced Six Sigma tools such as FMEA, benchmarking and marketing surveys can be effectively used in KDSS. These tools are helpful in analyzing data, obtaining information, statistical process evaluation and generating knowledge. We can say that KDSS is based on these scientific and statistical methods.

We believe that DT, KM and Six Sigma can be combined in a highly satisfactory way so that companies can adopt them and build an excellent management system.

12.2.2 ALTERNATIVE PROCESS FLOW IN KDSS

For implementation of KDSS, we want to propose that a good alternative process flow to DMAIC is the DMAMP cycle. It means Define, Measure, Analyze, Model and Predict. In each phase, the major activities are as follows.

(1) Define phase:
  - Project identification related to DT and KM
  - Project definition (project range and objective)
  - Project approval (action plan and Champion approval)

(2) Measure phase:
  - Identification of core Y's related to DT and KM
  - Finding of current sigma level, and setting up improvement level
  - Finding of potential X's

(3) Analyze phase:
  - Collection of data related to DT and KM
  - Analysis of data to find out a few vital X's
  - Ranking of the X's

(4) Model phase:
  - Modeling the function to explain the relationship between X and Y
  - Finding a model to predict the future of Y for given set of X's
  - Checking the predictability of the chosen model

(5) Predict phase:
  - Predicting the Y for a given condition of X's
  - Checking the goodness-of-fit of the predicted values of Y
  - Utilization of the predicted Y values

12.2.3 AN ILLUSTRATED EXAMPLE OF DMAMP CYCLE

A project team was formed in a petrochemical company in order to improve the color of a product called SW-kerosene which has been a major customer complaint. The color of the product is represented by a color saybolt number which is a "larger-the-better" characteristic. The process which produces the product is the kerosene sweetening process (U-1700 process) which is a key process of the company. The company adopted a Six Sigma strategy to improve its major quality problems, and formed several project teams to tackle some important quality problems. Under the guidance of authors, a project team decided to use the flow, DMADP, which is recommended in KDSS.

(1) Define Phase

The project team consisted of 5 members; team leader (black belt (BB)) and 4 green belts (GB). They decided to call the topic "Improvement of SW-kerosene color to satisfy customer requirements", and obtained Champion approval for project implementation. They started to investigate the current status of the problem. They set up the action plan, and planned to finish the project in 4 months.

(2) Measure Phase

The lower specification limit of the saybolt number for the product color is 20.0, and the sigma quality level of the saybolt number has been known to be about 3.2 sigma which is not at all satisfactory. They set up the goal to make the sigma quality level at least 4.5. For the core Y, which is the saybolt number, the team proposed 20 different possible independent variables for potential X's. However, by several brain-storming meetings, they finally decided to choose the following process variables which could influence the core Y.

X1 = Content of feed mecaptan (wt. ppm)

X2 = Air feed percentage (%)

X3 = Feed temperature (centigrade)

X4 = Feed endpoint (centigrade)

X5 = Input amount of catalyst (wt. ppm)

X6 = I/H content of petroleum (Vol. %)

The usual operating conditions for these input variables are X1 = 118, X2 = 129, X3 = 49, X4 = 265, X5 = 51, and X6 = 23, which were believed to be optimal.

The project team decided to collect the related data from the past two months to find the functional relationship between Y and X. They collected a total of 246 cases as shown in Table 12.2.

Table 12.2 Raw data for product color.

Data number	X1	X2	X3	X4	X5	X6	Y
1	101.6	130.5	46	247	62	100.0	21.1
2	91.5	135.1	50	251	70	94.2	20.3
3	90.8	135.4	49	245	68	94.5	19.8
4	80.8	156.0	45	248	66	99.6	22.4
5	82.8	145.5	48	253	64	98.0	21.5
.	.	.	.	.	.	.	.
.	.	.	.	.	.	.	.
.	.	.	.	.	.	.	.
243	119.5	119.0	51	271	55	88.3	20.0
244	105.9	133.5	52	260	47	78.9	22.8
245	110.0	133.2	50	269	54	80.1	23.0
246	101.6	133.8	51	273	60	77.8	22.3

(3) Analyze Phase

In order to find out the correlation relationship between the variables, the correlation matrix in Table 12.3 and the mean and standard deviation in Table 12.4 were obtained.

First of all, they found that the normality assumption for the color data was not rejected. From Table 12.3, they found that the variable X2 has a significantly big negative correlation with Y, and the variables X3, X4, X5 and X6 also have significant correlation with Y at 5% significant level. They also found in Table 12.4 that, since the lower specification limit of Y is 20.0, the defective rate is about 3.75% which indicates that the sigma level is 3.28.

Table 12.3 Correlation matrix.

	X1	X2	X3	X4	X5	X6	Y
X1	1.0	-0.175	0.532	0.632	-0.553	0.154	-0.075
X2		1.0	-0.077	-0.343	0.251	0.393	-0.748
X3			1.0	0.553	-0.562	0.166	-0.188
X4				1.0	-0.763	-0.340	0.171
X5					1.0	0.285	-0.264
X6						1.0	-0.360
Y							1.0

Table 12.4 Mean and standard deviation.

	Mean	Standard deviation
X1	117.29	22.81
X2	128.27	16.31
X3	49.33	5.02
X4	264.00	8.81
X5	50.95	15.62
X6	23.04	34.12
Y	22.80	1.57

(4) Model Phase

In order to find a good model to explain the functional relationship between Y and X, the stepwise regression procedure was used, and stepwise models were selected as shown in Table 12.5. Note that the order of inclusion in the models for the input variables is X2, X3, X5, X1 and X6. Note also that the variable X4 was not statistically entered in the stepwise regression procedure.

Among the 5 regression equations, listed in Table 12.5, the team decided to choose the third equation

$$\hat{y} = 55.952 - 0.136\ X2 - 0.269\ X3 - 0.067\ X5. \qquad (12.1)$$

Since the variables X1 and X6 are not easily controlled by engineers, and the coefficient of determination of the model (12.1) is 0.680, it is significant enough for a model.

Table 12.5 Results of stepwise regression.

Step	Input variables	Regression equation	R^2
1	X2	$\hat{y} = 40.538 - 0.145$ X2	0.560
2	X2, X3	$\hat{y} = 48.697 - 0.149$ X2 $- 0.156$ X3	0.620
3	X2, X3, X5	$\hat{y} = 55.952 - 0.136$ X2 $- 0.269$ X3 $- 0.067$ X5	0.680
4	X2, X3, X5, X1	$\hat{y} = 58.682 - 0.139$ X2 $- 0.218$ X3 -0.084 X5 $- 0.339$ X1	0.725
5	X2, X3, X5, X1, X6	$\hat{y} = 65.046 - 0.153$ X2 $- 0.265$ X3 -0.111 X5 $- 0.045$ X1 $- 0.021$ X6	0.755

(5) Predict Phase

Since the response variable Y is a "larger-the-better" characteristic, we want to find the operating condition of (X2, X3, X5) which maximizes Y-hat under some possible operating ranges of X2, X3 and X5. Because of operability and economic reasons, the possible operating ranges of these variables are $120 \le X2 \le 160$, $45 \le X3 \le 55$, and $40 \le X5 \le 70$. The value of $\hat{y}$ is maximized when X2, X3 and X5 take the lowest possible values. These values are X2 = 120, X3 = 45, and X5 = 40. If these values are inserted into the equation (1), the estimated value of Y is $\hat{y} = 24.85$.

The new optimal operating condition for the input variables are X1 = 118, X2 = 120, X3 = 45, X4 = 265, X5 = 40, and X6 = 23. The team tested the product color 5 times in this new condition, and found that the sample mean is about 25 with standard deviation 1.50. If this result is correct, the defective rate for the product color is 0.000434, which is equal to 4.83 sigma quality level. The quality level of 4.83 is greater than 4.5, which was the goal for this project. The team achieved the goal.

12.3 SIX SIGMA FOR SERVICE INDUSTRY

12.3.1 TRANSACTIONAL/SERVICE SIX SIGMA

As mentioned earlier, Six Sigma in a big manufacturing company is composed of three parts: DFSS, manufacturing Six Sigma and Transactional Six Sigma (TSS). However, there are many service companies that deal only with all kinds of service works. In this section, TSS, including service Six Sigma, will be discussed.

Although the Six Sigma approach to quality and process improvement has been used predominantly by manufacturing organizations, currently the popularity of Six Sigma in service organizations is growing exponentially, especially in banks, hospitals, financial services, the air line industry and utility services, to just name a few. The objective of a Six Sigma strategy in service processes is to understand how defects occur and then to devise process improvements to reduce the occurrence of such defects which improve the overall customer experience and thereby enhance customer satisfaction. Traditional manufacturing companies are taking their Six Sigma experiences and moving the applications of Six Sigma to their service operations. Service industries adopting a Six sigma business strategy will have the following benefits (Antony, 2005):

- Improved cross-functional teamwork throughout the organization.
- Transformation of the organizational culture from fire-fighting mode to prevention mode.
- Increased employee morale.
- Reduced cost of poor quality.
- Improved consistency of service through systematic reduction of variability in processes.
- Effective management decisions due to reliance on data and facts rather than assumptions and gut-feelings.

12.3.2 WHY IS SIX SIGMA REQUIRED IN THE SERVICE INDUSTRY?

Research has shown that most service processes such as shipping, billing, invoicing, payroll, customer order entry, etc. are performing at less than a 3.5 sigma quality level, with a defect rate of 23,000 ppm or 97.7% yield. If we improve the sigma quality level of the above processes to a 4 sigma quality level, the defect rate will drop to 6210 ppm. This clearly indicates a 3.7 fold improvement in process performance. The process yield will be increased to 99.38%. This would bring the significant financial returns to the bottom line of any organization engaged in continuous improvement programs such as Six Sigma. Service processes create scrap and rework in the form of cost of poor quality just like manufacturing processes (Bisgaard and Freiesleben, 2004). Six Sigma in service processes can be used to reduce the costs of poor quality so that a more consistent process for service delivery can be achieved.

Another important reason for the introduction of the Six Sigma strategy in many service companies is that customers of today feel "process variability" in the delivery of the service provided and not just on "process mean". The objective of a Six Sigma strategy is to reduce process variability around the acceptable target service performance.

It is quite common view among many people engaged in service processes that Six Sigma requires complicated statistical tools and techniques. The truth is that Six Sigma is not a collection of tools and techniques. Service organizations need more basic tools of Six Sigma. The majority of process related problems in service organizations can be tackled using process mapping, Pareto analysis, cause and effect analysis, control charts, etc.

12.3.3 COMMON CHALLENGES AND POTENTIAL IMPEDIMENTS FOR SERVICE SIX SIGMA

There are various challenges when applying Six Sigma in the service sector (especially in healthcare industry and financial services). Some of the potential difficulties and challenges encountered when implementing Six Sigma in a service environment include the following.

(1) Data not readily available for analysis or plenty of data available but not sure as what to do with the gathered data. The accuracy and completeness of the data is another major challenge in the context of service processes.

(2) In most service processes, the customers are themselves the supplier and thus the organization has low bargaining power. Therefore service organizations need to demonstrate to their clients "what is in there for them" in order to get them involved in the data collection process.

(3) The measurement of customer satisfaction in a service environment is more difficult due to the human behavioral interaction associated with the delivery of service. Human behavioral characteristics such as eagerness to help, honesty, friendliness, etc. are thought to have major influences on service processes that determine the quality of service provided to customers.

(4) It is often difficult in service organizations to identify processes which can be measured in terms of defects per million opportunities or mistakes per million opportunities, etc.

(5) The cultural change in a service-focused environment is more difficult to achieve than many manufacturing companies due to the high involvement of soft issues such as leadership, communication, teamwork, etc.

(6) The data collection strategy in a service-focused organization is quite weak and in many cases the data collection process takes more time due to the need for a large sample size for statistical robustness reasons.

(7) Service processes are subject to more noise or uncontrollable factors (psychological factors, sociological factors, etc.) as compared to manufacturing processes.

(8) In service industry, we do not describe activities in process terms. This is a major problem in many service environments. The use of process mapping and value stream mapping are quite uncommon in many service processes.

(9) A transfer function is not explicitly defined for many service-oriented processes and hence the application of advanced statistical techniques such as design of experiments is limited.

(10) Service processes in general are much more dependant on human and organizational change than the changes to manufacturing processes. Changing a machine parameter on a machine is far different from adjusting work procedures or tasks.

12.3.4 SIX SIGMA FOR IT/SOFTWARE BUSINESS

Gillies (1997) pointed out that software is fundamentally different from other processes, particularly manufacturing, due to various reasons:

- Software has no physical existence.
- Software process uses fault tolerance as opposed to design tolerance used in manufacturing.
- The process variation within software development can never be eliminated or even reduced below a moderate level. No two modules are alike so process performance always includes an intrinsic degree of variability.
- There are significant differences in skills and experience from one developer to another.
- Lack of knowledge about clients' needs at the beginning.

The behavior of a software process is difficult to predict compared to the constrained, limited and highly predictable behavior of a milling machine. Software cannot be simply measured like any physical units (e.g. length, thickness, etc.). In other words, meaningful measurements such as what constitute a defect in software development process is often a difficult task compared to manufacturing process. Software development processes can be fully characterized by three simple measurements (Janieszewski and George, 2003):

Time — the time required to perform a specific task.
Size — the size of the product produced.
Defects — the number of defects, the type of defects, time to eliminate defects, etc.

Some barriers to successful application of Six Sigma to software are:

- Standard Six Sigma training does not directly relate to software development.
- Software is different and software developers cannot bridge the gap on their own.
- Too many competing initiatives — total quality management, ISO 9000, Six Sigma, etc.
- Six Sigma projects are not considered integral to software development.
- Business as usual after training — no time for continuous improvement.

12.3.5 MEASUREMENT AND PROJECT TEAM ACTIVITIES

Many companies have learned a key lesson in their implementation of Six Sigma: successful outcomes are very often produced in transactional processes such as sales, purchasing, after-service and financing. However, arriving at a meaningful definition of defects and collecting insightful metrics are often the biggest challenges in transactional and service processes. Projects involving these processes sometimes lack objective data. When the data do exist, the practitioner is usually forced to work with attribute data such as pass/fail requirements or number of defects. Teams should strive for continuous data over attribute data whenever possible, since continuous data provide more options in terms of the available statistical tools and yield more information about the process with a given sample size.

In transactional/service projects, a process may be defined and a goal can be set but frequently without a set of known specification limits. Setting a goal or soft target as a specification limit for the purpose of determining the process capability/performance indices can yield only questionable results. It requires persistence and creativity to define the process metrics that yield true insight into transactional/service processes. However, many of the "low-hanging fruit" projects can be successfully attacked with some of the 7 QC tools: cause-and-effect analysis, histogram, Pareto diagram, scatter-diagram or simple graphs. These tools can help teams determine where to focus their efforts initially while establishing the data collection system to determine the root cause of the more difficult aspects of a project.

The correlation/regression or DOE (design of experiments) techniques are frequently associated with manufacturing processes, but they can provide significant benefits to transactional/service projects as well. A well-designed DOE can help establish process parameters to improve a company's efficiency and service quality. The techniques offer a structured, efficient approach to experimentation that can provide valuable process improvement information.

As mentioned in Chapter 11, the suggested process of the project team activities in Six Sigma is DMAIC, and often DMAD(O)V process is often used. Both flows of DMAIC and DMAD(O)V could be used in service industry. However, from the authors' experience, a modified flow, which is called DMARIC, is also very powerful in service industry. In Table 12.6, the suggested actions in each phase of DMARIC are listed.

Table 12.6 Suggested actions in each step of DMARIC project team activities.

Phase	Actions
Define (D)	1. Define the scope and surrounding conditions of the project. 2. Identify critical customer requirements and CTQ*y*'s. 3. Check the competitiveness of the CTQ*y*'s by benchmarking. 4. Describe the business impact of the project.
Measure (M)	1. Identify the project metrics for the CTQ*y*'s. 2. Measure the project metrics, and start compiling them in time series format by reflecting the long-term variabilities. 3. Address financial measurement issues of project.
Analyze (A)	1. Create a process flowchart/process map of the current process at a level of detail that can give insight into what should be done differently. 2. Create a cause-and-effect diagram or matrix to identify input variables, CTQ*x*'s, that can affect the process output, CTQ*y*. 3. Rank importance of input variables using a Pareto diagram. 4. Conduct correlation, regression and analysis of variance studies to gain insight into how input variables can impact output variables.
Redesign (R)	1. Consider using DOEs to assess the impact of process change considerations within a process. 2. Consider changing work standards or process flow to improve process quality or productivity. 3. Determine optimum operating windows of input variables from DOEs and other tools.
Implement (I)	1. Set-up the best work standards or process flow. 2. Test whether the optimum operating windows of input variables are suitable, and implement them. 3. Verify process improvements, stability, and performance using run charts.
Control (C)	1. Update control plan. Implement control charts to check important output and input variables. 2. Create a final project report stating the benefits of the project. 3. Make the project report available to others within the organization. 4. Monitor results at the end of 3 and 6 months after project completion to ensure that project improvements are maintained.

12.3.6 SIX SIGMA FOR E-BUSINESS

Recently, e-business (electronic business) has been rapidly increasing and it is of great interest to consider Six Sigma for e-business. The author would like to call Six Sigma for e-business "e-Sigma". It is clear that the ultimate management concept of e-Sigma should be customer satisfaction. There are 4 ingredients for customer satisfaction management. They are labeled CQCD, which stands for convenience, quality, cost and delivery. To have an excellent e-Sigma system for providing convenient, high-quality, low-cost products, and accurate and speedy delivery, the following e-Sigma model should be established in e-business companies.

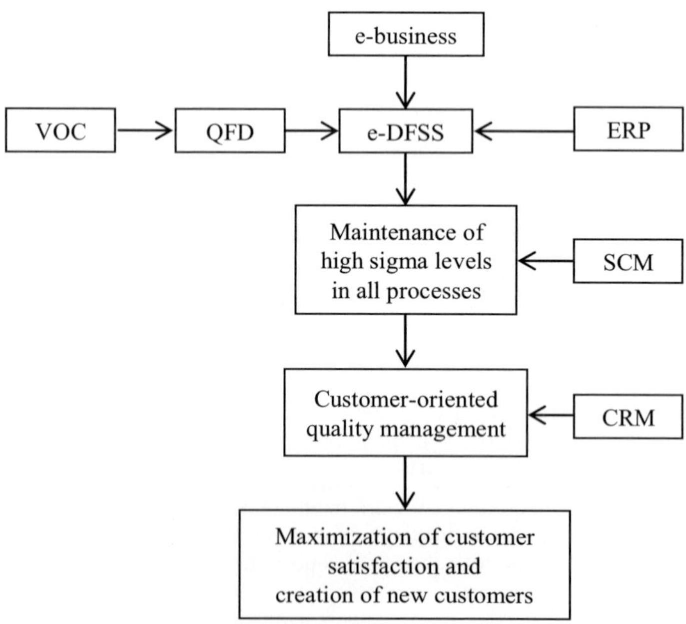

Figure 12.5 e-Sigma model.

The voice of customer (VOC) should be input into DFSS by using QFD, which converts VOC to technical requirements. These technical requirements are reflected in design aspects for Six

Sigma, which may be called e-DFSS (electronic design for Six Sigma). An ERP scheme which is suitable for the e-business should be employed to manage necessary resources. Also an efficient SCM is required for systematic acquisition, handling, storage and transportation of products. In all processes of an e-business, the sigma level of each process should be evaluated and improved to assure high-quality performance of each process. For customer-oriented quality management, CRM is required in e-business. Eventually, such e-Sigma flow will guarantee a high-level customer satisfaction and simultaneous creation of new customers.

12.4 BLACK BELT TRAINING

12.4.1 A GOOD BLACK BELT TRAINING PROGRAM

Depending upon each company, the content and duration of a black belt course could be different. Most Korean companies take four five-day sessions and one final graduation day. The duration is usually four months; one week for one session and three weeks for the practice period in each month. Hence, it takes four months. Usually a project is carried out during the four-month period, and a certified examination is conducted before graduation. Also a homework assignment is given after each session. On the final graduation day, the project is presented and the black belt certification is awarded. The following are the major contents of the four sessions.

(1) **First session** (Focus on Define & Measure in DMAIC)
- Introduction to Six Sigma: The history, definition, philosophy and major strategies of Six Sigma.
- Basic statistics: Basic descriptive statistics, PPM, DPMO, DPO, DPU, continuous data, normal distribution, Z-transform.
- The 7 QC tools.

- Six Sigma statistics: Sigma quality level, process capability, rolled throughput yield, attribute data, Poisson and binomial distributions.
- Advanced statistics: Concept of statistical estimation and hypothesis testing, t-test, confidence interval, F-test, case studies and exercises.
- Correlation and regression analysis: Theories and case studies.
- Benchmarking.
- Costs of poor quality (COPQ): Quality costs, hidden factory.
- Long-term quality management: Measure process performance and case studies.
- Homework (or project) assignment (between first and second session): Several homework exercises can be assigned to make use of the above methodologies. For example:
  1. Select a process with a chronic problem which has been awaiting a solution for a long time where a certain economic advantage is to be gained by improvement. Run a project, first using the 7 QC tools and show an economic advantage.
  2. Measure process performance of at least three different characteristics and compute the sigma quality level for each one and the combination of the three characteristics.
  3. Run a regression analysis for a process, find significant factors and suggest improvements with a cost reduction potential.

(2) Second session (Focus on Analyze in DMAIC)

- Review of homework assignment.
- Understanding variation, quality and cycle time.
- Process management: Principles and process flow charts.
- Measurement evaluation analysis.
- Introduction to design of experiments (DOE): Full factorial design and fractional factorial design.

- DOE, introduction and software: Exercises with Minitab, JMP and others.
- Quality function deployment (QFD).
- Reliability analysis: FMEA (failure mode and effects analysis).
- Homework assignment (between second and third session):
  1. Find a process where a certain economic advantage is to be gained by improvement. Run a full factorial with two or three factors.
  2. Collect VOCs (voice of customers) and, using QFD, find CTQs which you should handle in your process.

(3) **Third session** (Focus on Improve in DMAIC)

- Review of homework assignments.
- DOE: ANOVA, p-value, Robust design (parameter design, tolerance design).
- Response surface design: Central composite designs, mixture designs.
- Gage R&R test.
- Six Sigma deployment.
- Six Sigma in non-manufacturing processes: Transactional Six Sigma methodologies.
- Homework assignment (between third and fourth session): Select a process with a chronic problem in CTQ deployment. Screen important factors by regression analysis, optimize the process by using a robust design or a response surface design.

(4) **Fourth session** (Focus on Control in DMAIC)

- Review of homework assignments.
- Control charts.
- Statistical process control.
- DFSS (design for Six Sigma).
- Black belt roles: Job description of BBs.

- Six Sigma and other management strategies: The relationship of Six Sigma to ISO 9000, TQC, TQM, ERP, and other management strategies.
- Six Sigma in a global perspective.
- Group work (evening program): Why is Six Sigma necessary for our company?
- Homework assignment (between fourth and graduation): Take a project where the economic potential is at least $50,000 in annual cost reduction and complete the project.

(5) Graduation

- Review of homework assignments.
- BB certified test.
- Presentation of the projects completed.
- Graduation ceremony.

12.4.2 JOB DESCRIPTION OF A BLACK BELT

The role of a BB is very critical for the success of Six Sigma. The job description of a BB could be different from company to company, but the following is a general guideline for the job description of a BB.

- Lead a project improvement team, and also lead a focused effort to systematically assess the performance of business systems and processes (measure DPMO).
- Apply the Six Sigma skills and tools to analyze inadequate processes and recommend solutions for improvement.
- Communicate the plans, methods and the results achieved in a documented fashion in regularly scheduled meetings.
- Provide training and consultation to local personnel in Six Sigma strategies and tools.

12.5 A PRACTICAL FRAMEWORK FOR SIX SIGMA IMPLEMENTATION

This section of the chapter focuses on a practical framework which can assist people in organizations with little expertise and knowledge in the implementation of Six Sigma business strategy. If Six Sigma is poorly launched, it will be very difficult to reorganize and regain momentum. As defined by *Cambridge Advanced Learner Dictionary* (2005) framework is "a supporting structure around which something can be built; a system of rules, ideas or beliefs that is used to plan or decide something". A sound framework should first assess the current state of the organization, i.e. what an organization does or tries to do, followed by analyzing the steps taken to do it in a correct sequence (Struebing and Klaus, 1997). Adopting Six Sigma as a business improvement methodology is not sufficient to emulate the success of the key players. Evidence indicates that failure to select a change management program based on a structured approach or framework can lead to malpractices, fire-fighting and sub-optimization of resources (Davies and Kochhar, 2000). Some of the reasons why an implementation framework is needed are:

- It provides a focus on the objective to be achieved and link the change management initiative to business objectives.
- It minimizes fire fighting and provides a guideline or structure that are easy to understand, efficient and can be implemented at reasonable costs and time.
- It forces the management to address a substantial list of key issues which otherwise might not be addressed.
- It also provides an avenue for managers who are interested and enthusiastic in quality improvement to provide essential leadership for the change initiative agenda.
- It facilitates the unification of several disciplines in the change process.

The framework recommended in this chapter encompasses *four phases*:

12.5.1 PHASE 1 — MANAGING THE CHANGE

The preliminary stage of understanding the need for change and commitment from the top management are vital steps which forms the foundation of whole Six Sigma framework. The framework begins with the objective or identification of the need for Six Sigma and exploring its potential impact on the company. Conducting the business review and gap analysis would facilitate in identifying the strengths and weaknesses of the organization and also determining the potential change needed. It is imperative to note that if a business is going through restructuring that requires significant layoffs or a merger that creates uncertainty and too little executive attention, then Six Sigma is not a way forward.

Six Sigma works best with a top-down approach when the CEO and senior management team own it, support it, and drive it (Pande et al., 2000; Antony and Banuelas, 2002). This is the very important aspect to start with because many companies have failed in their attempt to implement Six Sigma either due to lack of commitment at the managerial level or due to lack of strong leaders to drive the initiative. It is essential that senior management approves the Six Sigma initiative, defines the purpose and scope of Six Sigma, and links it to the mission and vision of the organization. Leadership from top of the organization is necessary to define, prioritize and construct the quality and process management thinking in the organization.

To constantly maintain competitive edge and effectively transfer knowledge throughout the organization, it is important to start training at the top of the organization and should then be cascaded down through the organizational hierarchy (Antony and Taner, 2003). Education and awareness sessions will secure the senior management commitment and faith in the initiative. The action plan for this step is listed below:

- Educate senior management team with Six Sigma executive or champion training.
- Identify and select executive leaders to drive the Six Sigma initiative.

- Provide an awareness to all senior management team about the benefits realized by world-class companies (similar to their businesses) after Six Sigma implementation.

12.5.2 PHASE 2 — INITIALIZE

It is indispensable to attract the best people with good leadership skills to be involved in the company's first wave of training on Six Sigma (Pande et al., 2000; Pyzdek, 2003;). There are three major reasons why top talent is so important to the Six Sigma effort (Snee and Hoerl, 2003):

- The better the talent, the better the result.
- Top talent attracts more top talent.
- Top talent becomes the next organizational leader.

Start planning at the process level to identify potential starting points for Six Sigma Black Belts. Identification of core processes is an area of potential bottleneck for many organizations because, if the core processes are not known, the rest of the framework is difficult to implement (Oakland, 2003). It is preferable to launch Six Sigma by focusing on a few strategic areas (Snee and Hoerl, 2003). It is equally important to establish the process performance measures, which could be used to define what's important for success and are used to select project.

Projects are the core activity driving the change in Six Sigma organizations. As quoted by Snee and Hoerl (2003), "*Project selection is a critical component of success. Often the battle is lost before we have even begun due to selection of poor projects*". Some early wins are crucial and necessary to buy-in management and company-wide commitment for the initiative. Primarily, initial pilot projects should focus on key problem areas with strategic alignment in terms of high customer satisfaction and critical to business success in terms of faster or larger financial return (Jackenthal, 2004). The project should also be completed with reasonable time span of less than 4 months. The success of pilot projects would act as a model for the rest of the organization to follow. It is desirable that finance department is involved from the beginning of project to ensure that the cost-benefit analysis is carried out for each Six Sigma project and to ensure that the savings generated to the bottom-line are realistic.

12.5.3 PHASE 3 — INSTITUTIONALIZE

All financial savings generated from pilot projects should be reported upon project completion. Awareness and recognition of all employees is achieved through the communication of Six Sigma savings generated from projects. Senior Management team should communicate to the organization at all levels as why they chose to deploy Six Sigma, what they hope to get out of this initiative and where it will take the organization to (Snee and Hoerl, 2003). Different communication strategies or media such as intranet, newsletter, bulletin board, etc. can be used.

A sound organizational infrastructure for each of the Six Sigma roles (Six Sigma Deployment Champion, Master Black Belt (MBB), Black Belt (BB), Green Belt (GB), Yellow Belt (YB)) is a key element of institutionalizing the Six Sigma initiative. It is not only important to allocate resources for a wave of mass training, but also to identify all the training needs of all the roles, and put together a sustained, ongoing system to continuously satisfy these needs in the most efficient way possible (Snee and Hoerl, 2003). Training should not just be short-term but should involve educating on the long-term basis, with regular training follow-ups and briefings (Antony and Taner, 2003). Training should be provided across the department and also to the suppliers, who play the key role as a supply chain partner of the larger counterpart.

12.5.4 PHASE 4 — SUSTAIN

Sustaining the improvement is a challenging task, where many organizations struggle due to change in business focus or fluctuation in economy. To keep the momentum going, it is necessary to develop generations of managers, who not only understand but are dedicated to the pursuit of continuous improvement in meeting external and internal customer needs (Oakland, 2003). Management should believe in the power of "intrinsic motivation" (self-motivation) rather than solely relying on "extrinsic motivation" (coerced or bribed to do it).

A progression towards becoming a learning organization would ensure Six Sigma companies sustain benefits in long-term. Some of the key elements that an organization should focus on to become a learning organization are:

- Regular project reviews meeting and briefings to enable both management and employees to share experiences and progress on projects, and factors critical for its success and failure (Senge *et al.*, 1999).
- Individual and organizational learning should be actively encouraged (Kaye and Anderson, 1999).
- Regular review of employees' training needs, evaluation of their performance and feedback for improvement.

It has been observed that communication and leadership commitment play key roles in all the above four phases. A sound strategy for implementation is always necessary and decisions about rolling the program in segment or as a whole is often determined by the culture within the company, time available and the current needs of the organization. In implementing the framework, it is important that communication of information is effective and that accountability is explicit and supported by committed senior management team and leaders of the organisation. Top leaders should make themselves available for staff, keep in touch with staff and communicate all managers about the long-term strategies of the organization. Middle Managers play a key role in the success of any change program (Snee and Hoerl, 2003; Kaye and Anderson, 1999) and their confidence in the initiative is equally important.

12.6 KEYS FOR SIX SIGMA SUCCESS

From the authors' consulting experience for Six Sigma, it is believed that the following points are the keys for a Six Sigma success. The points could be slightly different depending on the type of company business. However, the general ideas remain applicable to all types of businesses.

(1) Get the top managers involved.

Until senior managers of the corporation or business unit really accept Six Sigma as part of their jobs and as the company's management strategy, the true importance of the initiative will be in doubt and the energy behind it will be weakened.

(2) Keep the message simple and clear, and request the participation of all employees.

Since Six Sigma is a new management strategy, the core of the system and the vision for Six Sigma should be simple, clear, meaningful and accessible to everyone. While new vocabulary and skills are obviously part of the Six Sigma discipline, beware of the possibility of alienating some people by the strange terms and jargons that could create "classes" in a Six Sigma environment. Participation of all employees in the Six Sigma efforts is essential for a Six Sigma success.

(3) Select the right Six Sigma projects and train an adequate number of full-time BBs to concentrate on project team efforts.

Project selection is perhaps the most critical activity in launching a Six Sigma project. Well-selected and well-defined improvement projects equal better and faster results. In selecting project themes, a top-down approach based on the company's CTQ deployment is often used. In running the project teams, it is recommended that the BBs become the full-time leaders who can concentrate their entire efforts to the team project for a success.

(4) Focus on short-term results and long-term growth.

It is very stimulating to have initial achievements in the first four to six months. Hence, focusing on short-term results at the beginning is a good strategy. However, it is also important to balance the push for short-term results with the recognition that those gains must lay the foundation for the real power of Six Sigma. Creation of a more responsive, customer-focused, and successful company for the long-term is the major source of Six Sigma success.

(5) Publicize and award results, and admit setbacks.

Recognize and celebrate successes, but pay equal attention to challenges and disappointments. Don't expect that Six Sigma will work perfectly in your company. Be ready to continuously improve and even redesign your Six Sigma processes as you progress.

(6) Develop your own style toward Six Sigma.

Your themes, priorities, projects, training, and structure — all should be decided based on what works best for you. Develop your own style toward Six Sigma based on your company's culture and habits, if there are any. Setting up a "Six Sigma day" each month to evaluate the progress of Six Sigma, and to publicize and reward results is a good idea, if your company's culture suits this.

(7) Link customers and your processes.

Customer satisfaction is one of the core elements of the Six Sigma approach. To ensure customer satisfaction, there must be a way to link customers and your processes efficiently to build an excellent Six Sigma system.

(8) Make learning an ongoing activity, and make an investment to make it happen.

A few months of training, however intensive, won't cement all the new knowledge and skills needed to sustain Six Sigma. Making learning a continuous and ongoing activity is necessary. Without time, support and money, the habits and existing processes in your business won't change much. You have to make an investment to make it happen.

(9) Use Six Sigma tools wisely.
There are many tools available in Six Sigma. However, very often, no single tool can create happier customers or improve profits. Statistics can answer questions, but can't solve all possible problems. Your success with Six Sigma will depend on applying all the methods wisely, in the right balance, to maximize your results. In general, using the simplest tool that works — not the most complex — should be highly valued.

13

Design for Six Sigma

13.1 DFSS FRAMEWORK

13.1.1 WHAT IS DFSS?

Design for Six Sigma (DFSS) is the part of Six Sigma which is employed for R&D activities. DFSS is a systematic methodology that uses project management discipline and tools to optimize the design process of products, processes or services in order to achieve excellent designs to maximize customer value with high quality and low vulnerability.

The major objective of DFSS is to "design it right the first time" to avoid lengthy R&D time and costly downstream experiences. In general, two major design vulnerabilities may affect the quality of a design entity:

(1) System design vulnerabilities due to violations of design principles such as customer value, innovative idea generation, and cost effectiveness.
(2) Parameter and tolerance design vulnerabilities due to lack of robustness in the noises. Elimination or reduction of operational vulnerabilities through parameter and tolerance designs is the objective of quality initiative in Six Sigma as well as in quality engineering.

The DFSS objective is to attack the design vulnerabilities, both conceptual (system design) and operational (parameter and tolerance designs), by deriving and integrating tools and methods for their elimination and reduction. Operational vulnerabilities takes variability reduction and mean adjustment of the critical-to-quality (CTQ) requirements, and have been the subject of parameter designs described in Chapters 5 through 8, and the subject of tolerance design in Chapter 9. Note that tolerance design is also important in operational vulnerabilities as it deals with the assignment of tolerances in the design parameters and process variables, the assessment and control of manufacturing processes, the metrological issues, as well as the geometric and cost models.

Organizations that have adopted the principles and concepts of the Six Sigma methodology become aware that, once they have achieved high sigma capability level (generally about 5 sigma capability), the only way to progress further is to redesign their products, processes or services. This has led to the development of what today is called DFSS. DFSS is a systematic methodology that uses tools, strategies, tactics, project management, to optimize the design process during the designing of products, services and processes, in order to achieve superior designs to maximize customer value at Six Sigma quality levels (Yang and EI-Haik, 2004). DFSS is a process to define, design and deliver innovative products, processes and services that provide competitively attractive value to customers in a manner that achieves the critical-to-quality characteristics for all the significant functions (Watson, 2005).

Design for Six Sigma has its roots in systems engineering. Much of the learning that underpins systems engineering evolved under the guidance of the Department of Defence and NASA. DFSS is the application of Six Sigma principles to the design of products and their manufacturing and support processes. It is a powerful approach to designing products, processes or services in a cost-effective and simple manner to meet the needs and expectations of the customer while driving down costs. It is important to note that DFSS is not a replacement for New Product Development Process (NPDP), rather it is a methodology to make the introduction of products, processes and services more efficient, reliable and capable of meeting customer expectations and requirements.

DFSS has the potential to simplify design configuration, eliminate non-value added processes in the design of a product or service and hence reduce labour costs, material costs and overhead costs. It provides a systematic integration of tools, techniques, methods and team members throughout product, process and service design. The DFSS initiatives vary dramatically from company to company, however, it often starts with a charter linked to the organization's strategic plan, followed by an assessment of customer needs, functional analysis, identification of critical-to-quality (CTQ) characteristics, concept selection, detailed design of products, and so on. Table 13.1 shows some of the fundamental differences between Six Sigma for manufacturing or service parts and DFSS.

Table 13.1 Fundamental differences between Six Sigma and DFSS.

Six Sigma for manufacturing/service parts	Design for Six Sigma (DFSS)
Focus is on detecting and resolving problems in processes or products or services.	Focus is on preventing problems.
Cultural change is fairly greater.	DFSS requires much greater cultural change than Six Sigma.
More reactive.	More proactive.
Savings generated from Six Sigma projects can be quantified rather quickly.	Benefits from DFSS projects are more difficult to quantify and tend to be much more long term.
Looks at existing processes and fixes problems.	Focuses on the up-front design of the product and process.

13.1.2 DFSS PROCESSES

The phases which are used for DFSS processes are basically as follows.

(1) DMAIC: Define, Measure, Analyze, Improve, Control
(2) DMAD(O)V: Define, Measure, Analyze, (Optimize), Verify
(3) DIDOV: Define, Identify, Design, Optimize, Verify (or Validate)

(4) DMEDI: Define, Measure, Explore, Develop, Implement
(5) DCOV: Define, Characterize, Optimize, Verify
(6) DIDES: Define, Initiate, Design, Execute, Sustain

The first process, DMAIC, has been fully explained before. This process can be used widely whether the process is manufacturing, R&D, or non-manufacturing (service). In principle, DMAIC is used when we need some improvement to the already existing system or process. However, all the other processes are used when we need some new design, new product or new system.

Based on the authors' consulting experiences, it is not easy for a company to adopt DFSS. However, once it is fully adopted, the net effect and cost savings can be enormous. Figure 13.1 shows DMADOV or DIDOV process which is quite effective in a research institute. Many excellent companies including Samsung and LG Electronics are using this process. Note that in the "Define" phase, project planning and approval is the major step with defining relevant product and technology contents. Very often feasibility study is necessary at this phase. In the "Identify" phase, which is in fact "Measure and Analyze" phase, to identify the necessary CTQs, gather data and analyze/obtain necessary information are the things to be done. In the "Design" phase, initial design (system design), product and process design (parameter design) are the major things. And in the "Optimize" phase, optimization of product or process is done through tolerance design or simulation. Finally in the "Verify" phase, the previous DMADO phases are verified through reliability test, and a process control system is established for mass production.

Note that DIDOV is very much similar to DMADOV. In fact, "I" in DIDOV implies "MA" in DMADOV, and the other initials (D, D, O, V) are the same. According to Atwood (Atwood, 2005), if DMAIC is excluded, DMAD(O)V is the most used DFSS process, and the share of companies which use this process is about 62%.

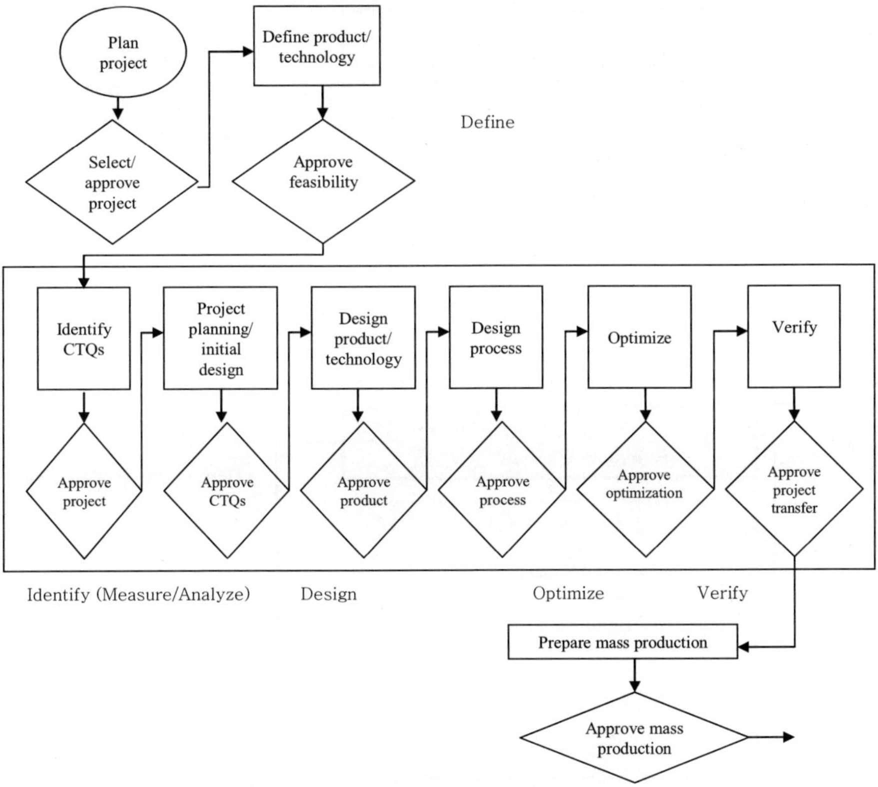

Figure 13.1 A typical DFSS process.

13.1.3 MAJOR ACTIVITIES OF DMADOV PROCESS

Figure 13.2 shows major activities and methods in each phase of DMADOV which are quite effective in a research institute. The major activities are listed in the center column of Figure 13.2, and the related methods are listed at the right hand side of this figure. Since there are many methods available in each activity, just a few methods are mentioned in this figure.

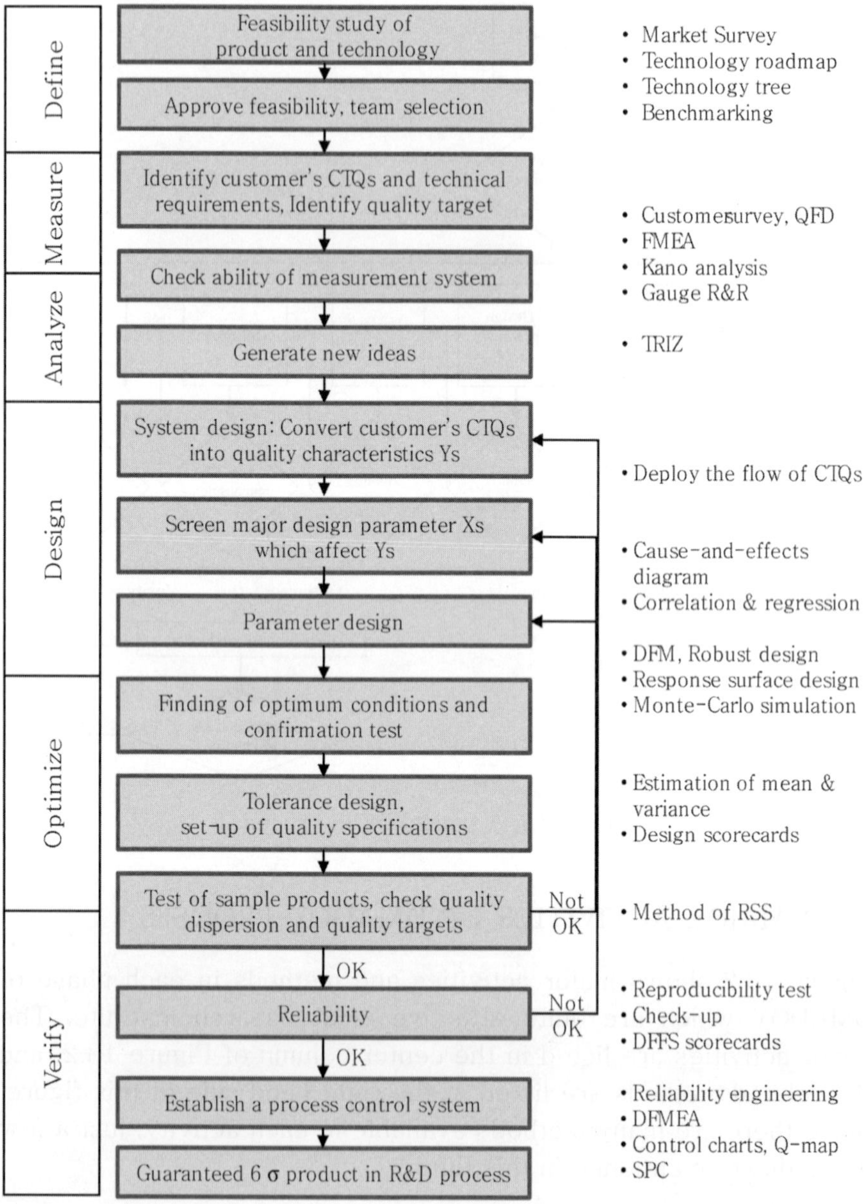

Figure 13.2 Major activities and methods in each phase of DMADOV process.

The major activities and methodologies used in each phase can be explained as follows.

(1) Define

This is the same as "Define" phase of the DMAIC process. This phase is concerned with selection of a project theme, and the team to tackle the selected theme. For selection of a theme, relevant technology roadmap and technology tree are considered, and a feasibility study is conducted if necessary. It is also concerned with benchmarking of key product or process characteristics of other world-class companies.

(2) Measure and Analyze (Identify)

This phase is to identify customer and business requirements, to translate customer requirements to product/process functional requirements, and to collect necessary data. The following is the list of tasks in this phase.

- Obtain customer needs and wants and transform them into the voice-of-customer (VOC) list.
- Translate the VOC list into product/process functional and measurable requirements.
- Clear definition of the design requirements for the product/process.
- Planning of functional and engineering requirements.
- Identify important customer's critical-to-quality (CTQ) with quality target, critical-to-delivery (CTD), critical-to-cost (CTC), and check ability of measurement system.
- Generate new ideas.

DFSS tools used in this phase include market/customer survey, quality function deployment (QFD), benchmarking, Kano analysis, feasibility study and risk analysis, Gauge R&R, and TRIZ.

(3) Design

This phase is to generate and evaluate design alternatives. After the identification of the functional requirements for the new design product/process, we need to develop design product/process alternatives that will be able to deliver those functional requirements. Then we need to evaluate the design alternatives, and make a final decision on which concept will be used.

Many methods can be used in design evaluation, including the Pugh concept selection technique, robust design, design reviews, design for manufacturability (DFM), design vulnerability analysis (El-Haik 1996, Yang and Trewin 1999), and FMEA. If a process is designed, process management techniques are used as an evaluation tool.

Some of the important tasks which should be involved in this phase are as follows.

- Select the design concept — select the best design concept from many alternatives.
- Utilization of concurrent engineering practice.
- Study of the relation of design parameters to CTQs at sub-levels in complex processes or systems.
- Perform the risk analysis — identify the risks involved and typical failures using design failure mode and effect analysis.
- Perform the engineering analysis and select materials.
- Establish the DFSS score card — record the CTQs, specifications, design capability data, measurement system capability, design parameters, transfer functions Y = f(X), where Y's = customer requirements or CTQs and X's = design parameters.

(4) Optimize

This phase is to optimize design product/process with all functional requirements released at the Six Sigma performance quality level. As the system design is finalized, there are still a lot of design parameters that can be adjusted and changed. With the help of computer simulation and hardware testing, DOE modeling, Taguchi's robust design methods, and response surface methodology, the optimal parameter settings will be determined. Usually the parameter optimization is followed by a tolerance optimization, which is to provide a logical and objective basis for setting manufacturing tolerances.

Some of the important tasks which should be involved in this phase are as follows.

- Identification of sources of variability.
- Minimizing product performance sensitivity to all sources of variation using robust design.
- Application of tolerance design for critical design parameters identified from the parameter design.
- Optimizing the design for assembly, manufacturability, reliability, serviceability, etc.
- Determination of design capability and comparison with design specifications.

(5) Verify

This phase is to verify and validate the design after the parameter and tolerance design is completed. For product/service that goes to market, pilot test and refining is necessary. Here design failure mode-effect analysis (DFMEA), as well as pilot and small-scale implementations, can be used to test and evaluate real-life performance.

In this phase the process controls in manufacturing and production are established in order to ensure that critical characteristics are always produced to specification of the optimization phase. As the design product/process is validated and process control is established, full-scale commercial rollout can be handed over to process owners with requirement settings and statistical process control (SPC) and monitoring systems.

13.1.4 BENEFITS FROM DFSS

The following benefits can be gained from the application of DFSS principles to a new product or service development process.

(1) Superior customer value and robust performance of the product/service can be intrinsically determined in the early design stage.
(2) Reduced time to market for new or revised products/services.
(3) Reduced life cycle costs associated with products/services.
(4) Poor design concepts adopted at early design stage is easy to correct at early stage of product/service development cycle, but is very costly to correct at later stage. DFSS ensures the designers select the best design concepts at the early stages of the design.
(5) Enhanced ability to manage risk in design processes of products/services.
(6) Reduced warranty costs.
(7) Improved communication among functions such as marketing, design, product/service development and manufacturing.
(8) Reduced number of design changes or iterations and hence reduced number of prototypes made during the design stage.

13.1.5 WHEN DO WE CHOOSE DMAD(O)V?

In research world, the choice between DFSS and Six Sigma is still unclear to many practitioners and academics. Many organizations use the rule of thumb promoted and advocated by Harry and Schroeder (2000) which clearly states that the only way to surpass the 5 sigma quality level is to re-design products, processes or services by means of DFSS. There is not enough data or evidence to date to support this claim. The decision to choose DMAD(O)V is fundamentally dependent upon a number of variables such as complexity, risk involved, time, cost, customer demands, etc.

Two projects in two different industry sectors have been examined using action research as a method of investigation. It can be concluded that the decision as to when to employ a redesign approach in a Six Sigma project is characterized by the following issues:

- The "5 sigma quality wall".

- A number of factors affect the decision to improve or re-design.
- Multi-criteria decision analysis technique such as AHP.
- Some tools and techniques in DMAD(O)V are very useful.

13.1.6 SOME PROBLEMS FOR DFSS IMPLEMENTATION

There are several problems to be tackled for DFSS implementation. These problems must be solved for a smooth introduction of DFSS. Some of the problems are as follows.

(1) Researchers tend to resist introduction of any new scientific methodology into their research activities. Hence, their understanding and cooperation or approval should be sought before introducing the DFSS into their activity.
(2) GB or BB education/training is especially necessary, since there are many scientific tools for R&D including QFD, DOE, simulation techniques, robust designs and regression analysis. For such education/training, textbooks which contain real and practical examples should be carefully prepared in order to make researchers understand why DFSS is a very useful tool.
(3) Project team activities are usually not popular in R&D departments. In this case, BB should be assigned as full-time project leaders. It is desirable that the company gives time, space and necessary financial support to the BB to solve the projects.

13.2 CASE STUDY OF DMADOV PROCESS

Here CPT means Color Picture Tube. Samsung SDI is one of the two initiators of Six Sigma in Korea. When the company applied for a National Six Sigma Quality Award In 2000, it submitted a book entitled "Six Sigma case studies for quality innovation". This book contains the 10 most remarkable results obtained by Six Sigma project teams. One DFSS (R&D Six Sigma) case study is presented here. The team consisted of 8 persons (one is a champion, and the other 7 members are all BBs). The duration of this study was from January to June of 2000. The team basically used the DMADOV process. Table 13.1 shows the project implementation steps used by this team.

Table 13.1 Project implementation steps of a DFSS team.

DFSS steps	Detailed steps	Tools used	Design review
D (Define)	• Analysis of CPT market trends • Preparation of customer value map • Organization of team • Theme selection of CPM flow-down	• Customer review • Business planning • QFD • Concept engineering	DR1
M (Measure)	• Selection of project CFR • Selection of Omega CPT CFR • Measurement analysis	• CPM, FMEA • MSA	
A (Analyze)	• Failure analysis • List of all input variables	• Benchmarking & gap analysis • Cause & effect matrix	
D (Design)	• Design of basic shape and Decision of prototype • Tolerance analysis for yield improvement	• Simulation, Capability study • Tolerance design	DR2
O (Optimize)	• Determination of big Xs which influence Y • Determination of optimal levels of big Xs • Quality check through pilot study • Completion of paper design	• DOE & ANOVA • Robust design • DFM	DR3
V (Verify)	• Verification for mass production • Analysis of process capability • Determination of final product quality	• Process mapping • Capability study • Reliability study	DR4

Abreviation: CPM = Critical parameter method, QFD = Quality function deployment, CFR = Critical functional responses, FMEA = Failure mode and effect analysis, MSA = Measurement system analysis, DR = Design review, DOE = Design of experiments, ANOVA = Analysis of variance, DFM = Design for manufacturability.

(1) Define

The current management strategy of Samsung SDI is to have four number 1 products in the world. In order to have the world's best CRT, customer needs must be met. The major customer demands for a new CRT are as follows.

- Slim (short back length)
- Larger scale and really flat
- High-quality screen performance
- HD resolution
- Long life and quick start

To meet the above customer demands, it was necessary to develop a new product, called Omega CPT. The key problems to be solved for the above demands are as follows.

Slim — The short length increases deflection angle and decreases I/S (inner shield) height. The Omega CPT is sensitive to external magnetic field. Hence, the key issue is to minimize the influence of the external magnetic field.

High resolution — High resolution decreases spaces between stripes, which makes the pitch small. The small pitch makes the landing shift of the electron beams large. Hence, the key issue is to minimize the landing shift.

To solve this problem, a cross-functional team (CFT) was formed as described above.

(2) Measure

The technology relating to magnetic shields for solving the above is to consider the design of the inner shield material and inner shield shape as shown below.

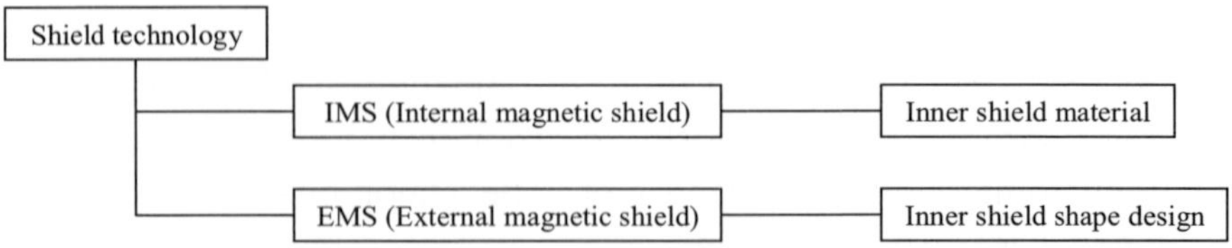

In order to determine the critical parameters, the following critical parameter method (CPM) was used, and the inner shield was identified as the major critical parameter.

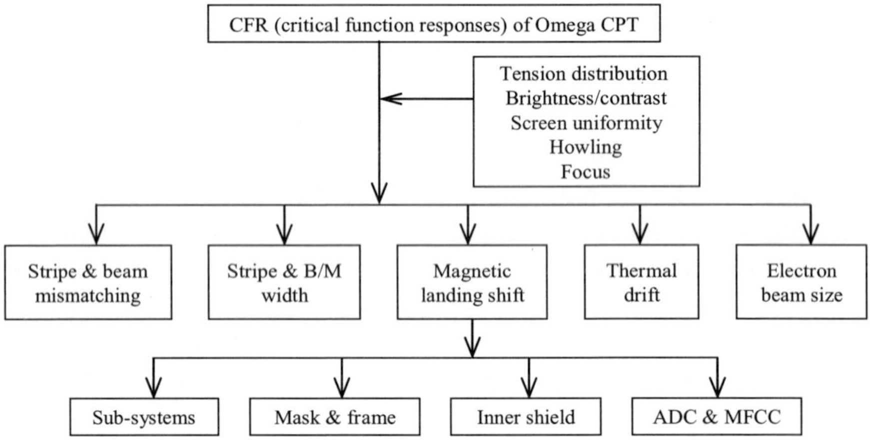

(3) Analyze

In order to control the inner shield, the magnetic landing shift had to be minimized. However, the landing shift was directly related to CFRs of the design of the inner shield. The goals for the magnetic landing shift are as follows.

	Magnetic shift	Yield	Sigma level
Current level	C	C	1.25
First goal	B	B	4.38
Final goal	A	A	6.00

(4) Design

What are the key parameters of the inner shield for minimizing the magnetic landing shift?

To determine the parameters, the flowchart of the design process (Figure 13.3) for the inner shield was sketched.

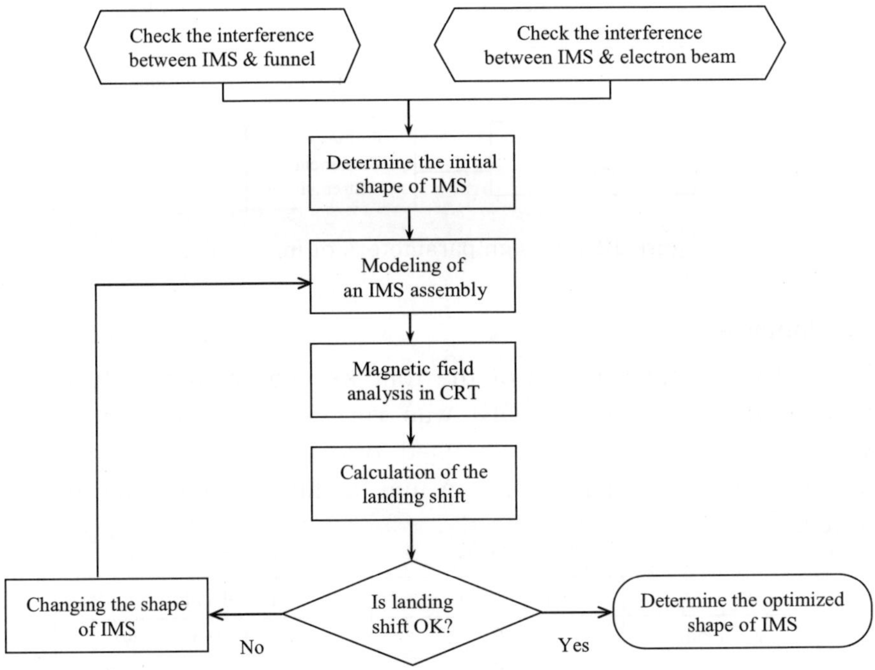

Figure 13.3 Design process of inner shield.

The design parameters of the inner shield are listed in Figure 13.4 according to the sub-system level CFRs, shape, and material.

A cause-and-effect matrix and an engineering simulation study were made to select the critical parameters. As a result, four parameters (material, hole size, height, V-notch) were selected.

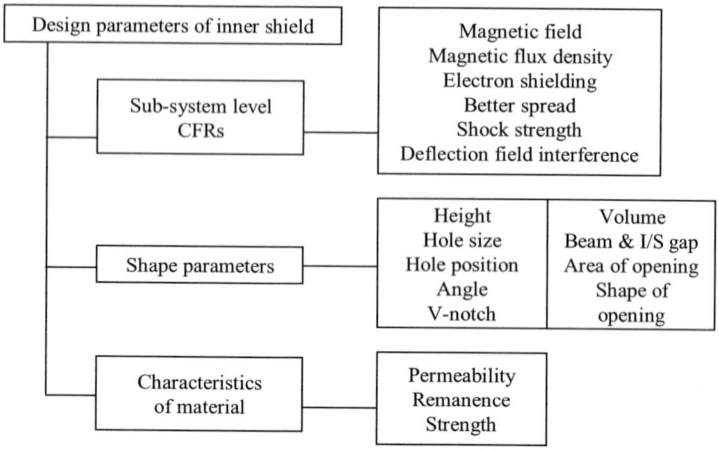

Figure 13.4 Design parameters of inner shield.

(5) Optimize

To find the optimal levels of the four key parameters selected, a design of experiments (DOE) was run. The levels investigated were as follows. The levels used originally were IV (old) for material, medium for hole size, A mm for height, and B mm for V-notch.

Factors	Number of levels	Level values
Material	2	IV (old), POS (new)
Hole size	3	Large, medium, small
Height	3	A mm, B mm
V-notch	3	A mm, B mm, C mm

The total number of treatment (factor) combinations could be as many as $2 \times 3 \times 3 \times 3 = 54$, which is too many in practice. Hence, $L_{18}(2^1 \times 3^7)$, which is an orthogonal array was used and a total of 18 treatment combinations were run. The experimental results and the analysis are not given here. However, the optimal levels were found to be POS (new) for material, small for hole size, A mm for height, and C mm for V-notch.

(6) Verify

A confirmation test was attempted to validate the results of DOE and the optimality was confirmed. Finally, a cost/benefit analysis was made and the manufacturability and productivity were studied to prove all were satisfactory. Thus, the first goal of this project (magnetic shift B, yield B, and Sigma level 4.38) was achieved. From this, the cost reduction was estimated to be $0.2/each, which is equivalent to $0.25 million per year.

13.3 CASE STUDY OF DMAIC PROCESS

13.3.1 DESIGN FOR AN EFFICIENT COMPUTERIZED CONTROL SYSTEM

(1) Define

There were many materials necessitating storage on a shelf for some period of time to be subsequently used to create various products. Each material had its own specified shelf life-time depending upon whether it was stored in a refrigerator or not. Some frequently used materials and their specifications are listed in Table 13.2. The shelf life-time was counted from the manufactured date.

However, due to poor storage conditions and other reasons, the shelf life-times became short, and they could not be used in good conditions. Such SSLM resulted in some COPQ, environmental pollution and additional test expenses.

(2) Measure

During the period of July to December, 2001, scrap materials were found during the process of manufacturing many products. Table 13.3 shows the scrap materials for the product, stator bar and connecting ring.

Table 13.2 Stored materials and their specified shelf life-times.

Name of material	Storage in a refrigerator		Storage in a storeroom	
	Shelf life-time	Storage condition	Shelf life-time	Storage condition
Mica paper				
tape(#77865)				
Mica paper	6 month	below 7°C	3 month	below 23°C
tape(#77906)	6 month	2-10°C	3 month	18-32°C
Gl yarn flat tape	1 year	below 5°C	3 month	18-32°C
prepreg	6 month	2-10°C	2 month	18-32°C
Mica M tape(#77921)	6 month	below 10°C	2 month	18-32°C
Modified epoxy varnish	1 year	2-10°C	6 month	18-32°C
Polyester resin-35%	6 month	2-10°C	1 month	18-32°C
Epoxy impreg fiber	10 month	2-10°C	3 month	18-32°C
cloth	10 month	2-10°C	3 month	18-32°C
Pa-polyster resin	1 year	2-10°C	3 month	18-32°C
Pb-catalyst	1 year	2-10°C	3 month	below 25°C
Polyester comp	1 year	2-10°C	3 month	18-23°C
Glass cloth & tape				
Transposition filler				
......				

Table 13.3 Scrap materials in the stator bar & connecting ring.

Name of material	Purchase quantity	Scrap			Cause of scrap			
		Quantity	Unit	Number of times	Change of manufacturing schedule	Earlier purchase	No control of storage	Etc.
Modified epoxy varnish	92	31	GL	2				
Epoxy impeg fiber cloth	7,890	120	SH	2	1	1	2	
Glass cloth & tape	958	84	RL	5	2	1	1	1
Transposition filler	4,118	55	LB	1			1	
....								

The products/processes which were of concern are listed in Table 13.4 and their current process capabilities at the time are shown in the same table.

Table 13.4 Current process capabilities.

Product/Process	Defect	Unit	Opportunity	Total opportunity	DPU	DPO	DPMO	Process capability (sigma level)
Stator bar & connecting ring	61	12	50	600	5.083	0.101	101,667	2.77
Stator W'g ass'y	148	13	71	923	11.385	0.160	160,347	2.49
Lower frame A.	31	14	8	112	2.214	0.277	276,786	2.09
Rotor coil A.	4	17	5	85	0.235	0.047	47,059	3.17
Total	244			1,720		0.142	141,860	2.57

Defect: Over shelf life-time of SSLM
Unit: 4 items categorized in the processing using SSLM
Opportunity: Quantities of SSLM used in unit
Total opportunity = Unit * Opportunity
DPU = Defects/Unit
DPO = Defects/Total opportunity
DPMO = DPO/1,000,000
Short-term capability = Long-term capability + 1.5

(3) Analyze

In order to discover the sources of defects and variation, a cause-and-effect diagram was sketched by the team as shown in Figure 13.5.

In the past six month period, the total defect count on the materials was 244, and the Pareto diagram for the types of defects is shown in Figure 13.6.

Figure 13.6 shows that the insufficient control of SSLM in the storehouse accounted for 42.2% of the total and the unexpected change of manufacturing schedule was responsible for 39.8% of the total defects.

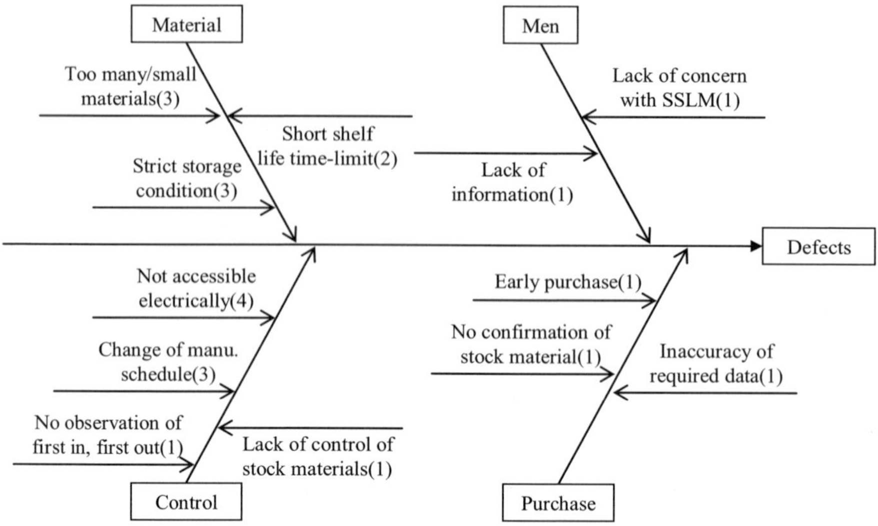

Figure 13.5 Cause-and-effect diagram for SSLM.

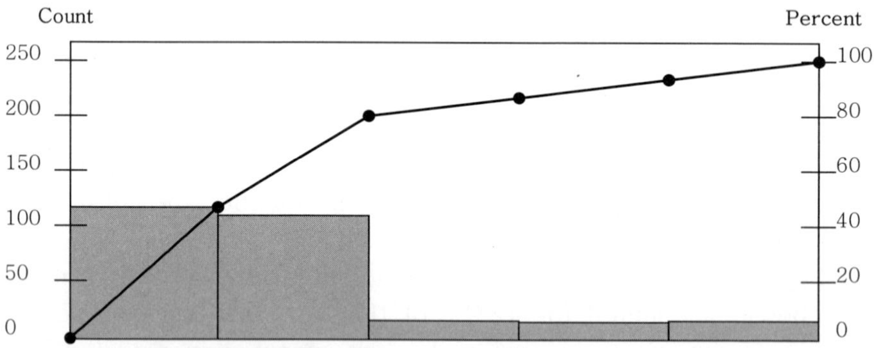

Defect pattern	Lack of control of stock materials	Change in manufacturing schedule	Strict storage condition	Too many/ small storage materials	Others
Count	103	97	15	14	15
Percent	42.2	39.8	6.1	5.7	6.2
Cum. %	42.2	82.0	88.1	93.8	100.0

Figure 13.6 Patterns of defects.

(4) Improve

In order to reduce the defects of SSLM, the computerized inventory control system was redesigned to increase the control efficiency of SSLM. The current process after the redesign looks as follows.

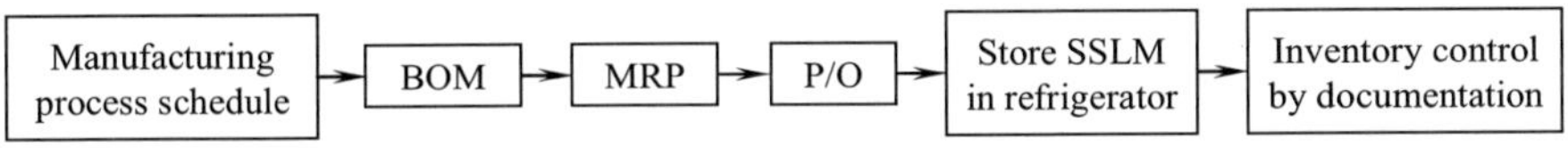

In this current process, there is no tool for checking and monitoring SSLM, and no one is assigned for checking the defects. The redesigned and improved process (Figure 13.7) makes cross checking of the manufacturing schedule in advance possible. Also, the related departments can monitor and control SSLMs through the on-line system.

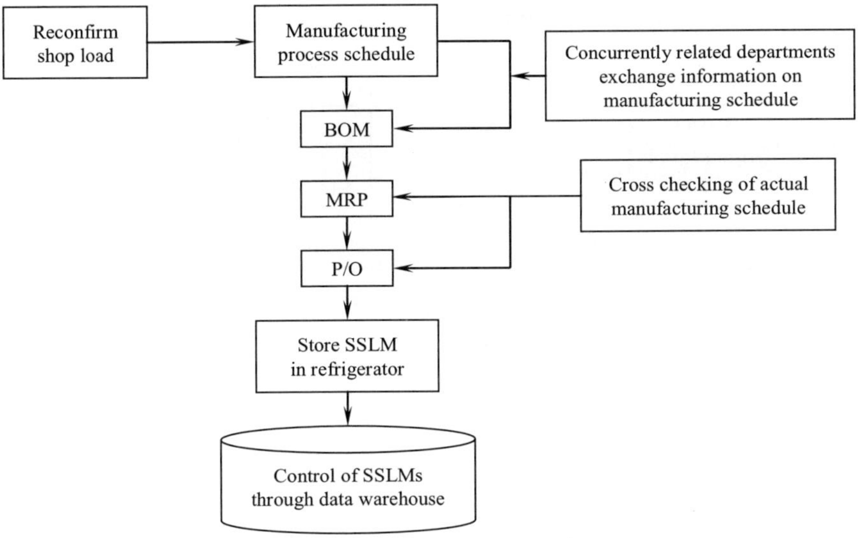

Figure 13.7 Redesigned process for SSLMs.

By practicing the improved process, they could obtain the following data for the first three months after the start of its use for SSLMs.

Product/ Process	Defect	Unit	Opportunity	Total opportunity	DPU	DPO	DPMO	Process capability (sigma level)
Stator bar & connecting ring	4	13	24	312	0.308	0.01282	12,820	3.73
Stator W'g ass'y	3	7	45	315	0.429	0.06667	66,670	3.00
Lower frame A.	1	6	3	18	0.167	0.05556	55,560	3.09
Rotor coil A.	0	10	5	50	0	0	0	6.00 (estimated)
Total	8			695		0.01151	11,510	3.77

The quality performances of the old and newly improved processes were compared as follows, clearly showing the impact of the Six Sigma team activities.

	Before improvement	After improvement
DPMO	141,860	11,510
Sigma level	2.57	3.77
COPQ	$190,000/year	$15,400/year (estimated)

The estimated savings by this project is about $174,600/year. However, the impact was bigger on the management side rather than the financial side.

(5) Control

In order to maintain the benefits, the team decided to follow the following control procedures:

- Update the SSLM instruction manual, and check the manual every six months.

- Educate the workers on SSLM information every month.
- Monitor related data through the on-line computer system every other month.

13.3.2 CONCLUSION OF CASE STUDY

Doosan is now one of the best companies in Korea, which is employing Six Sigma for their management innovation initiative. Many project leaders in manufacturing and non-manufacturing areas are using DMAIC for their project solving. However, in Doosan, many leaders in R&D parts are using DMADOV (Define, Measure, Analyze, Design, Optimize, Verify). It seems that both processes are working well in most project problem solving.

By the big help of Six Sigma, Doosan is emerging as a super power in heavy industries. Doosan's recent achievements include the LWR (Light Water Reactor) Project in North Korea, the nuclear power project for Qinshan of China, the Al-Shoaiba Desalination Plant in Saudi Arabia, the large-scale desalination plants in Al-Taweelah of the United Arab Emirates, and the 2002 World Cup Stadiums in Seoul.

13.4 CASE STUDY OF PRODUCT DESIGN THROUGH RSM

13.4.1 PRELIMINARY STUDY BY FACTORIAL DESIGN

This section presents a simple case study in the domestic appliance industry (named Company ABC from now onwards) in which designers employed sequential experimentation to find the design parameters that provide the "most desirable" response. The first part of the case study shows a full factorial 2^k design of experiments for product characterization. After that, a response surface methodology (RSM) is employed to a more general approximation of the response than the linear model. Finally, desirability function is employed to identify the combination of design parameter settings that jointly optimize the two quality characteristics.

In this case study, a sequential experimentation was carried out to investigate three design parameters potentially affecting the energy consumption of a domestic appliance. Sequential experimentation allows building a transfer function incrementally without wasting experimental resources. Accordingly, at each step of the sequential experimentations the model is evaluated statistically in order to know how well it models the behavior of the system under study.

In Company ABC, design and development department proposed the use of a new technology which has the potential of reduced energy consumption. This technology is composed by three main design parameters which might have some impact on the reduction of energy consumption. For confidentiality reasons, these design parameters are not disclosed and will be called A, B and C. A team of designers and product engineers decided to build a transfer function in order to finely tune and optimize the system's performance using sequential experimentation.

According to Montgomery (1997), the 2^k factorial design is particularly useful in the early stages of the experimental work, when many factors are likely to be investigated. Kiemele *et al.* (1997) recommend full factorial DOE when no more than four factors at two levels are studied. Following the above recommendations, the team has decided to run a full factorial 2^3 experiment (8 trials). Each trial condition was replicated twice to measure variation due to noise or lurking variables. The quality characteristic which is of interest to the team was energy consumption expressed in kWh/Kg. The results of the experiment are shown in Table 13.5.

Having collected the data from the experiment, next was to identify those design parameters (and their interactions) which have significant influence on energy consumption. Table 13.6 shows the estimates of effects, regression coefficients and P-vale associated with each effect (both main and interaction effects of design parameters).

Table 13.5 Results of full factorial experiment.

Run	A	B	C	Energy consumption
1	60	60	14	0.6830, 0.7056
2	60	60	42	0.7140, 0.7123
3	60	80	14	0.7120, 0.7165
4	60	80	42	0.7160, 0.7170
5	80	60	14	0.7060, 0.7120
6	80	60	42	0.6920, 0.7036
7	80	80	14	0.7170, 0.7245
8	80	80	42	0.7212, 0.7211

Table 13.6 Estimates of effects and regression coefficients.

Design parameters/interactions	Effect estimates	Coefficients	P-values
A	0.002637	0.001318	0.466
B	0.014612	0.007306	0.003 **
C	0.002587	0.001293	0.474
AB	0.002949	0.001475	0.417
AC	-0.007955	-0.003978	0.050 *
BC	-0.001259	-0.000630	0.724
ABC	0.007050	0.003525	0.075

** statistically significant at 1% significance level.
* statistically significant at 5% significance level.

Table 13.6 shows that the main effect of design parameter B is statistically significant at 1% significance level. Similarly the interaction between A and C is statistically significant at 5% significance level. The interaction between A and B as well as B and C are not statistically significant.

13.4.2 OPTIMIZATION BY RSM

Response Surface Methodology (RSM) can be employed to determine the settings of the design parameters that produce the optimal performance within the defined range of parameter settings. RSM can answer the following questions:

(1) What settings of the design parameters will yield a maximum or minimum response or quality characteristic?
(2) What is the local geography of the response surface near this maximal or minimal value?
(3) Will slight changes in the design parameters produce large changes in the response or quality characteristic?

The practical deployment of a response surface model often arises through sequential experimentation, i.e. a full or fractional factorial experiment to fit a first-order model, and axial and centre points are then added to allow the quadratic terms to be incorporated into the model. Kiemele et al. (1997) recommend Central Composite Design (CCD) or Box Behnken design when three factors are studied. In such designs, experimenters augment a factorial design by the addition of further runs involving "star" or "axial" points and centre points. The centre points are located in the midway between the low and high level settings of design or process parameters. The distance of the axial points (α) from the centre of the design cube can be calculated using the equation (Montgomery 1997):

$$\alpha = \sqrt[4]{n_F} \,,$$

where n_F = number of points in the factorial design.

For 3 design parameters (A, B and C) at 2-levels, there are 8 design points. Table 13.7 shows the results of a CCD with two quality characteristics (energy consumption and time).

The first step in this case study was to find a suitable approximation of the true functional relationship between the quality characteristic of interest and the set of design parameters. In this study, the team needed to determine the levels of design parameters that result in lower energy consumption at a reasonable operational time. Since each property is important in determining the quality of the appliance, the team considered these properties (quality characteristics) simultaneously. Optimal settings of design parameters for one quality characteristic may be far from optimal or even physically impossible for another quality characteristic. The team obtained the first order models for energy consumption and time.

Table 13.7 Results of a Central Composite Design.

Run order	A	B	C	Energy consumption	Time
1	70	70	4	0.699	114
2	70	70	28	0.703	115
3	60	80	14	0.712	111
4	70	70	28	0.709	117
5	80	80	14	0.717	112
6	80	60	42	0.692	120
7	70	70	28	0.709	116
8	80	60	14	0.706	123
9	53	70	28	0.701	115
10	87	70	28	0.698	115
11	60	80	42	0.716	112
12	70	53	28	0.693	126
13	80	80	42	0.721	112
14	70	70	28	0.696	114
15	70	70	28	0.712	117
16	70	70	28	0.699	115
17	60	60	42	0.714	125
18	70	87	28	0.720	110
19	60	60	14	0.683	120
20	70	70	52	0.707	118

The least square regression for "Energy Consumption" is:

$$\hat{y} = 0.704514 + 0.0004x_A + 0.0085x_B + 0.0028x_C - 0.0008x_A^2 + 0.001x_B^2 + 0.0004x_C^2 + 0.0011x_Ax_B - 0.0056x_Ax_C - 0.0011x_Bx_C. \tag{13.1}$$

The surface analysis is most of the times done in terms of the fitted surface. If the fitted surface is adequate approximation of the true response function, analysis of the fitted surface will be approximately equivalent to analysis of the actual system (Montgomery and Runger, 2003). Table 13.8 presents the estimated regression coefficients and the corresponding P-values for the energy consumption. The R^2 value for the model shown in equation (13.1) is 72.4%, which indicates that over 70% of the variability can be explained by the regression model. There was no real problem with the lack of fit. Moreover, a large difference in R^2 and adjusted R^2 value has been observed. This implies that some terms were not adding any value and if these terms were eliminated, a more meaningful model can be constructed.

Table 13.8 Estimated regression coefficients and corresponding P-values.

Terms	Regression coefficients	P-value
Constant	0.7045	0.000
A	0.000436	0.832
B	0.008524	0.002 *
C	0.002816	0.190
A * A	0.000829	0.679
B * B	0.001646	0.418
C * C	0.000408	0.838
AB	0.001125	0.676
AC	-0.005625	0.057 **
BC	-0.001125	0.676

* design parameter (or its interaction) is significant at 1% significance level.
** design parameter (or its interaction) is significant at 10% significance level

A simplified regression model for energy consumption is:

$$\hat{y} = 0.704514 + 0.0004x_A + 0.0085x_B + 0.0028x_C + 0.0056\, x_A x_C.$$

Similarly, a regression model for quality characteristic (time) has been developed (see below).

$$\hat{y} = 115.72 - 0.073\, x_A - 4.972\, x_B + 0.712\, x_C + 0.922 x_B^2 - 1.125\, x_A\, x_C. \tag{13.2}$$

The R^2 value for the model shown in equation (13.2) is 95% and the adjusted R^2 value was over 90%. This implies that we have captured all the important terms in the regression model. There was no real problem with the lack of fit.

In response surface, the response is usually represented graphically (Montgomery, 1997). Response surface plots such as contour and surface plots are useful for establishing desired response values and design parameters. In a contour plot, the response surface is viewed as a 2-dimensional plane where all points that have the same response are connected to produce contour lines of constant responses (Antony, 2003; Montgomery and Runger, 2003). Figure 13.8 shows a response surface plot for interaction AC with energy consumption as the quality characteristic of interest.

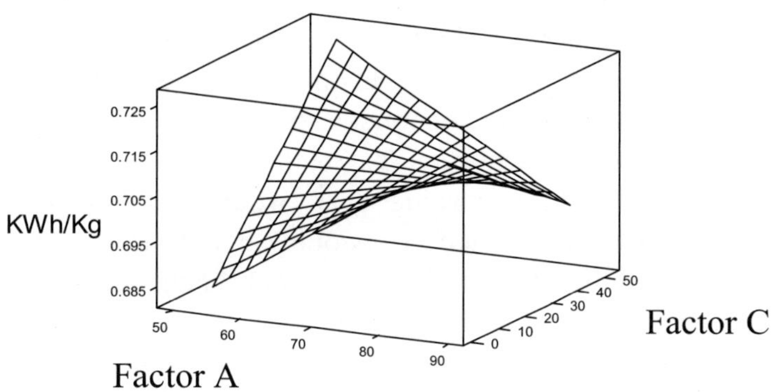

Figure 13.8 Response surface plot showing interaction AC.

The above figure suggests that both factors or design parameters A and C should be kept at low levels for minimizing energy consumption. A response surface plot displays a three dimensional view that may provide a clearer picture of the response than the contour plots (Antony, 2003). The next step was to optimize the quality characteristics energy consumption and time simultaneously. Here the authors proposed a desirability function approach for simultaneous optimization of these characteristics.

13.4.3 OPTIMIZATION OF MULTIPLE QUALITY CHARACTERISTICS

The desirability function approach was explained in detail in Section 10.6. The first step on the desirability function approach consists of conducting experiments and fitting response models for all k quality characteristics or responses. The second step involves defining individual desirability functions for each quality characteristic. Depending on whether a particular response or quality characteristic Y_i is to be maximized, minimized, or assigned a target value, different desirability functions $d_i(Y_i)$ can be used.

In this case, a smaller response is desired because if the responses are below a particular value, the situation is ideal and the desirability is 1 (i.e. the lower energy consumption the better). Decreasing cannot significantly enhance quality once the situation is ideal. Meanwhile, if a response exceeds the maximum acceptable level, the desirability is 0. The value of Y_i* gives the maximum acceptable value of $\hat{Y}_i$. In this case the team assigned a value of Y_i* for each of the responses, time and energy. $\hat{Y}_{time}$ has a $Y_{Time}*$ of 120 minutes, whereas $\hat{Y}_{energy}$ has a maximum $Y_{energy}*$ of 0.704. Therefore, a higher value of $\hat{Y}_i$ would result in an overall unacceptable product, since it would make $d_i = 0$ and thus composite desirability $D = 0$ (Derringer and Suich, 1980).

By employing a modified simplex algorithm (which maximizes the composite desirability to determine the numerical optimal solution (optimal factor settings)), the desirability function is calculated. After the individual desirability for each response is estimated, they are combined to provide a measure of the composite, or overall, desirability of the multi-response system (D). This measure of composite desirability (D) is the weighted geometric average of the individual desirabilities for the responses. Derringer and Suich (1980) used geometric mean of the individual desirabilities to estimate the overall desirability D:

$$D = (d_1 \times d_2 \times d_3 \times \ldots d_k)^{1/k},$$

with k denoting the number of responses. This function has values inside the interval (0, 1) and approaches unity as the desirabilities of the responses increases. The geometric mean is used because it vanishes whenever any d_i is equal to zero (Derringer and Suich, 1980 and Khuri and Cornell, 1996). The optimal solution (optimal operating conditions) can then be determined by maximizing the composite desirability.

However, the two quality characteristics are not equally important for the customer. This information can be incorporated into the optimal solution by setting unequal important values. Larger values correspond to more important quality characteristics and smaller values to less important quality characteristics. The team assessed the importance of each quality characteristic in order to assign appropriate values for importance and assigned a value of (2) for the energy consumption response and a (1) for the time response (values must be between 0.1 and 10). The composite desirability is then the geometric mean of the individual desirabilities (refer to Figure 13.9).

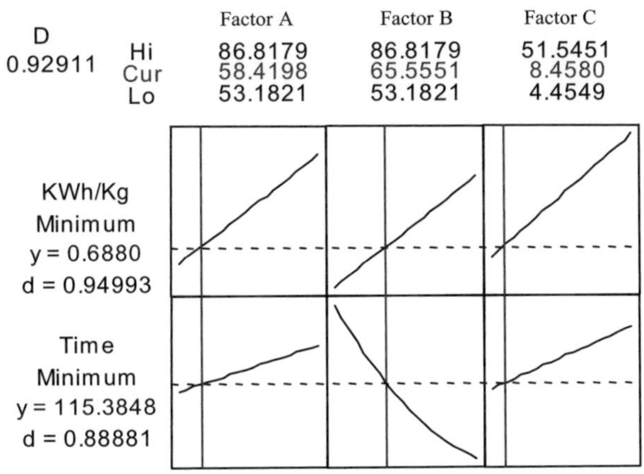

Figure 13.9 Individual desirabilities for quality characteristics.

As it can be seen from Figure 13.9, the overall desirability function is 0.929, which indicates a good multiple optimization of both quality characteristics. According to the experimentation carried out the design parameters should be set as follows:

Design Parameter A — 58
Design Parameter B — 65
Design Parameter C — 8

The above combination of design parameter settings has produced an energy consumption of 0.6880 kWh/Kg at 115 units of time. This represents an improvement of 6.7% (0.734/0.6880) on energy consumption due to the optimization of the new technology.

14

Robust Design and Implementation of Six Sigma

14.1 BARRIERS AND BENEFITS OF ROBUST DESIGN

The first part of this section will provide the readers with a good introduction to Robust Design (RD) and its role in general. The second part will be looking into how Robust Design (RD) can be integrated into the Six Sigma and Design for Six Sigma methodologies. RD is a powerful methodology intended as a cost-effective approach for improving the quality of products and processes. The methodology of RD was popularized by the Japanese quality engineering consultant Genichi Taguchi in Japan in the 1970s and the USA in the mid 1980s. In RD experiments, the purpose of the experimenters is to determine the optimal levels of controllable parameters of the process (or design) that will make the process performance (or design performance) insensitive to the variation caused by the uncontrolled process parameters called noise parameters. RD means systematic efforts to achieve insensitivity to noise factors. These efforts are founded on an awareness of variation and can be applied to all stages of product design (Arvidsson and Gremyr, 2006). For example, when designing an automobile engine, it is important that its performance is insensitive to the noise factors such as ambient temperature, relative humidity, etc.

Creating a robust product or process is like putting shock absorbers on a car. Shock absorbers enable the driver to get a smooth ride regardless of the conditions of the road. Achieving robustness is critical to minimizing quality loss advocated by Taguchi. Box and Jones (1992) provide an excellent example for RD by seeking the best recipe for a cake mix, so that the taste of the cake is insensitive to two uncontrollable factors; baking time and baking temperature. The idea of RD is to determine the best settings of the control factors (flour, egg powder, etc.) which would yield the best taste, regardless of the variations of uncontrollable factors.

Although the concept of RD has been used for many years, it is not widely accepted by many organizations in the European manufacturing organizations. Experimenters and scientists in the UK are far more likely to use the traditional one-factor-at-a-time (OFAT) approach to experimentation. Unlike European manufacturers, Japanese manufacturers make extensive use of RD techniques during product design/development phases. Nearly 80% of Japanese quality programs use experimental design methods (such as Robust Design) to optimize process performance (Studt, 1993). Although many engineers and managers in the UK manufacturing companies are familiar with experimental design, very few of them are actually applying this powerful methodology for problem solving and process optimization problems.

14.1.1 BARRIERS IN THE SUCCESSFUL APPLICATION OF ROBUST DESIGN

The following are some of the typical barriers in the successful application of RD methodology in the UK manufacturing organizations.

(1) Educational barriers

The educational barrier stems from the current statistical education received by the engineering community in the UK engineering and business schools. The courses in engineering statistics currently offered to engineering students are primarily focused on the mathematical aspects of probability and probability distributions. Very little attention is paid to the practicality and proven applied statistical methodologies such as design of experiments (DOE), RD, etc. Moreover, even if applied statistical methods are taught in the engineering schools, very little thought has been given to Six Sigma, Design for Six Sigma and Lean Thinking. A good introduction along with case studies will ensure engineering and business students have a better understanding of the application of various quality and process improvement tools and techniques before they enter the job market.

(2) Management barriers

Managers often do not understand the importance of RD in Six Sigma or Design for Six Sigma (DFSS) or appreciate the competitive value it brings. If managers view RD as a technical tool for professionals, then it will be relegated to an optional "tool box" status. On the other hand, if they view it as a management tool, then their leadership takes on a stronger role. In many manufacturing organizations, managers encourage the engineers to use their "home-grown" solutions for process and product related quality problems. These home-grown solutions are consistent with OFAT approach to experimentation as managers are always after fire-fighting and quick fix solutions which yield short-term benefits to their organizations.

(3) Cultural barriers

This is possibly one of the principal reasons for the uncommon applications of RD in the European industrial world. The management must be prepared to address all cultural barrier issues that might be present within the organization, plus any fear of training or reluctance to embrace the application of RD. Many companies are not culturally ready for the introduction and implementation of advanced quality improvement techniques such as RD.

(4) Communication barriers

Research has indicated that there is very little communication between the academic and industrial communities. Moreover, the communication among engineers, managers and statisticians in many organizations is not effective. For the successful introduction and development of any quality improvement initiative, these communities should work together and make this barrier less formidable. For example, lack of statistical knowledge for engineers could lead to problems such as misinterpretation of historical data or misunderstanding of the nature of interactions among process or design parameters. Similarly, statisticians lack of engineering knowledge of the process could lead to problems such as improper selection of design parameters for the experiment.

14.1.2 BENEFITS OF ROBUST DESIGN

Robust Design (RD) is extremely useful and valuable to any company, small or large, that must innovate, experiment, compete and change in order to survive. It is a powerful method for improving the competitiveness of any organization which is engaged in the journey of continuous improvement of product and service quality. The following benefits can be gained from the effective application of RD in organizations (Antony, 2002):

- Develop new products to markets faster than their competitors.
- Guarantee an increased market share due to shortened product-introduction cycle.
- Achieve better product definition characteristics.
- Minimize product sensitivity to everyday variations in manufacturing conditions.
- Obtain improved product design and thereby ensuring better product quality.
- Reduce the number of design changes in NPDP.
- Maximize benefit-cost ratio of experimental effort.
- Reduce the number of prototypes to be tested in a minimum number of experimental trials.

- Obtain a greater insight into the way a product design and development process works.
- Understand the most critical design parameters which affect the product performance or critical-to-quality characteristics.
- Reduce warranty and maintenance costs by improving product reliability substantially.
- Practically eliminating “design-trial-debug-modify” cycle.

14.2 CASE STUDY OF ROBUST DESIGN IN FIBER OPTIC SENSOR DEVELOPMENT

14.2.1 INTRODUCTION

Communications grade optical fibers, developed in the early 1970s, were born out of the demand to carry large amounts of information quickly and economically. Fiber optic sensor is a growing field spawned by the tremendous advances in optical fiber fabrication. Fiber optic sensors are being deployed to measure pressure, temperature, strain, acoustic waves, electromagnetic fields, and various chemical and biological agents, in a variety of industrial environments including oil wells at depths of over ten thousand feet; in jet engines to monitor both test and in-flight parameters; in power transformers to monitor dielectric breakdown; in civil structures to monitor bridges and buildings and along railroad systems to monitor track integrity, just to name a few.

Much of the work currently being performed on these optical fiber sensors is in the research and development stages at universities or small companies. Regardless of the type of optical fiber sensor or the application area, very little work has been done to design these sensors for robust manufacturing. Utilization of experimental design techniques for optimization of fiber optic sensors to ensure robustness against noise both in the manufacturing environment and the customer application environment has not been widely published. In this section, we report on the initial application of experimental design techniques in the research and development stage of optical fiber sensors to understand sources of variation and to factor these into the design changes. This case study is extracted from Chen, Pickrell and Antony (2004).

Micro-fabrication techniques have been developed that allow differential etching of the core with respect to the cladding of the fiber. The fiber optic sensor used in this study has been fabricated by chemically etching a pit in the core region of a single mode optical fiber and fusion splicing this fiber with another single mode fiber. The small air-gap produced in the fibers and the cleaved fiber end face form a Fabry-Perot interferometric cavity. Deposition of the polymer film onto the fiber end will cause an increase of the cavity length due to the similarity of the refractive index of the materials. The total optical cavity length is the sum of the fiber cavity length and the thin film thickness. The resolution of this sensor can be as high as 0.1 nm in determination of the cavity length. Figure 14.1 shows the optical sensor system and enlarged view of the sensor head.

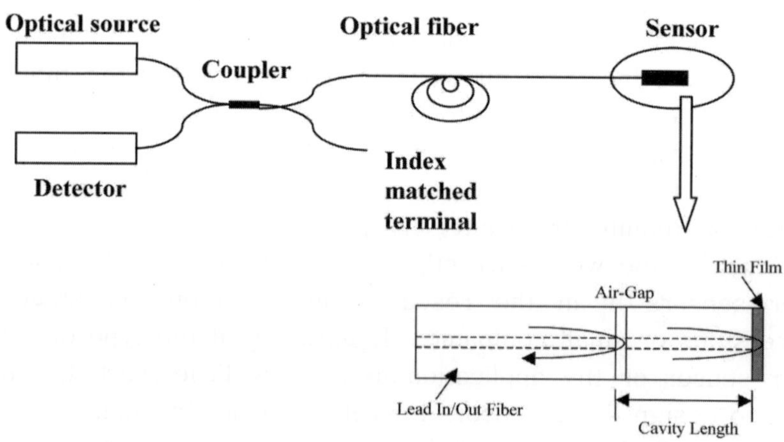

Figure 14.1 Scheme diagram of the optical system.

In order to evaluate the growth of the thin film on the surface of the fiber end with the required resolution, it was necessary to design the system to be robust against noise factors present during the measurement so that the thickness measurement would not be adversely affected. A large amount of variation due to the measurement process would be manifest in lower confidence levels for factor significance determination in the ANOVA (Analysis of Variance) analysis in subsequent sensor development. Therefore, understanding the sources causing variation of the measurement was the first step in optimization of the sensor design. The maximum variation of the cavity length due to uncontrolled factors was determined to be around 5nm during the pre-experiments. This was deemed to be an unacceptable amount of scatter in the measurement process due to unknown factors. Therefore an experiment was designed to identify the major sources of variation in the system as a prelude to subsequent testing.

14.2.2 DESIGN OF EXPERIMENTS BY USING TAGUCHI METHODS

There are at least five steps to follow in the experimental design process (Lausby, 1994). They include:

- Planning,
- Selection of factors, factor levels, interactions to study, and orthogonal array,
- Conducting the experiment,
- Analysis of the data,
- Confirmation experiment.

(1) Planning

In this experiment, the parameter of interest is the measured thin film thickness (L_f), but it cannot be measured directly. So first, we measured the fiber cavity length without the thin film (L_0). Then we measure the optical length (L), which is the sum of the cavity length, L_0 and the thickness of the thin film (L_f). The thin film thickness can be obtained by subtracting L_0 from L. In order to get a more accurate value of the thin film thickness, the variation of the cavity length should be as small as possible.

By brainstorming and some pre-experiments, it was concluded that the variation of the cavity length was most likely due to some combination of the following four factors: (1) the change of refractive index of the solutions (which is required when depositing the thin film on the fiber end); (2) fiber bending effects (from which the spectrum of the single mode fiber can be deformed significantly); (3) temperature of the environment; and, (4) vibrations during the measurement process. These were thought to be the most important factors influencing the performance of the sensor. We segregated these factors depending on whether they were control or noise factors. The vibration caused by moving the optical fiber during the experiments is not easy to control and was therefore selected as a noise factor. The levels of control factors and noise factors are shown in Tables 14.1 and 14.2.

Table 14.1 Control factors and levels.

	Factors	Level 0	Level 1	Level 2	Level 3
1	Temperature (°C)	RT*	35	45	50
2	Refractive index**	1	1.3330	1.3586	1.3772
3	Fiber bending (diameter, cm)***	None	5	3	2

*RT: Room temperature is around 25°C.

**Refractive Index change was induced by placing the sensor in air (refractive index = 1), water (refractive index = 1.33), acetone (refractive index = 1.3586) and isopropanol alcohol (refractive index = 1.3772).

***Fiber bending effect was induced by making the lead in fiber a circle with diameter of 5cm, 3cm, 2cm and no bending.

Table 14.2 Noise factors.

Factors	N0	N1
Vibration	Without	With

Per our knowledge, refractive index of liquids is temperature dependent. As the temperature of a liquid rises, the refractive index usually decreases. For example, for water, the refractive index at a wavelength of 1550 nm changes 0.0001 per degree C (Weber, 2003). Since in this experiment, the temperature differences between levels are less than 15°C, we ignored the interaction between the temperature and the refractive index. In this experiment, it was determined that no interaction needed to be studied and that only main factor effects would be considered.

(2) Selection of the orthogonal array

In this case, a full factorial experimental design array with three factors and at four levels would have required 64 experiments. Adding in the two level noise factor with two repetitions would have taken the total number of observations over two hundred, an unacceptable number under the constraints of the experimentation. A fraction factorial array was selected to study the main factor effects and the levels adjusted accordingly in order to reduce the number of experiments. Although more than two levels in the initial screening array may not be an optimum strategy, the information was required to make the necessary decisions. In addition, four different kinds of media with different refractive indexes were required based on operational constraints. Also, calibration data over four temperatures was required for future software algorithm development. Four levels of fiber bending were chosen because of past experience and data published in the literature (Xiao, 2000) indicating a highly nonlinear response of the optical signal deformation with extent of fiber bending. The experimental design methodology utilized one of the Taguchi fractional factorial arrays.

From the orthogonal arrays popularized by Taguchi, an L_{16} array (shown in Table 14.3) was chosen based on the factors and levels selected in the brainstorming sessions. The L_{16} array was initially designed for 2-level experiments. In order to use it for four-level experiments, modifications can be made by combining levels (maintaining, of course, the number of degrees of freedom of the combined levels and matching this to the number of degrees of freedom required by the factor levels). In the 2-level orthogonal array, each column has 1 degree of freedom. Since the 4-level factor requires 3 degrees of freedom, we merged three mutually interactive columns in L_{16} to accommodate the four level factor. Three sets of mutually interacting columns in L_{16} array can be columns 2, 4 and 6, or columns 1, 10 and 11, or columns 7, 8 and 15. The merging of mutually interactive columns minimizes confounding of interactions as much as possible. The original L_{16} array and modified L_{16} array are shown in Table 14.3. In this manner, 16 experiments were performed based on the conditions specified in the array.

Table 14.3 L_{16} orthogonal array for 15 factors of 2-level.

	Column no.														
Trial no.	2	4	6	1	10	11	8	15	7	3	5	9	12	13	14
1	0	0	0	0	0	0	0	0	0	0	0	0	0	0	0
2	0	0	0	0	1	1	1	1	0	0	0	1	1	1	1
3	0	1	1	0	0	0	0	1	1	0	1	0	1	1	1
4	0	1	1	0	1	1	1	0	1	0	1	1	0	0	0
5	1	0	1	0	1	1	0	1	1	1	0	0	0	0	1
6	1	0	1	0	0	0	1	0	1	1	0	1	1	1	0
7	1	1	0	0	1	1	0	0	0	1	1	0	1	1	0
8	1	1	0	0	0	0	1	1	0	1	1	1	0	0	1
9	0	0	0	1	0	1	0	1	1	1	1	1	0	1	0
10	0	0	0	1	1	0	1	0	1	1	1	0	1	0	1
11	0	1	1	1	0	1	0	0	0	1	0	1	1	0	1
12	0	1	1	1	1	0	1	1	0	1	0	0	0	1	0
13	1	0	1	1	1	0	0	0	0	0	1	1	0	1	1
14	1	0	1	1	0	1	1	1	0	0	1	0	1	0	0
15	1	1	0	1	1	0	0	1	1	0	0	1	1	0	0
16	1	1	0	1	0	1	1	0	1	0	0	0	0	1	1

Table 14.4 Column-merged L_{16} orthogonal array for three factors of 4-level.

Trial no.	Column no.								
	(2,4,6) 1	(1,10,11) 2	(8,15,7) 3	(3) 4	(5) 5	(9) 6	(12) 7	(13) 8	(14) 9
1	0	0	0	0	0	0	0	0	0
2	0	1	3	0	0	1	1	1	1
3	1	0	1	0	1	0	1	1	1
4	1	1	2	0	1	1	0	0	0
5	2	1	1	1	0	0	0	0	1
6	2	0	2	1	0	1	1	1	0
7	3	1	0	1	1	0	1	1	0
8	3	0	3	1	1	1	0	0	1
9	0	2	1	1	1	1	0	1	0
10	0	3	2	1	1	0	1	0	1
11	1	2	0	1	0	1	1	0	1
12	1	3	3	1	0	0	0	1	0
13	2	3	0	0	1	1	0	1	1
14	2	2	3	0	1	0	1	0	0
15	3	3	1	0	0	1	1	0	0
16	3	2	2	0	0	0	0	1	1

Three control factors (Temperature, Refractive Index, Fiber Bending) can be assigned to the columns 1 and 2 and 3 of the column-merged L_{16} orthogonal array for three factors of 4-level. The noise factors are assigned to a full factorial outer array. The experimental layout showing all the factors and interactions along with the test results is shown in Table 14.4.

Trial no. 8 in Table 14.4 is used as an example of the procedure followed in the testing of the sensors. In this trial, the sensor was placed in the alcohol (refractive index = 1.3772, level 3), and the temperature of alcohol was raised from room temperature to 35°C (temperature level 1).

The fiber was bent to a diameter of 3cm (fiber bending level 2) and the sensor was tested with the vibration of the fiber (noise factor N0). The sensor was again tested without vibration of the fiber (noise factor N1). Such tests were repeated twice and the test results were recorded and shown in the noise factor columns in Table 14.5. The cavity length was measured in units of microns in this experiment.

Table 14.5 Experiment results (before temperature compensation).

Trial no.	Temp	Refractive index	Fiber bending	Noise factors				SN
				N0	N1	N0	N1	
1	RT	1	100	266.7027	266.7035	266.7021	266.7048	57.89815
2	RT	1.3330	5	266.7105	266.71	266.7103	266.7104	42.15561
3	RT	1.3586	3	266.7115	266.7111	266.7116	266.7104	41.24678
4	RT	1.3772	2	266.7117	266.7104	266.7113	266.71	41.53756
5	35	1	5	266.7233	266.724	266.7237	266.7245	33.40005
6	35	1.3330	100	266.7274	266.7284	266.7271	266.7285	31.91792
7	35	1.3586	2	266.7243	266.7299	266.7259	266.7305	31.94234
8	35	1.3772	3	266.7315	266.7297	266.731	266.7291	31.10607
9	45	1	3	266.737	266.7411	266.7461	266.739	28.30221
10	45	1.3330	2	266.7439	266.744	266.7447	266.7441	27.60225
11	45	1.3586	100	266.7437	266.7456	266.7432	266.7453	27.54279
12	45	1.3772	5	266.7455	266.7463	266.7456	266.7463	27.24488
13	50	1	2	266.7627	266.7596	266.7607	266.758	24.76516
14	50	1.3330	3	266.7575	266.7575	266.758	266.7574	25.17689
15	50	1.3586	5	266.7534	266.7537	266.7545	266.754	25.78047
16	50	1.3772	100	266.7529	266.7539	266.7533	266.7535	25.86543
								28.77959

(3) Results Analysis

Signal-to-noise (SN) ratio, advocated in Taguchi approach, is to maximize the performance of a system or product by minimizing the effect of noise. As the objective of this research was to control the cavity length of the optical fiber, it was decided that the cavity length is a nominal-is-best characteristic.

To obtain the SN ratio, equation (2.24) based on the loss function approach was used, which is

$$\mathrm{SN} = -10\log\left[\frac{1}{n}\sum_{i=1}^{n}(y_i - m)^2\right].$$

The target value m is set to be 266.7025 here. The results are shown in Table 14.4.

The ANOVA table (Table 14.6.) was obtained based on the following equations.

$$S_A = \frac{A_0^2}{4} + \frac{A_1^2}{4} + \frac{A_2^2}{4} + \frac{A_3^2}{4} - CT,$$

$$V_A = \frac{S_A}{3},$$

$$F_A = \frac{V_A}{V_e},$$

$$S'_A = S_A - (V_e \times 3),$$

$$S_T = \sum_{i=1}^{N} y_i^2 - \frac{T^2}{N},$$

where A_0, A_1, A_2 and A_3 are the sum of the results of 4 levels of factor A, CT is the correction factor ($=T^2/N$), T is the sum of all results, N is the total number of all results, S_T is total sum of squares of all factors, S_A is the factor sum of squares of factor A, V_e is variance of error term, and S'_A is sum of squares of factor A.

Table 14.6 ANOVA table before temperature compensation.

Source of variation	D.F.	Sum of squares (S)	Variance (V)	F-ratio (F)	Pure sum (S')	Contribution ratio (%)
Temperature	3	995.573	331.857	22.101	950.528	79.269
Refractive index	3	61.109	20.369	1.356	16.064	1.339
Fiber bending	3	52.332	17.444	1.161	7.287	0.607
Other/Error	6	90.088	15.014			18.785
Total:	15	1199.104				100%

Since the F ratio of temperature, 22.101, is great enough, we can conclude that the temperature is a significant factor affecting the response at this confidence level, and we can conclude that the temperature was the most significant factor from the resulting data. The other factors were determined to be not significant. Although bending will deform the spectrum, there was no significant effect on the cavity length calculation even with a bending diameter as small as 2 cm which was somewhat of a surprise based on the initial information from the brainstorming session.

(4) Confirmation

Since temperature was determined to be the major source of the variation in the previous testing and confirmed in the confirmation experiment, a temperature compensation technology was developed to reduce this temperature effect. Another set of tests could have been designed and conducted to determine which other factors might be significant to reduce the susceptibility of the sensor design to temperature fluctuations in the environment, however, it was decided that controlling development of a temperature compensation system within the sensor itself would be the most beneficial. Figure 14.2 shows the structure of the thin-film sensor with temperature compensation. The original thin-film sensor (Sensor 1) was multiplexed with another temperature sensor (Sensor 2).

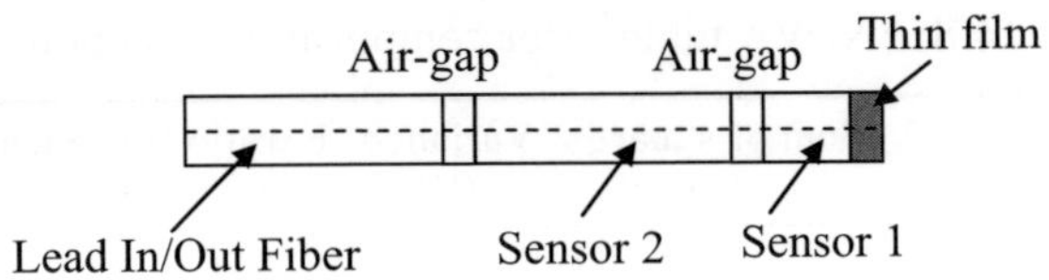

Figure 14.2 Structure for the temperature compensation of thin-film sensor.

As shown in Table 14.7, we repeated the experiment again, but with the temperature compensation. ANOVA calculation results in Table 14.8 which shows the improvement of the temperature compensation. The temperature influence was reduced from F-ratio 158 to 0.3 while the signal to noise ratio (SN) was increased from 28.78 dB to 45.898 dB. The temperature effect has been fully compensated by this multiplexed sensor structure.

Table 14.7 Experiment results (after temperature compensation).

Trial no.	Temp	Refractive index	Fiber bending	Noise factors				SN
				N0	N1	N0	N1	
1	RT	1	100	266.7017	266.7016	266.7018	266.7017	61.9044
2	RT	1.3330	5	266.707	266.7068	266.707	266.7064	47.31656
3	RT	1.3586	3	266.708	266.7079	266.7079	266.7073	45.54357
4	RT	1.3772	2	266.7095	266.7091	266.7096	266.709	43.34372
5	35	1	5	266.7048	266.7046	266.7048	266.7047	53.04737
6	35	1.3330	100	266.7068	266.7064	266.7068	266.7064	47.734
7	35	1.3586	2	266.7052	266.7098	266.7065	266.7105	44.54136
8	35	1.3772	3	266.7089	266.7088	266.7089	266.7082	44.14257
9	45	1	3	266.7035	266.7033	266.7035	266.7032	61.06516
10	45	1.3330	2	266.7089	266.708	266.7091	266.7082	44.33975
11	45	1.3586	100	266.7073	266.7077	266.7072	266.7072	46.27733
12	45	1.3772	5	266.7083	266.7077	266.7082	266.708	45.10674
13	50	1	2	266.7028	266.7014	266.7026	266.7013	61.62727
14	50	1.3330	3	266.7117	266.7114	266.7116	266.7112	40.93732
15	50	1.3586	5	266.7072	266.7073	266.7074	266.7074	46.32877
16	50	1.3772	100	266.7076	266.7078	266.7075	266.7075	45.84609
								45.89766

Table 14.8 ANOVA table after temperature compensation.

Source of variation	D.F.	Sum of squares (S)	Variance (V)	F-ratio (F)	Pure sum (S′)	Contribution (%)
Temperature	3	10.587	3.529	0.341	0	0
Refractive index	3	611.017	203.672	19.707	580.013	82.881
Fiber bending	3	16.194	5.398	0.522	0	0
Other/Error	6	62.006	10.334			17.119
Total:	15	699.806				100%

We can see from Table 14.8, after the temperature compensation, refractive index became the next most significant factor causing the variation. But in order to grow a thin film on the fiber end, putting the sensor into different solutions is necessary. At this stage, the total variation due to the other factors was deemed to be within the acceptable limits of variation to allow the design and development process to proceed on this sensor.

(5) Conclusion and future work

Through signal-to-noise ratio based ANOVA analysis, we found temperature had the most significant effect on the variations in the cavity length measurement. We also found that the refractive index of the surrounding media contributed to the variation for this sensor structure. Compared to the effect of refractive index, bending of the lead in/out fiber had much smaller effect. Unless we bend the fiber to be a circle with diameter less than 2 cm, the bending effect can be ignored. Vibration of the sensor could not be controlled easily in practice to date and was used as a noise factor to design robustness in the process. The experimental results and conclusions are not only suitable for this specific sensor structure, but also for other fiber optic sensors.

A confirmation experiment was done in order to ensure that our understanding of the effect of the temperature is correct. The results from the confirmation experiment showed that this structure worked very well. The signal-to-noise ratio was increased with the incorporation of the temperature compensation.

This multiplex sensor structure is going to be useful for any other temperature sensitive fiber optic sensors for temperature compensation.

14.3 A NEW DIMENSION OF SIX SIGMA: SAMSUNG DFSS

The usual DFSS (Design for Six Sigma) was defined and explained in detail in Chapter 13. However, the usual DFSS is not enough for a good company to be a world-class competitive company. Then what is a better system of DFSS to be an innovative, creative and leading company? There could be several solutions for this question. The authors would like to suggest a solution for a new version of DFSS, which has been adopted by Samsung. Since this new version of DFSS has been employed by SAIT (Samsung Advanced Institute of Technology) and has been quite successful, this new dimension will be called "Samsung DFSS".

The general characteristics of Samsung DFSS can be sketched as follows.

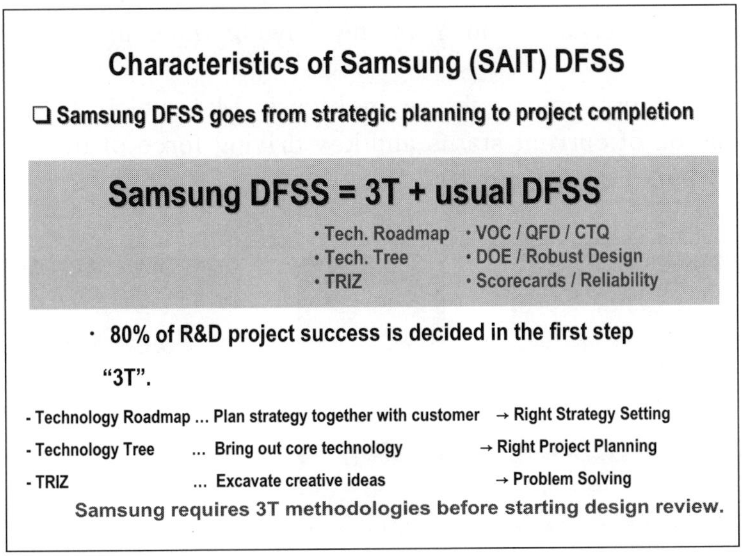

Figure 14.3 Important characteristics of Samsung DFSS.

Note that the usual DFSS employs the DMADOV (DIDOV or DMAIC) process which was explained in Chapter 13, and the major tools are gathering of VOC (voice of customer), QFD (quality function deployment), deployment of CTQ (critical-to-quality), DOE (design of experiments), robust design, balanced scorecards, reliability, etc.

Samsung realized that the usual DFSS is not enough for creative and innovative product development. They experienced that, if technology roadmap, technology tree and TRIZ (these are called "3T") are added in front of the usual DFSS, the new version of DFSS would be much powerful and effective. The "3T" is explained below.

14.3.1 TECHNOLOGY ROADMAP

Technology roadmap (TR) is a roadmap which shows strategic technology development along the coming years. In fact, TR takes a long-range look at technology to examine possible development plans. TR is updated annually, since technology changes in a very fast way. TR looks like an X-Y matrix, in which X contains major products or markets, and Y is the flowing time, in general, the years.

The TR process is shown in Figure 14.4 which starts from examination of current status and key driving force of the relevant industry and our company.

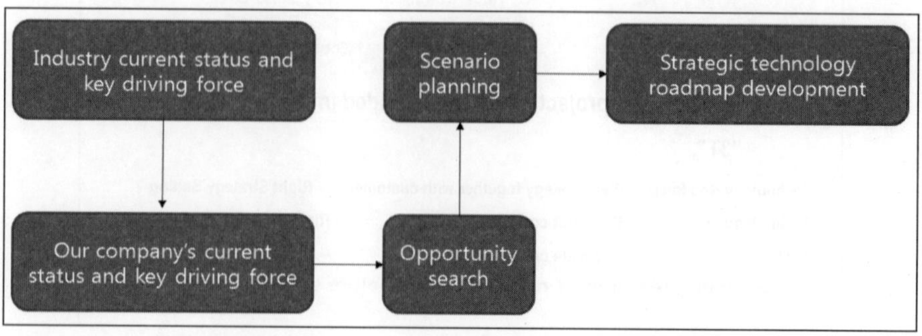

Figure 14.4 Technology roadmap process.

In the scenario planning, several scenarios are checked by many questions such as:

- What would be the threats and opportunities in this scenario?
- What are the business needs and leverage points in this scenario?
- What technology strategies and actions would we implement?

In the scenario planning, an opportunity search based on our company's current status and key driving force is necessary, and its information is transmitted to scenario planning.

The process of developing a strategic technology roadmap based on scenario planning has three steps as follows.

Step 1: Industry-level technology roadmaps are sketched to show how technology development will play out in each scenario.
- Start with a roadmap template and segment the X space by major products or markets.
- Identify currently existing mainstream technologies for these products or markets and enter them on the technology roadmap.
- Convert the technology advances described in scenarios from text to graphic form.

Step 2: The current company strategy based on observations of the industry-level technology roadmaps is adjusted.

Step 3: The company-specific strategic technology roadmap is developed. The strategic technology roadmap is a focused picture of technology advances that relate specifically to the company's product and market focuses. Figure 14.5 shows a general look at a strategic technology roadmap.

When the strategic technology roadmap is ready, the next phase is to develop the R&D project plan and roadmap in support of the strategic technology roadmap. This phase is to show the specific R&D projects and their timing relative to the product or technology milestones for the strategic technology roadmap.

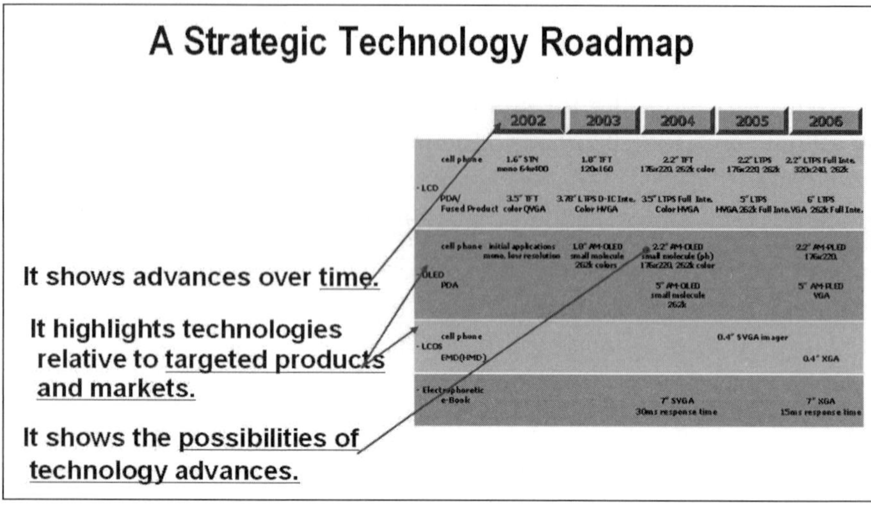

Figure 14.5 A look at strategic technology roadmap.

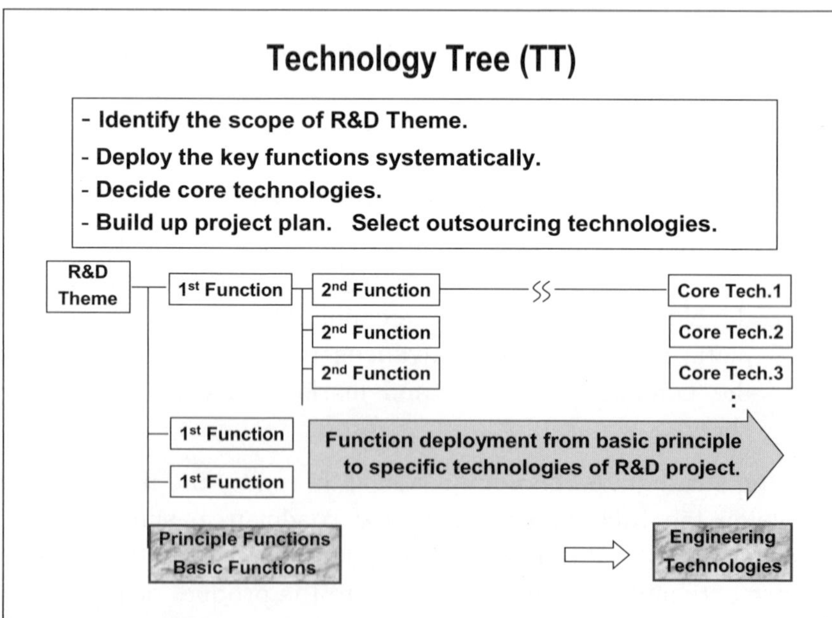

Figure 14.6 The major functions of TT and its general look.

14.3.2 TECHNOLOGY TREE

When an R&D project (or theme) is determined from TR, technology tree (TT) is then used to identify the key functions, secondary functions and core technologies of the R&D theme. Since the deployment of the functions and core technologies looks like a tree, it is called technology tree. The major role of TT and its general look is shown in Figure 14.6.

When a TT is established, the core technologies are tackled first by the R&D project team (DFSS team). And then the team checks whether the secondary functions are satisfied, and finally checks the major functions (1st functions).

14.3.3 TRIZ

(1) Introduction of TRIZ

TRIZ (Teoriya Resheniya Izobreatatelskikh Zadatch) in Russian is the Theory of Inventive Problem Solving (TIPS) developed in the Soviet Union starting in the late 1940s. TRIZ is a combination of methods, tools, and a way of thinking. The goal of TRIZ is to achieve absolute excellence and effectiveness in design and innovation. Genrich S. Altshuller, the creator of TRIZ, initiated the investigation on invention in 1946. After initially reviewing 200,000 former Soviet Union patent abstracts, Altshuller selected 40,000 as representatives of inventive solutions. He separated the patents' different degrees of invention into five levels, ranging from level 1, the lowest, to level 5, the highest, as follows.

1. Level 1: Apparent or conventional solution — solution by methods well-known within specialty (32% of total patent inventions), which is called "company knowledge".
2. Level 2: Small invention inside paradigm — improvement of an existing system, usually with some compromise (45% of total patent inventions), which is called "industry knowledge".

3. Level 3: Substantial invention inside technology — essential improvement of an existing system (18% of total patent inventions), which is called "new industry".
4. Level 4: Invention outside technology — new generation of design using science and technology (4% of total patent inventions), which is called "new technology".
5. Level 5: Discovery — major discovery and new science (1% of total patent inventions), which is called "new science".

Since 1946, TRIZ has developed based on more than 1,500 person-years of research over many of the world's most successful solutions of problems from science and engineering, and systematic analysis of successful patents from around the world, as well as the study of the psychological aspects of human creativity (Mann, 2002). TRIZ has five key philosophical elements. They are as follows.

- Ideality. Ideality is the ultimate criterion for system excellence.
- Functionality. Functionality is the fundamental building block of system analysis. It is used to build models showing how a system works, as well as how a system creates benefits, harmful effects, and costs.
- Resource. Maximum utilization of resource is the key to achieve maximum ideality.
- Contradiction. Contradiction is a common inhibitor for increasing functionality. Altshuller found that almost all invention problems contain at least one contradiction. A contradiction is defined as a situation in which an attempt to improve one feature of the system detracts from another feature.
- Evolution. The evolution trend of the development of technological systems is highly predictable, and can be used to guide further development.

(2) 40 principles to resolve contradictions

Altshuller analyzed more than 40,000 patents and identified about 1,250 typical technical contradictions. These contradictions are further expressed into a matrix of 39 × 39 "engineering parameters". To resolve these contradictions, Altshuller compiled 40 principles such as segmentation, taking out, local quality, asymmetry, merging, universality, anti-weight, preliminary anti-action, preliminary action, and so on. Each of the 40 principles contains a few sub-principles, totaling up to 86 sub-principles.

For instance, for the first principle "segmentation", it has the following sub-principles.

- Divide an object into independent parts.
- Make an object easy to disassemble.
- Increase the degree of fragmentation (or segmentation) of an object.

For more details on these principles and on TRIZ, Yang and El-Haik (2003) is an excellent reference.

(3) Four step process of TRIZ problem solving logic

In the context of DFSS, TRIZ can be used in concept (or idea) generation and solving problems. TRIZ may suggest different system design, design parameters, and tolerance parameters. Based on the five key elements described above, TRIZ developed a system of methods, which is a complete problem definition and solving process. The TRIZ has four steps for problem solving logic as follows.

1. Problem definition (real problem). This is a very important step in TRIZ. If you define the right problem and do it right, then that is 80% of the solution. This step includes the following tasks.
   - Project definition. This is the same as D in DMADOV or DMAIC.
   - Functional analysis. This is the functional modeling of the system and analysis.
   - Technological evolution analysis. This looks into the relative maturity in technology development of all subsystems and parts.

- Ideal final result. This result is the virtually hopeful limit of the system, which may never be achieved but provides us with an "ultimate dream".

2. Problem classification and tool selection (model problem). TRIZ has a wide array of tools for inventive problem solving. However, we must first classify the problem type and then select the tools accordingly.
3. Solution generation (model solution). We apply TRIZ tools to generate solutions for the problem. Because TRIZ has a rich array of tools, it is possible to generate many solutions.
4. Evaluation (real solution). We need to evaluate the soundness of the solutions in order to choose the final best solution.

This four step problem solving logic of TRIZ may be summarized as in Figure 14.7.

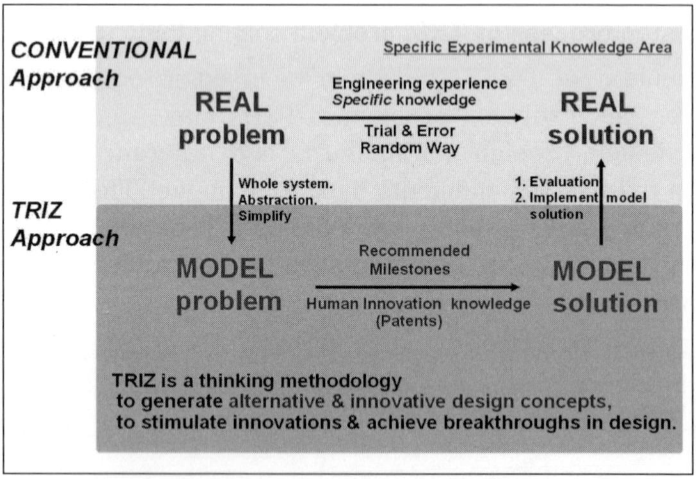

Figure 14.7 TRIZ problem solving logic.

14.3.4 SAMSUNG DFSS

(1) TRIZ project process

Samsung started to use TRIZ extensively in the late 1990s with emphasis on DFSS, as Samsung introduced Six Sigma as its management strategy. A TRIZ champion at SAIT (Samsung Advanced Institute of Technology), whose name is H. J. Kim, mentioned at the 4th Japanese IMC User's Conference that, "The economic benefit after applying TRIZ ideas is impossible to count." He mentioned that 180 million dollars gained by 63 TRIZ projects in 2003, and total 1 billion dollars were cumulatively achieved until 2003 by R&D projects applying TRIZ.

TRIZ has four steps for problem solving logic as explained in Figure 14.7. Samsung uses 5 step process (DAGEV) for its TRIZ project process as shown in Figure 14.8. Note that the 5 step process is Define – Analyze – Generate – Evaluate – Verify (DAGEV) which encompasses the four steps for problem solving logic by TRIZ. The "Real Solution" after evaluation is very often "Concept Solution" obtained by using TRIZ ideas.

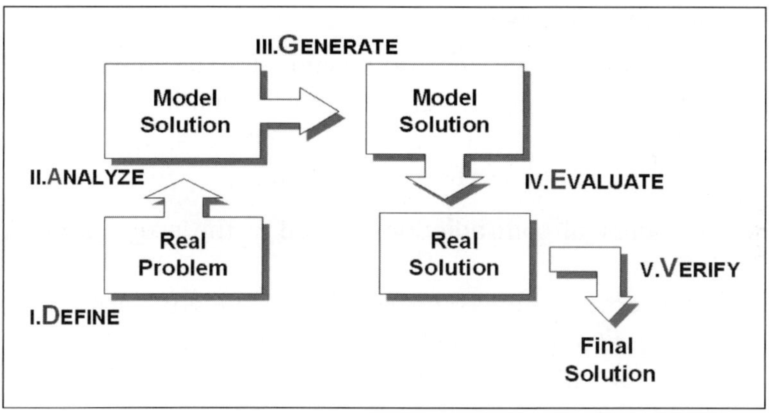

Figure 14.8 5 Steps of TRIZ project process (DAGEV).

(2) DMADOV process using TRIZ

The DMADOV process, which is the most typical process used in DFSS, includes TRIZ in "Design" phase, which utilizes the five steps of TRIZ project process (DAGEV). In Figure 14.9, note that DOE (design of experiments) and Taguchi (robust design) are extensively used in "Optimization" phase.

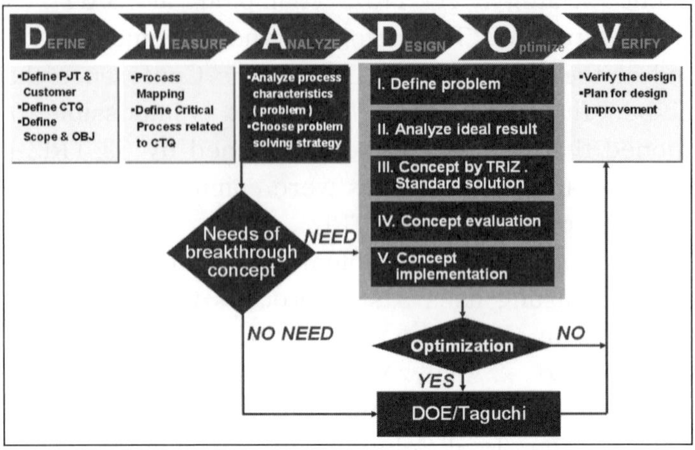

Figure 14.9 DMADOV process with utilization of TRIZ.

(3) Applying TRIZ in robust design

For using robust design (RD) or design of experiments (DOE) in general, TRIZ can be effectively utilized to identify critical factors and their experimental levels. The RD process by applying the concept of TRIZ can be sketched as follows.

1. Define technical contradictions raised in the problem definition of TRIZ.
2. Excavate creative ideas through TRIZ. Identify critical design (or process) factors, and some noise factors involved.
3. Establish an RD (or DOE) experimental planning such as parameter design and/or tolerance design.
4. Optimize the relevant parameters through RD (or DOE).
5. Real solutions are generated, and through evaluation process, a final solution is obtained.

(4) Overall picture of Samsung DFSS

The Samsung DFSS, "3T + usual DFSS", has been explained, which, we believe, is an excellent methodology for Six Sigma in the area of R&D. The overall picture of Samsung DFSS can be sketched as in Figure 14.10.

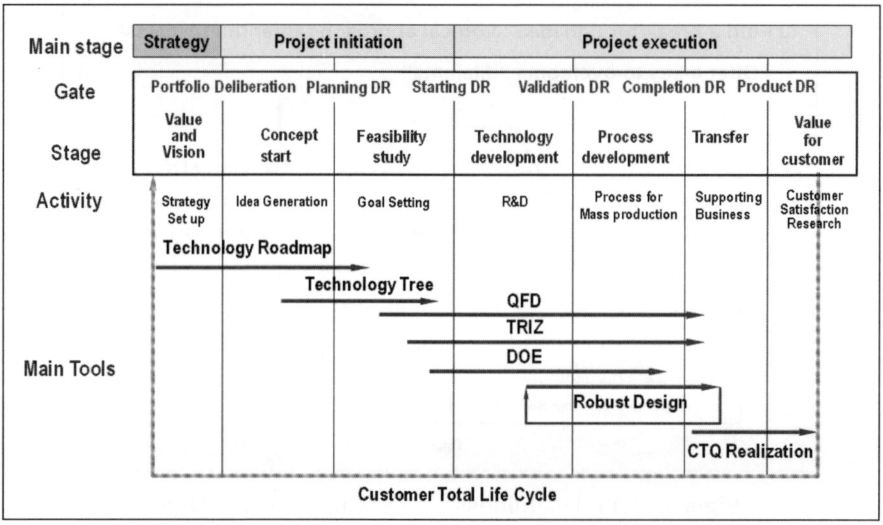

Figure 14.10 The overall R&D process of Samsung DFSS methodologies.

Note that the overall R&D process consists of 3 main stages; strategy, project initialization and project execution. Also notice that the whole process has 6 gates starting from "portfolio deliberation" to "product DR (design review)". The process is composed of 7 stages from "value and vision" to "value for customer", which means that creation of value for customer is the major object of R&D activities. Also the R&D activities are divided into 7 categories starting from "strategy set up" to "customer satisfaction research". For the main tools, 3 Ts (technology roadmap (TR), technology tree (TT) and TRIZ) play the major roles in the early stages of R&D, and then quality function deployment (QFD), design of experiments (DOE), robust design (RD) and critical to quality (CTQ) realization play their roles in the later stages of R&D.

(5) Usefulness of TRIZ in Samsung DFSS

The usefulness of TRIZ in addition to TR and TT can be sketched as shown in Figure 14.11.

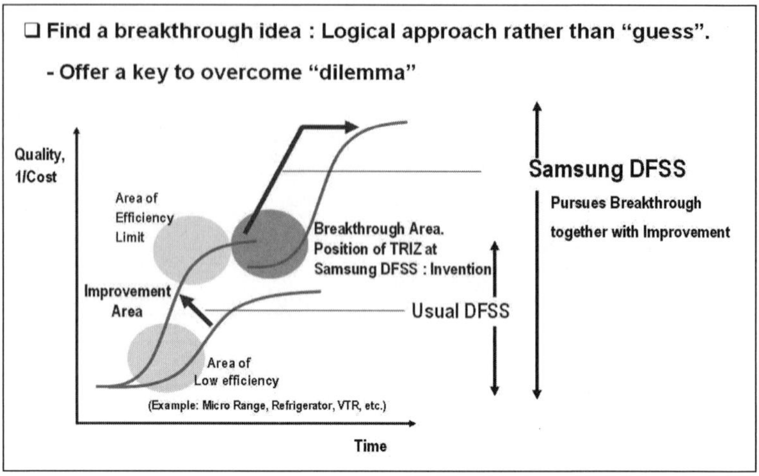

Figure 14.11 Usefulness of TRIZ in Samsung DFSS.

Note that the area of low efficiency can be moved to the improvement area and then to the breakthrough area by utilizing TRIZ in addition to TR and TT. Also note that the area of efficiency limit can be broken by TRIZ.

It has been proved in Samsung that TR maximizes business efficiency of R&D activities. The business strategy and R&D goal can be shared by TR, and thus duplicated R&D activities can be avoided. Through TT, technology flow-down and project planning can be simultaneously sketched. In general. TT can be done in parallel with QFD, and it helps systematic development of R&D project from basic principles to technology.

14.4 CASE STUDY OF SIX SIGMA IMPLEMENTATION

Samsung SDI was the first introducer of Six Sigma to Korean industry in 1996. Since then, it has implemented Six Sigma including DFSS quite well. The First National Quality Prize of Six Sigma was given to two companies. One is Samsung SDI and the other is LG Electronics, which are virtually the leaders of Six Sigma in Korea. Samsung SDI was founded in 1970 as a producer of the black/white braun tube. It began to produce the color braun tube from 1980, and now it is the number one company for braun tubes in the world. The company's achievements are truly wide-ranging. It puts in place the world's first PDP (plasma display panels) mass-production line. It unveiled new concepts for a small-sized LCD (liquid crystal display) called UFS (ultra fine and high speed) LCD as well as the world's highest resolution AM (active matrix) OLED (organic light-emitting diodes) for high-end mobile phones. Besides, other major products are CDT (color display tube), CPT (color picture tube), VFD (vacuum fluorescent display), and C/F (color filter). The total sales volume is more than $6 billion and the total number of employees is about 18,000, including 8,000 domestic employees. It has 6 overseas subsidiaries in Mexico, China, Germany, Malaysia and Brazil.

The typical characteristics of Six Sigma in Samsung SDI is explained below.

14.4.1 EVOLUTION OF QM STRATEGIES

Since its founding in 1970, Samsung SDI has employed several quality management strategies such as QC, TQC/TPM, TQM/ISO9000, and PI as shown in Figure 14.12. In 1996, it began PI as the beginning stage of Six Sigma. Note that the direction of evolution in management strategies is from manufacturing areas to all areas of the company, and from product quality/small group activities to process innovation and redesign.

Product quality/small group activity ⇒ Process innovation & redesign

The necessity of PI and Six Sigma stems from the problems of the company as shown in Figure 14.13. The problems were in the large quality variations in many products, repeated occurrences of the same defects, high quality costs (in particular, high failure costs), insufficient unified information for quality and productivity, manufacturing-oriented small group activities, and infrequent use of advanced scientific methods. The company concluded that the directions for solving these problems lay in scientific and statistical approaches for product quality, elimination of waste elements for process innovation, and continuous learning system for people. These directions in turn demanded a firm strategy for a complete overhaul, implying a new paradigm shift to Six Sigma.

Scope of participation	Product quality	Small group activities, policy deployment	Standard management	Customer-oriented process redesign	Participation of all areas
Transactional areas					
Marketing, R&D sales, Purchasing					
Accounting				PI	Six Sigma
Material, Facility			TQM/ISO9000		
Manufacturing areas	QC	TQC/TPM			
Beginning	1970	1984	1992	1996	1999

QC=quality control, TQC=total quality control, TPM=total productive maintenance, TQM=total quality management, ISO=international organization for standardization, PI=process innovation.

Figure 14.12 Evolution of quality management strategies in Samsung SDI.

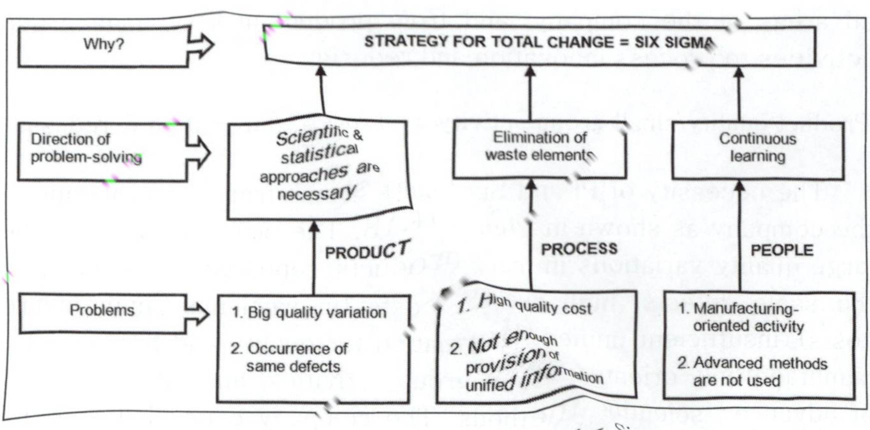

14.4.2 VISION OF SIX SIGMA

The CEO of Samsung SDI, Son Wook, declared the slogan "True leader in digital world" as the Six Sigma vision at the end of 1996. The definition of Six Sigma in the company is as follows.

"Six Sigma is the management philosophy, strategy and tool which achieves innovative process quality and development of world number one products, and which cultivates global professional manpower, and a way of thinking and working from the viewpoint of customer satisfaction."

The company demonstrates its vision as seen in Figure 14.14. In this figure, "7 values" indicates vision, customer, quality, innovation, communication, competency and integrity. These values are in fact "the principles of action behavior" by which the employees are working in the company.

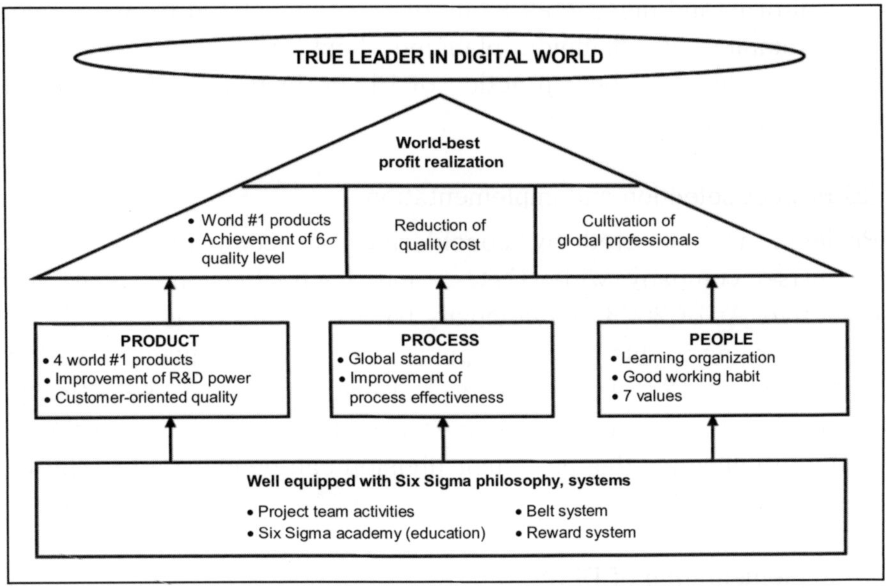

Figure 14.14 The vision of Six Sigma.

14.4.3 MAJOR IMPLEMETATION OF SIX SIGMA

(1) Realization of champion leadership

Six Sigma is basically a top-down management tool. For implementation of Six Sigma, executive officers (i.e. champions) should be the leaders of Six Sigma. In Samsung SDI, the following points have been implemented for champion leadership.

- Champion education: all champions take the champion education course of 4 days, and they obtain the GB certification.
- Champion planning: Each champion is supposed to plan a "Six Sigma roadmap" for his division twice a year. The champion selects the themes of projects, and he supervises the Six Sigma plan for his division.
- Champion day: One day each month is designated as the champion day. On this day, the champions wear Six Sigma uniform, and discuss all kinds of subjects related to Six Sigma. Examples of champion planning, best practice of champion leadership, and best practice of BB projects are presented on this day.

(2) Project selection and implementation

Projects are selected by considering the company 6Y, which comprise company-wide CTQs, and each division's goal and objective. As of 2003, the company 6Y are as shown in Table 14.9.

According to this matrix mapping, the customer quality gets the highest mark, 12.0, hence the first priority for project selection is given to the company Y, customer-oriented quality. Then several project themes for this particular division can be chosen to achieve this company Y.

(3) Implementation of DFSS

The development system of Samsung SDI is based on E-CIM (Engineering Computer Integrated Manufacturing). E-CIM is a tool for maximizing the company's competitiveness from the viewpoint of customer demand through efficient development process,

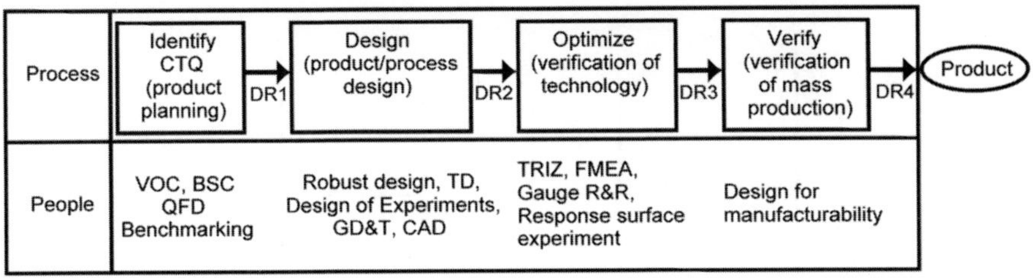

Figure 14.15 DFSS process.

technology standardization, PDM (product data management) and DR (design review). The DFSS process of Samsung SDI follows the IDOV (identify, design, optimize, verify) process, and after each step, DR helps to validate the process as shown in Figure 14.15.

There are 4 different types of design review (DR). Each one reviews and validates the previous immediate step. For instance, DR1 reviews the product planning and decides whether DFSS process can flow to the next step or not.

(4) Manpower cultivation

Six Sigma education really began from 1999, when 1228 GBs, 30 BBs, and 9 MBBs were cultivated. However, in 2000, 62 champions, 44 MBBs, 192 BBs, 1385 GBs and WBs out of a total of 7,818 employees were educated. In 2004, 384 BBs and 915 GBs were cultivated, so that almost all possible GB candidates were educated. The belt system ran as shown in Table 14.10.

Full-time BBs are the backbone of Six Sigma management. As soon as a BB completes the BB education course, he/she becomes a "nominated BB". When he completes two BB projects, he becomes a "certified BB" or "full-time BB" depending on his division's situation. If a BB becomes a full-time BB, he is supposed to spend all of his time working on his project with several GBs. Usually, his mission as a full-time BB lasts one year. After one year, his performance is evaluated. If he meets the maintenance standard, he can be a full-time BB once again for the next year. However, if he cannot meet the maintenance standard, he should be a certified BB or nominated BB for the next year.

Table 14.9 Matrix mapping for project selection.

Division's goal and object	Rate of importance	Company 6Y					
		Improvement of R&D	Improvement of marketing	Customer quality	Global management	30% improvement of effectiveness	SDI's 7 values
Customer satisfaction	2	3	1	3	3	1	1
Process	1	2	1	2	3	2	1
Learning	0.5	1	3	2	1	1	3
Financial achievement	1	1	2	3	1	1	1
Sum		9.5	6.5	12.0	10.5	5.5	5.5

Table 14.10 Belt system: qualification and maintenance.

Belt	Class	Qualification	Maintenance standard	Effective period
MBB	Full-time	• Full-time project supervisor • Graduate of BB and MBB courses • Supervision of at least 3 BB projects	• Full-time Six Sigma supervision • Supervision of at least 6 BB projects	One year
	Certified	• Part-time project supervisor • Graduate of BB and MBB courses • Supervision of at least 3 BB projects	• Part-time Six Sigma supervision • Supervision of at least 3 BB projects	One year
	Nominated	• Graduate of BB and MBB courses • Supervision of less than 3 BB projects		Permanent
BB	Full-time	• Full-time project leader • Graduate of BB course • Completion of 2 BB projects	• Full-time project leader • Completion of 2 projects • Supervision of at least 12 GB projects	One year
	Certified	• Part-time project leader • Graduate of BB course • Completion of 2 BB projects	• Part-time project leader • Completion of 1 project • Supervision of at least 4 GB projects	One year
	Nominated	• Graduate of BB course, completion of less than 2 BB projects	• None	Permanent
GB	No class	• Graduate of GB course • Completion of 1 project	• Completion of 1 project	One year
WB	No class	• Graduate of WB course		Permanent

(5) Major results of Six Sigma

The pre-tax profits for the Six Sigma beginning years 1998, 1999, 2000 were $51.7 million, $166.7 million, and $600 million, respectively, exhibiting dramatic yearly increases. The excellent Six Sigma programs contributed to the sharp increases. In 2004, the total sales became more than $4 billion. The financial benefits from the project teams in 2004 were $46 million from BB projects and $45 million from GB projects.

The quality cost (prevention cost, appraisal cost and failure cost combined) in 1999 was $0.38 billion, or 11.3% of total sales. However, due to intensive project activities to reduce the quality cost, the quality cost for 2003 was about $0.30 billion, or roughly 7.5% of the total sales. This remarkable gain in sales and profit, together with reduction of quality costs, attests to the positive effects of Six Sigma projects in the early stages of introduction of Six Sigma in Samsung SDI.

14.5 PRACTICAL QUESTIONS IN IMPLEMENTING SIX SIGMA

14.5.1 IS SIX SIGMA RIGHT FOR US NOW?

(1) Key questions to answer

Many commercial firms are wondering whether Six Sigma is right for them now. Embarking on a Six Sigma initiative begins with a decision to change — specifically, to learn and adopt methods that can boost the performance of your organization. The starting point in gearing up for Six Sigma is to verify that an organization is ready or is required to embrace a big change. There are several essential questions and facts an organization has to consider in making its readiness assessment.

- Is change a critical business need now, based on bottom-line, cultural, or competitive needs?

- Can we come up with a strong strategic rationale for applying Six Sigma to our business?
- Will our existing improvement systems and methods be capable of achieving the degree of change needed to keep us a successful, competitive organization?

If the answers are Yes, Yes and No, an organization may well be ready to explore further how to adopt Six Sigma in its organization. However, if an organization is in one or more of the following situations, it probably would be best to say "No thanks for now" to Six Sigma adoption.

- An organization already has a strong, effective performance and process improvement effort.
- Recent changes are already overwhelming employees and resources.
- The potential gains are not expected to be much.

(2) Cost/benefit perspective

Without investment, we cannot expect a big change or a big gain. Some of the most important Six Sigma investment budget items include the following:

- Direct payroll: Individuals dedicated to the effort full-time such as BBs.
- Indirect payroll: The time devoted by executives, team members, process owners, and others to such activities as measurement, data gathering for VOC (voice of customer), and improvement projects.
- Training and consulting: Teaching people Six Sigma skills and obtaining advice on how to make the effort successful.
- Improvement implementation costs: Expenses related to installation of new solutions or process designs proposed by project teams.
- Other expenses such as travel and lodging, facilities for training, and meeting space for teams.

Estimating potential benefits is not an easy task. There is no way to accurately estimate the gains without examining the improvement opportunities present in the business, and without planning the implementation to see what the relative payoff will be. However, the following benefits could be expected:

- The total quality costs (prevention cost, appraisal cost and failure cost) can be reduced. Eventually, the costs of poor quality (COPQ) can be reduced substantially, and the company's profits can soar.
- By improving quality and productivity through process evaluations and project team efforts, the total sales and profits can dramatically increase.
- Through a sound Six Sigma initiative, better strategic management, more systematic data collection and analysis, and efforts directed toward customer satisfaction will result in a better market image and customer loyalty.
- As a result of systematic education through belt systems, cultivation and efficient utilization of manpower becomes possible, which eventually fosters employees' pride in their company.

Based on the responses to key questions and the cost/benefit analysis, a company can decide whether it should take the Six Sigma initiative now or later. One important point to be kept in mind when a company prepares to embark on Six Sigma efforts is that the company should factor at least six to twelve months for the first wave of DMAIC projects to be completed and concrete results to be realized. The company can push teams for faster results. Giving them extra help or coaching as they work through their "learning curve" can be a good way to accelerate their efforts, although it may also boost your costs. It would be a mistake to forecast achievement of big tangible gains sooner than a period of six months. The company must have patience and make consistent efforts at the embarkation stage.

14.5.2 HOW SHOULD WE INITIATE OUR EFFORTS FOR SIX SIGMA?

When a company decides to start a Six Sigma initiative, the first important issue to resolve is "How and where should we embark on our efforts for Six Sigma?" Since Six Sigma is basically a top-down approach, the first action needed is a declaration of commitment of top-level management. For making the management decisions, it is best to look at the criteria impacting the scale and urgency of their efforts which will strengthen the company by removing the weaknesses. Based on these facts, he should decide his Objective, Scope and Time-frame for Six Sigma engagement.

(1) Determination of objective

Every business desires "good results" from a Six Sigma effort, but the type of the results and the magnitude of the changes may vary a great deal. For example, Six Sigma may be attractive as a mean to solve nagging problems associated with product failures or gaps in customer service, or as a way to create a responsive management culture for future growth. Each of these objectives could lead to different types of Six Sigma efforts. It is possible to define three broad levels of Objectives — Management innovation, Statistical measurement and process evaluation, and Strategic improvement by problem solving (see Table 14.11).

(2) Assessing the feasibility scope

What segments of the organization can or should be involved in the initial Six Sigma efforts? Scope is very important in the initial stage of Six Sigma. Usually we divide the whole company into 3 segments; the R&D part, manufacturing part, and transactional (or non-manufacturing) part. Mostly the manufacturing section is the target for initial Six Sigma efforts. However, the author is aware of some companies in Korea that began their efforts from the transactional section. It would be desirable to consider the following three factors in determining the scope of the initial Six Sigma efforts.

Table 14.11 Three levels of Six Sigma Objectives.

Objective	Description
Management innovation	A major shift in how the organization works through cultural change. • Creating customer-focused management • Abandoning old structures or ways of doing business • Creating a top-level world-beating quality company
Statistical measurement and process evaluation	All processes are statistically measured, and the sigma quality levels are evaluated. • The sigma quality level of each process is evaluated. • Poor processes are designated for improvement. • A good system of statistical process control is recommended for each process.
Quality and productivity improvement by problem solving	Key strategic and operational weaknesses and opportunities become the targets for improvement. • Speeding up product development • Enhancing supply chain efficiencies or e-commerce capabilities • Shortening processing/cycle time • Project team efforts for key quality and productivity problems

- Resources: Who are the best candidates to participate in the effort? How much time can people spend on Six Sigma efforts? What budget can be devoted to the start-up?
- Attention: Can the business focus on start-up efforts? Are they willing to listen to new ideas for management innovation?
- Acceptance: If people in a certain area (function, business unit, division, etc.) are likely to resist, for whatever reasons, it may be best to involve them later. It is wise to start from the section which accepts the new Six Sigma efforts.

(3) Defining time frame

How long are you willing to wait to get results? A long lead-time for a payoff can be frustrating. The time factor has the strongest influence on most Six Sigma start-up efforts. The top management should define the time-frame for Six Sigma implementation.

14.5.3 WHAT ARE THE MAIN CRITICISMS OF SIX SIGMA?

Since Six Sigma itself is only 15 years old, and its historical development has been one of dynamic changes, there are some criticisms on Six Sigma. The major criticisms are as follows.

(1) Six Sigma is nothing new. It is just old tools in new clothing.

Critics of Six Sigma have often said that it contains nothing new. The proponents acknowledge that the tools applied in Six sigma are not new — they are proven statistical tools. However, Six Sigma is new in many aspects. It has provided a powerful corporate framework for these tools to become effective and enabled the important link between the bottom line and top line. The following points which differentiate Six Sigma from earlier attempts can be highlighted.

- Its strategic involvement of top management in the companywide improvement process.
- Its customer focus.
- Its focus on project team efforts and financial results.
- Its focus on education and training through belt systems.
- Its formalized improvement methodology such as DMAIC, DMADOV, DIDOV, etc.

(2) The expected benefits are unrealistic.

A criticism is that publicized expected results are unrealistic. This criticism is rejected by annual reports from many Six Sigma companies. GE alone achieved Six Sigma benefits of about \$1.2 billion on a \$450 million investment in 1998, for 1999 the savings were in the plus \$2 billion range. ABB, Motorola, Samsung SDI, LG and others report that Six Sigma is delivering what is promised.

(3) Other business improvement initiatives will soon replace Six Sigma. It is just another management fad.

The argument that other strategic initiatives will replace Six Sigma is not very controversial and applies to all strategic initiatives in the business world, be they widely deployed or not. Some argues that Six Sigma will disappear soon from corporate agendas, which means that it is a fad.

We believe that Six Sigma is more than just a fad. The Six Sigma concept has survived for more than a decade already and is a way beyond the point where it could become a management fad lasting just a few years. One reason could be that Six Sigma was developed by the industry and for the industry — with a deployment based on merit. Another reason could be that Six Sigma is a more systematic, pragmatic, statistical and smarter approach compared to other past initiatives.

(4) Key Six Sigma practices are based on faulty statistical assumptions.

Some of the assumptions often employed in Six Sigma practices are said to be faulty by many opponents. Most criticized are assumptions relating to:

- The Normality assumptions
- The acceptance of a $\pm 1.5\sigma$ long-term shift
- Predictability of the future outcome

A common answer to these concerns is that the assumptions are made for pragmatic reasons to make matters simple and easily understood by all in the company. Even though the Normal distribution assumption may not always be completely correct, the procedures based on the Normal distribution assumption are often very robust, i.e. the consequences in terms of the errors are almost negligible.

The use of the 1.5σ shift is criticized for being unrealistic and without foothold in reality. Of course there is no natural law telling us that all processes have this much long-term shift in average value. However, each process has its own variations arising from several sources, and it can be assumed that the sum of all acceptable sources of variation may add up to 1.5σ. In industrial practice, this has been confirmed to be reasonable. Of course, it would be possible to utilize for each process its own special long-term shift; however, that would not have been very practical. A pragmatic approach is to use the 1.5σ shift of the process average in either direction.

14.5.4 DOES SIX SIGMA APPLY WELL TO SERVICE INDUSTRY?

Many service industries such as banking, insurance, postal office and public administration often ask, "Does Six Sigma apply well to service industries?" Despite the successful application of Six Sigma in companies such as AIG Insurance, American Express, Citibank, GE Capital Services, NBC and the US Postal Service, executives and managers from the service industry very often wonder if Six Sigma is applicable to their type of business.

The primary response to this question is that Six Sigma has the potential to be successful in almost any industry. Since Six Sigma mainly focuses on customer satisfaction, variation reduction, quality improvement and reduction of COPQ, the results enjoyed by Six Sigma companies in the service industry are just as impressive as their counterparts in the manufacturing industries.

Let's take the example of GE Capital Services. Three years after the launch of Six Sigma (1995 was the beginning year), they reported: "In 1998, GE Capital generated over a third of a billion dollars in net income from Six Sigma quality improvements — double that of 1997. Some 48,000 of our associates have already been extensively trained in this complex process improvement methodology — and they have completed more than 28,000 projects."

The framework in Six Sigma for ensuring and measuring that customer requirements are met should also be attractive to most service organizations. In Six Sigma, the customers are asked to identify the critical characteristics of the services they consume and what constitutes a defect for each of the individual characteristics. Based on these, the Six Sigma measuring system is built up.

It is true that many service companies often find it difficult to measure their processes adequately. Compared to manufacturing processes, it is often more demanding to find appropriate characteristics to measure. Also it is difficult to measure the sigma quality level for a service process. In this case, a possible way to set up the quality level for a service process is as follows.

6σ level	The ideal level to be reached or the benchmark level of the best company in the world
3σ level	The current level of my company

According to the above levels, the company can achieve the levels of 4σ and 5σ. If the current level of the company is very poor, one can designate the company level as 2σ.

Appendices

Table of Acronyms

ABB	Asea Brown Boveri
ANOVA	Analysis of Variance
BB	Black Belt
BSC	Balanced Score Card
3C	Change, Customer and Competition in quality and productivity
CCD	Central Composite Design
CEO	Chief Executive Officer
CFR	Critical Functional Response
CL	Center Line
COPQ	Cost of Poor Quality
Cp, Cpk	Process Capability Index
CPL	Lower Capability Index
CPM	Critical Parameter Method
CPT	Color Picture Tube
CPU	Upper Capability Index
CRM	Customer Relationship Management
CST	Critical Success Theme
CSUE	Creating & Capturing, Storing & Sharing, Utilization and Evaluation
CTC	Critical-to-customer
CTQ	Critical-to-quality
DBMS	Data Base Management System
DCOV	Define-Characterize-Optimize-Verify

DFM	Design for Manufacturability
DFR	Design for Reliability
DFSS	Design for Six Sigma
DIDES	Define-Initiate-Design-Execute-Sustain
DIDOV	Define-Identify-Design-Optimize-Verify
DMAD(O)V	Define-Measure-Analyze-Design-(Optimize)-Verify
DMAIC	Define-Measure-Analyze-Improve-Control
DMARIC	Define-Measure-Analyze-Redesign-Implement-Control
DMEDI	Define-Measure-Explore-Develop-Implement
DOE	Design of Experiments
DPMO	Defects Per Million Opportunities
DPO	Defects Per Opportunity
DPU	Defects Per Unit
DR	Design Review
DT	Data Technology
ECIM	Engineering Computer Integrated Manufacturing
EPA	European Productivity Agency
ERP	Enterprise Resources Planning
E-CIM	Engineering Computer Integrated Manufacturing
FMEA	Failure Modes and Effects Analysis
Gage R&R	Gage repeatability and reproducibility
GB	Green Belt
GE	General Electric
IDOV	Identify-Design-Optimize-Verify (or Validate)
ISO	International Organization for Standardization
IT	Information Technology
JIT	Just-in-time
KBSS	Knowledge Based Six Sigma
KM	Knowledge Management
KPIV	Key Process Input Variable
KPOV	Key Process Output Variable
LCL	Lower Control Limit
LGE-DA	The Digital Appliance Company of LG Electronics
LSL	Lower Specification Limit
MAIC	Measure-Analyze-Improve-Control

MBB	Master Black Belt
MBNQA	Malcolm Baldrige National Quality Award
MRP	Material Requirement Planning
MSA	Measurement System Analysis
NPDP	New Product Development Process
OFAT	One-factor-at-a-time
PDM	Product Data Management
PI	Process Innovation
ppm	parts per million
QC	Quality Control
QFD	Quality Function Deployment
RD	Robust Design
R&D	Research and Development
RPN	Risk Priority Number
RSM	Response Surface Methodology
RSS	Root Sum of Squares
RTY	Rolled Throughput Yield
4S	Systematic, Scientific, Statistical and Smarter
SCM	Supply Chain Management
SPC	Statistical Process Control
SQC	Statistical Quality Control
TPC	Total Productivity Control
TPM	Total Productive Maintenance
TQC	Total Quality Control
TQM	Total Quality Management
TRIZ	Teoriya Resheniya Izobretatelskih Zadach (in Russian) Theory of Inventive Problem Solving (in English)
TSS	Transactional Six Sigma
UCL	Upper Control Limit
USL	Upper Specification Limit
VOC	Voice of Customer
WB	White Belt
YB	Yellow Belt

Appendix A Standard normal distribution table

$$P(Z \geq z) = \int_z^{\infty} \frac{1}{\sqrt{2\pi}} \exp(-t^2/2)dt$$

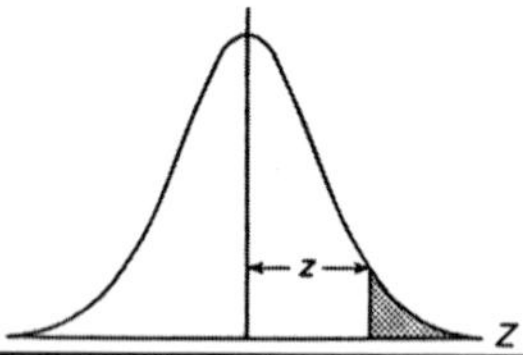

z	0.00	0.01	0.02	0.03	0.04	0.05	0.06	0.07	0.08	0.09
0.0	0.5000	0.4960	0.4920	0.4880	0.4840	0.4801	0.4761	0.4721	0.4681	0.4641
0.1	0.4602	0.4562	0.4522	0.4483	0.4443	0.4404	0.4364	0.4325	0.4286	0.4247
0.2	0.4207	0.4168	0.4129	0.4090	0.4052	0.4013	0.3974	0.3936	0.3897	0.3859
0.3	0.3821	0.3783	0.3745	0.3707	0.3669	0.3632	0.3594	0.3557	0.3520	0.3483
0.4	0.3446	0.3409	0.3372	0.3336	0.3300	0.3264	0.3228	0.3192	0.3156	0.3121
0.5	0.3085	0.3050	0.3015	0.2981	0.2946	0.2912	0.2877	0.2843	0.2810	0.2776
0.6	0.2743	0.2709	0.2676	0.2643	0.2611	0.2578	0.2546	0.2514	0.2483	0.2451
0.7	0.2420	0.2389	0.2358	0.2327	0.2296	0.2266	0.2236	0.2206	0.2177	0.2148
0.8	0.2119	0.2090	0.2061	0.2033	0.2005	0.1977	0.1949	0.1922	0.1894	0.1867
0.9	0.1841	0.1814	0.1788	0.1762	0.1736	0.1711	0.1685	0.1660	0.1635	0.1611
1.0	0.1587	0.1562	0.1539	0.1515	0.1492	0.1469	0.1446	0.1423	0.1401	0.1379
1.1	0.1357	0.1335	0.1314	0.1292	0.1271	0.1251	0.1230	0.1210	0.1190	0.1170
1.2	0.1151	0.1131	0.1112	0.1093	0.1075	0.1056	0.1038	0.1020	0.1003	0.0985
1.3	0.0968	0.0951	0.0934	0.0918	0.0901	0.0885	0.0869	0.0853	0.0838	0.0823
1.4	0.0808	0.0793	0.0778	0.0764	0.0749	0.0735	0.0721	0.0708	0.0694	0.0681
1.5	0.0668	0.0655	0.0643	0.0630	0.0618	0.0606	0.0594	0.0582	0.0571	0.0559
1.6	0.0548	0.0537	0.0526	0.0516	0.0505	0.0495	0.0485	0.0475	0.0465	0.0455
1.7	0.0446	0.0436	0.0427	0.0418	0.0409	0.0401	0.0392	0.0384	0.0375	0.0367
1.8	0.0359	0.0351	0.0344	0.0336	0.0329	0.0322	0.0314	0.0307	0.0301	0.0294
1.9	0.0287	0.0281	0.0274	0.0268	0.0262	0.0256	0.0250	0.0244	0.0239	0.0233
2.0	0.0228	0.0222	0.0217	0.0212	0.0207	0.0202	0.0197	0.0192	0.0188	0.0183
2.1	0.0179	0.0174	0.0170	0.0166	0.0162	0.0158	0.0154	0.0150	0.0146	0.0143
2.2	0.0139	0.0136	0.0132	0.0129	0.0125	0.0122	0.0119	0.0116	0.0113	0.0110
2.3	0.0107	0.0104	0.0102	0.0099	0.0096	0.0094	0.0091	0.0089	0.0087	0.0084
2.4	0.0082	0.0080	0.0078	0.0075	0.0073	0.0071	0.0069	0.0068	0.0066	0.0064
2.5	0.0062	0.0060	0.0059	0.0057	0.0055	0.0054	0.0052	0.0051	0.0049	0.0048
2.6	0.0047	0.0045	0.0044	0.0043	0.0041	0.0040	0.0039	0.0038	0.0037	0.0036
2.7	0.0035	0.0034	0.0033	0.0032	0.0031	0.0030	0.0029	0.0028	0.0027	0.0026
2.8	0.0026	0.0025	0.0024	0.0023	0.0023	0.0022	0.0021	0.0021	0.0020	0.0019
2.9	0.0019	0.0018	0.0018	0.0017	0.0016	0.0016	0.0015	0.0015	0.0014	0.0014
3.0	0.0013	0.0013	0.0013	0.0012	0.0012	0.0011	0.0011	0.0011	0.0010	0.0010
3.1	0.0010	0.0009	0.0009	0.0009	0.0008	0.0008	0.0008	0.0008	0.0007	0.0007
3.2	0.0007	0.0007	0.0006	0.0006	0.0006	0.0006	0.0006	0.0005	0.0005	0.0005
3.3	0.0005	0.0005	0.0005	0.0004	0.0004	0.0004	0.0004	0.0004	0.0004	0.0003
3.4	0.0003	0.0003	0.0003	0.0003	0.0003	0.0003	0.0003	0.0003	0.0003	0.0002
3.5	0.0002	0.0002	0.0002	0.0002	0.0002	0.0002	0.0002	0.0002	0.0002	0.0002
3.6	0.0002	0.0002	0.0001	0.0001	0.0001	0.0001	0.0001	0.0001	0.0001	0.0001
3.7	0.0001	0.0001	0.0001	0.0001	0.0001	0.0001	0.0001	0.0001	0.0001	0.0001
3.8	0.0001	0.0001	0.0001	0.0001	0.0001	0.0001	0.0001	0.0001	0.0001	0.0001

Appendix B t-distribution table of $t_{1-\alpha}(\phi)$

$$P[t(\phi) \leq t_{1-\alpha}(\phi)] = 1-\alpha$$

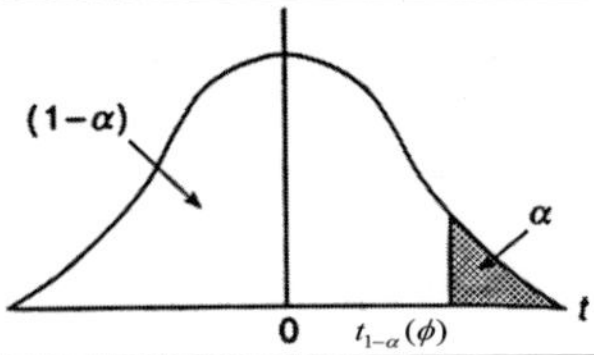

D.F. ϕ \ α	0.1	0.05	0.025	0.01	0.005	0.0005
1	3.078	6.314	12.706	31.821	63.657	636.619
2	1.886	2.920	4.303	6.965	9.925	31.599
3	1.638	2.353	3.182	4.541	5.841	12.924
4	1.533	2.132	2.776	3.747	4.604	8.610
5	1.476	2.015	2.571	3.365	4.032	6.869
6	1.440	1.943	2.447	3.143	3.707	5.959
7	1.415	1.895	2.365	2.998	3.499	5.408
8	1.397	1.860	2.306	2.896	3.355	5.041
9	1.383	1.833	2.262	2.821	3.250	4.781
10	1.372	1.812	2.228	2.764	3.169	4.587
11	1.363	1.796	2.201	2.718	3.106	4.437
12	1.356	1.782	2.179	2.681	3.055	4.318
13	1.350	1.771	2.160	2.650	3.012	4.221
14	1.345	1.761	2.145	2.624	2.977	4.140
15	1.341	1.753	2.131	2.602	2.947	4.073
16	1.337	1.746	2.120	2.583	2.921	4.015
17	1.333	1.740	2.110	2.567	2.898	3.965
18	1.330	1.734	2.101	2.552	2.878	3.922
19	1.328	1.729	2.093	2.539	2.861	3.883
20	1.325	1.725	2.086	2.528	2.845	3.850
21	1.323	1.721	2.080	2.518	2.831	3.819
22	1.321	1.717	2.074	2.508	2.819	3.792
23	1.319	1.714	2.069	2.500	2.807	3.768
24	1.318	1.711	2.064	2.492	2.797	3.745
25	1.316	1.708	2.060	2.485	2.787	3.725
26	1.315	1.706	2.056	2.479	2.779	3.707
27	1.314	1.703	2.052	2.473	2.771	3.690
28	1.313	1.701	2.048	2.467	2.763	3.674
29	1.311	1.699	2.045	2.462	2.756	3.659
30	1.310	1.697	2.042	2.457	2.750	3.646
40	1.303	1.684	2.021	2.423	2.704	3.551
50	1.299	1.676	2.009	2.403	2.678	3.496
60	1.296	1.671	2.000	2.390	2.660	3.460
70	1.294	1.667	1.994	2.381	2.648	3.435
80	1.292	1.664	1.990	2.374	2.639	3.416
90	1.291	1.662	1.987	2.368	2.632	3.402
100	1.290	1.660	1.984	2.364	2.626	3.390
110	1.289	1.659	1.982	2.361	2.621	3.381
120	1.289	1.658	1.980	2.358	2.617	3.373
∞	1.282	1.645	1.960	2.326	2.576	3.291

Appendix C χ^2-distribution table of $\chi^2_{1-\alpha}(\phi)$

$$P[\chi^2 \le \chi^2_{1-\alpha}(\phi)] = 1-\alpha$$

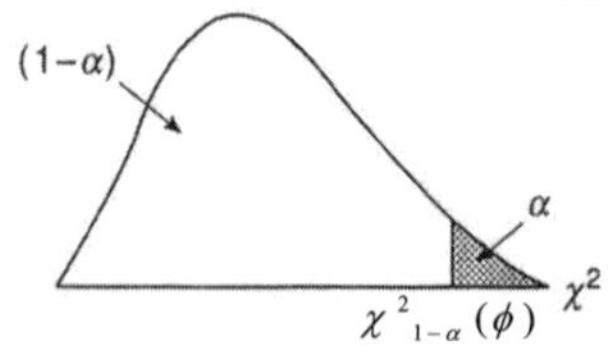

D.F. ϕ \ α	0.995	0.900	0.975	0.950	0.900	0.100	0.050	0.025	0.010	0.005
1	$0.0^4 393$	$0.0^3 157$	$0.0^3 982$	$0.0^2 393$	0.0158	2.71	3.84	5.02	6.63	7.88
2	0.0100	0.0201	0.0506	0.103	0.211	4.61	5.99	7.38	9.21	10.60
3	0.072	0.115	0.216	0.352	0.584	6.25	7.81	9.35	11.34	12.84
4	0.207	0.297	0.484	0.711	1.064	7.78	9.49	11.14	13.28	14.86
5	0.412	0.554	0.831	1.145	1.61	9.24	11.07	12.83	15.09	16.75
6	0.676	0.872	1.24	1.64	2.20	10.64	12.59	14.45	16.81	18.55
7	0.989	1.24	1.69	2.17	2.83	12.02	14.07	16.01	18.48	20.28
8	1.34	1.65	2.18	2.83	3.49	13.36	15.51	17.53	20.09	21.96
9	1.73	2.09	2.70	3.33	4.17	14.68	16.92	19.02	21.67	23.59
10	2.16	2.56	3.25	3.94	4.87	15.99	18.31	20.48	23.21	25.19
11	2.60	3.05	3.82	4.57	5.58	17.28	19.68	21.92	24.73	26.76
12	3.07	3.57	4.40	5.23	6.30	18.55	21.03	23.34	26.22	28.30
13	3.57	4.11	5.01	5.89	7.04	19.81	22.36	24.71	27.69	29.82
14	4.07	4.66	5.63	6.57	7.79	21.06	23.68	26.12	29.14	31.32
15	4.60	5.23	6.26	7.26	8.55	22.31	25.00	27.49	30.58	32.80
16	5.14	5.81	6.91	7.96	9.31	23.54	26.30	28.85	32.00	34.27
17	5.70	6.41	7.56	8.67	10.09	24.77	27.59	30.19	33.41	35.72
18	6.26	7.01	8.23	9.39	10.86	25.99	28.87	31.53	34.81	37.16
19	6.84	7.63	8.91	10.12	11.65	27.20	30.14	32.85	36.19	38.58
20	7.43	8.26	9.59	10.85	12.44	28.41	31.41	34.17	37.57	40.00
21	8.03	8.90	10.28	11.59	13.24	29.62	32.67	35.48	38.93	41.40
22	8.64	9.54	10.98	12.34	14.04	30.81	33.92	36.78	40.29	42.80
23	9.26	10.20	11.69	13.09	14.85	32.01	35.17	38.08	41.64	44.18
24	9.89	10.86	12.40	13.85	15.66	33.20	36.42	39.36	42.98	45.66
25	10.52	11.52	13.12	14.61	16.47	34.38	37.65	40.65	44.31	46.93
26	11.16	12.20	13.84	15.38	17.29	35.56	38.89	41.92	45.64	48.29
27	11.81	12.88	14.57	16.15	18.11	35.74	40.11	43.19	46.96	49.64
28	12.46	13.56	15.31	16.93	18.94	37.92	41.34	44.46	48.28	50.99
29	13.12	14.26	16.05	17.71	19.77	39.09	42.56	45.72	49.59	52.34
30	13.79	14.95	16.79	18.49	20.60	40.26	43.77	46.98	50.89	53.67
40	20.71	22.16	24.43	26.51	29.05	51.81	55.76	59.34	63.69	66.77
50	27.99	29.71	32.33	34.76	37.69	63.17	67.50	71.42	76.15	79.49
60	35.53	37.48	49.48	43.19	46.46	74.40	79.08	83.30	88.38	91.95
70	43.28	45.44	48.76	51.74	55.33	85.53	90.53	95.02	100.4	104.2
80	51.17	53.54	57.15	60.39	64.28	96.58	101.9	106.0	112.3	113.6
90	51.20	61.75	65.65	69.13	73.29	107.6	113.1	118.1	124.1	128.3
100	67.33	70.06	74.22	77.93	82.36	118.5	124.3	129.6	153.8	140.2

Appendix D F-distribution table of $F_{1-\alpha}(\phi_1,\phi_2)$

$$P[F(\phi_1,\phi_2) \le F_{1-\alpha}(\phi_1,\phi_2)] = 1-\alpha$$

$$F_{1-\alpha}(\varphi_1,\varphi_2) = \frac{1}{F_\alpha(\varphi_2,\varphi_1)}$$

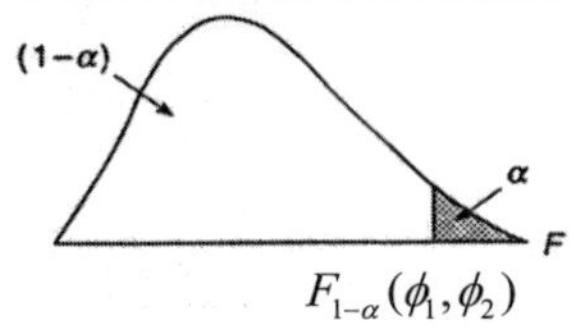

		ϕ_1								
ϕ_2	α	1	2	3	4	5	6	7	8	9
1	0.1	39.86	49.50	53.59	55.83	57.24	58.20	58.91	59.44	59.86
	0.05	161.45	199.50	215.71	224.58	230.16	233.99	236.77	238.88	240.54
	0.025	647.79	799.50	864.16	899.58	921.85	937.11	948.22	956.66	963.28
	0.01	4052.18	4999.50	5403.35	5624.58	5763.65	5858.99	5928.36	5981.07	6022.47
2	0.1	8.53	9.00	9.16	9.24	9.29	9.33	9.35	9.37	9.38
	0.05	18.51	19.00	19.16	19.25	19.30	19.33	19.35	19.37	19.38
	0.025	38.51	39.00	39.17	39.25	39.30	39.33	39.36	39.37	39.39
	0.01	98.50	99.00	99.17	99.25	99.30	99.33	99.36	99.37	99.39
3	0.1	5.54	5.46	5.39	5.34	5.31	5.28	5.27	5.25	5.24
	0.05	10.13	9.55	9.28	9.12	9.01	8.94	8.89	8.85	8.81
	0.025	17.44	16.04	15.44	15.10	14.88	14.73	14.62	14.54	14.47
	0.01	34.12	30.82	29.46	28.71	28.24	27.91	27.67	27.49	27.35
4	0.1	4.54	4.32	4.19	4.11	4.05	4.01	3.98	3.95	3.94
	0.05	7.71	6.94	6.59	6.39	6.26	6.16	6.09	6.04	6.00
	0.025	12.22	10.65	9.98	9.60	9.36	9.20	9.07	8.98	8.90
	0.01	21.20	18.00	16.69	15.98	15.52	15.21	14.98	14.80	14.66
5	0.1	4.06	3.78	3.62	3.52	3.45	3.40	3.37	3.34	3.32
	0.05	6.61	5.79	5.41	5.19	5.05	4.95	4.88	4.82	4.77
	0.025	10.01	8.43	7.76	7.39	7.15	6.98	6.85	6.76	6.68
	0.01	16.26	13.27	12.06	11.39	10.97	10.67	10.46	10.29	10.16
6	0.1	3.78	3.46	3.29	3.18	3.11	3.05	3.01	2.98	2.96
	0.05	5.99	5.14	4.76	4.53	4.39	4.28	4.21	4.15	4.10
	0.025	8.81	7.26	6.60	6.23	5.99	5.82	5.70	5.60	5.52
	0.01	13.75	10.92	9.78	9.15	8.75	8.47	8.26	8.10	7.98
7	0.1	3.59	3.26	3.07	2.96	2.88	2.83	2.78	2.75	2.72
	0.05	5.59	4.74	4.35	4.12	3.97	3.87	3.79	3.73	3.68
	0.025	8.07	6.54	5.89	5.52	5.29	5.12	4.99	4.90	4.82
	0.01	12.25	9.55	8.45	7.85	7.46	7.19	6.99	6.84	6.72
8	0.1	3.46	3.11	2.92	2.81	2.73	2.67	2.62	2.59	2.56
	0.05	5.32	4.46	4.07	3.84	3.69	3.58	3.50	3.44	3.39
	0.025	7.57	6.06	5.42	5.05	4.82	4.65	4.53	4.43	4.36
	0.01	11.26	8.65	7.59	7.01	6.63	6.37	6.18	6.03	5.91
9	0.1	3.36	3.01	2.81	2.69	2.61	2.55	2.51	2.47	2.44
	0.05	5.12	4.26	3.86	3.63	3.48	3.37	3.29	3.23	3.18
	0.025	7.21	5.71	5.08	4.72	4.48	4.32	4.20	4.10	4.03
	0.01	10.56	8.02	6.99	6.42	6.06	5.80	5.61	5.47	5.35
10	0.1	3.29	2.92	2.73	2.61	2.52	2.46	2.41	2.38	2.35
	0.05	4.96	4.10	3.71	3.48	3.33	3.22	3.14	3.07	3.02
	0.025	6.94	5.46	4.83	4.47	4.24	4.07	3.95	3.85	3.78
	0.01	10.04	7.56	6.55	5.99	5.64	5.39	5.20	5.06	4.94
11	0.1	3.23	2.86	2.66	2.54	2.45	2.39	2.34	2.30	2.27
	0.05	4.84	3.98	3.59	3.36	3.20	3.09	3.01	2.95	2.90
	0.025	6.72	5.26	4.63	4.28	4.04	3.88	3.76	3.66	3.59
	0.01	9.65	7.21	6.22	5.67	5.32	5.07	4.89	4.74	4.63

Appendix D (Continued).

		ϕ_1								
ϕ_2	α	1	2	3	4	5	6	7	8	9
12	0.1	3.18	2.81	2.61	2.48	2.39	2.33	2.28	2.24	2.21
	0.05	4.75	3.89	3.49	3.26	3.11	3.00	2.91	2.85	2.80
	0.025	6.55	5.10	4.47	4.12	3.89	3.73	3.61	3.51	3.44
	0.01	9.33	6.93	5.95	5.41	5.06	4.82	4.64	4.50	4.39
13	0.1	3.14	2.76	2.56	2.43	2.35	2.28	2.23	2.20	2.16
	0.05	4.67	3.81	3.41	3.18	3.03	2.92	2.83	2.77	2.71
	0.025	6.41	4.97	4.35	4.00	3.77	3.60	3.48	3.39	3.31
	0.01	9.07	6.70	5.74	5.21	4.86	4.62	4.44	4.30	4.19
14	0.1	3.10	2.73	2.52	2.39	2.31	2.24	2.19	2.15	2.12
	0.05	4.60	3.74	3.34	3.11	2.96	2.85	2.76	2.70	2.65
	0.025	6.30	4.86	4.24	3.89	3.66	3.50	3.38	3.29	3.21
	0.01	8.86	6.51	5.56	5.04	4.69	4.46	4.28	4.14	4.03
15	0.1	3.07	2.70	2.49	2.36	2.27	2.21	2.16	2.12	2.09
	0.05	4.54	3.68	3.29	3.06	2.90	2.79	2.71	2.64	2.59
	0.025	6.20	4.77	4.15	3.80	3.58	3.41	3.29	3.20	3.12
	0.01	8.68	6.36	5.42	4.89	4.56	4.32	4.14	4.00	3.89
16	0.1	3.05	2.67	2.46	2.33	2.24	2.18	2.13	2.09	2.06
	0.05	4.49	3.63	3.24	3.01	2.85	2.74	2.66	2.59	2.54
	0.025	6.12	4.69	4.08	3.73	3.50	3.34	3.22	3.12	3.05
	0.01	8.53	6.23	5.29	4.77	4.44	4.20	4.03	3.89	3.78
17	0.1	3.03	2.64	2.44	2.31	2.22	2.15	2.10	2.06	2.03
	0.05	4.45	3.59	3.20	2.96	2.81	2.70	2.61	2.55	2.49
	0.025	6.04	4.62	4.01	3.66	3.44	3.28	3.16	3.06	2.98
	0.01	8.40	6.11	5.18	4.67	4.34	4.10	3.93	3.79	3.68
18	0.1	3.01	2.62	2.42	2.29	2.20	2.13	2.08	2.04	2.00
	0.05	4.41	3.55	3.16	2.93	2.77	2.66	2.58	2.51	2.46
	0.025	5.98	4.56	3.95	3.61	3.38	3.22	3.10	3.01	2.93
	0.01	8.29	6.01	5.09	4.58	4.25	4.01	3.84	3.71	3.60
19	0.1	2.99	2.61	2.40	2.27	2.18	2.11	2.06	2.02	1.98
	0.05	4.38	3.52	3.13	2.90	2.74	2.63	2.54	2.48	2.42
	0.025	5.92	4.51	3.90	3.56	3.33	3.17	3.05	2.96	2.88
	0.01	8.18	5.93	5.01	4.50	4.17	3.94	3.77	3.63	3.52
20	0.1	2.97	2.59	2.38	2.25	2.16	2.09	2.04	2.00	1.96
	0.05	4.35	3.49	3.10	2.87	2.71	2.60	2.51	2.45	2.39
	0.025	5.87	4.46	3.86	3.51	3.29	3.13	3.01	2.91	2.84
	0.01	8.10	5.85	4.94	4.43	4.10	3.87	3.70	3.56	3.46
24	0.1	2.93	2.54	2.33	2.19	2.10	2.04	1.98	1.94	1.91
	0.05	4.26	3.40	3.01	2.78	2.62	2.51	2.42	2.36	2.30
	0.025	5.72	4.32	3.72	3.38	3.15	2.99	2.87	2.78	2.70
	0.01	7.82	5.61	4.72	4.22	3.90	3.67	3.50	3.36	3.26
30	0.1	2.88	2.49	2.28	2.14	2.05	1.98	1.93	1.88	1.85
	0.05	4.17	3.32	2.92	2.69	2.53	2.42	2.33	2.27	2.21
	0.025	5.57	4.18	3.59	3.25	3.03	2.87	2.75	2.65	2.57
	0.01	7.56	5.39	4.51	4.02	3.70	3.47	3.30	3.17	3.07
60	0.1	2.79	2.39	2.18	2.04	1.95	1.87	1.82	1.77	1.74
	0.05	4.00	3.15	2.76	2.53	2.37	2.25	2.17	2.10	2.04
	0.025	5.29	3.93	3.34	3.01	2.79	2.63	2.51	2.41	2.33
	0.01	7.08	4.98	4.13	3.65	3.34	3.12	2.95	2.82	2.72
120	0.1	2.75	2.35	2.13	1.99	1.90	1.82	1.77	1.72	1.68
	0.05	3.92	3.07	2.68	2.45	2.29	2.18	2.09	2.02	1.96
	0.025	5.15	3.80	3.23	2.89	2.67	2.52	2.39	2.30	2.22
	0.01	6.85	4.79	3.95	3.48	3.17	2.96	2.79	2.66	2.56
∞	0.1	2.71	2.30	2.08	1.95	1.85	1.77	1.72	1.67	1.63
	0.05	3.84	3.00	2.61	2.37	2.21	2.10	2.01	1.94	1.88
	0.025	5.03	3.69	3.12	2.79	2.57	2.41	2.29	2.19	2.11

Appendix D (Continued).

ϕ_2	α	ϕ_1 10	11	12	15	20	24	30	60	120	∞
1	0.100	60.19	60.47	60.71	61.22	61.74	62.00	62.26	62.79	63.06	63.32
	0.050	241.88	242.98	243.91	245.95	248.01	249.05	250.10	252.20	253.25	254.30
	0.025	968.63	973.03	976.71	984.87	993.10	997.25	1001.41	1009.80	1014.02	1018.21
	0.010	6055.85	6083.32	6106.32	6157.28	6208.73	6234.63	6260.65	6313.03	6339.39	6365.55
2	0.100	9.39	9.40	9.41	9.42	9.44	9.45	9.46	9.47	9.48	9.49
	0.050	19.40	19.40	19.41	19.43	19.45	19.45	19.46	19.48	19.49	19.50
	0.025	39.40	39.41	39.41	39.43	39.45	39.46	39.46	39.48	39.49	39.50
	0.010	99.40	99.41	99.42	99.43	99.45	99.46	99.47	99.48	99.49	99.50
3	0.100	5.23	5.22	5.22	5.20	5.18	5.18	5.17	5.15	5.14	5.13
	0.050	8.79	8.76	8.74	8.70	8.66	8.64	8.62	8.57	8.55	8.53
	0.025	14.42	14.37	14.34	14.25	14.17	14.12	14.08	13.99	13.95	13.90
	0.010	27.23	27.13	27.05	26.87	26.69	26.60	26.50	26.32	26.22	26.13
4	0.100	3.92	3.91	3.90	3.87	3.84	3.83	3.82	3.79	3.78	3.76
	0.050	5.96	5.94	5.91	5.86	5.80	5.77	5.75	5.69	5.66	5.63
	0.025	8.84	8.79	8.75	8.66	8.56	8.51	8.46	8.36	8.31	8.26
	0.010	14.55	14.45	14.37	14.20	14.02	13.93	13.84	13.65	13.56	13.46
5	0.100	3.30	3.28	3.27	3.24	3.21	3.19	3.17	3.14	3.12	3.11
	0.050	4.74	4.70	4.68	4.62	4.56	4.53	4.50	4.43	4.40	4.37
	0.025	6.62	6.57	6.52	6.43	6.33	6.28	6.23	6.12	6.07	6.02
	0.010	10.05	9.96	9.89	9.72	9.55	9.47	9.38	9.20	9.11	9.02
6	0.100	2.94	2.92	2.90	2.87	2.84	2.82	2.80	2.76	2.74	2.72
	0.050	4.06	4.03	4.00	3.94	3.87	3.84	3.81	3.74	3.70	3.67
	0.025	5.46	5.41	5.37	5.27	5.17	5.12	5.07	4.96	4.90	4.85
	0.010	7.87	7.79	7.72	7.56	7.40	7.31	7.23	7.06	6.97	6.88
7	0.100	2.70	2.68	2.67	2.63	2.59	2.58	2.56	2.51	2.49	2.47
	0.050	3.64	3.60	3.57	3.51	3.44	3.41	3.38	3.30	3.27	3.23
	0.025	4.76	4.71	4.67	4.57	4.47	4.41	4.36	4.25	4.20	4.14
	0.010	6.62	6.54	6.47	6.31	6.16	6.07	5.99	5.82	5.74	5.65
8	0.100	2.54	2.52	2.50	2.46	2.42	2.40	2.38	2.34	2.32	2.29
	0.050	3.35	3.31	3.28	3.22	3.15	3.12	3.08	3.01	2.97	2.93
	0.025	4.30	4.24	4.20	4.10	4.00	3.95	3.89	3.78	3.73	3.67
	0.010	5.81	5.73	5.67	5.52	5.36	5.28	5.20	5.03	4.95	4.86
9	0.100	2.42	2.40	2.38	2.34	2.30	2.28	2.25	2.21	2.18	2.16
	0.050	3.14	3.10	3.07	3.01	2.94	2.90	2.86	2.79	2.75	2.71
	0.025	3.96	3.91	3.87	3.77	3.67	3.61	3.56	3.45	3.39	3.33
	0.010	5.26	5.18	5.11	4.96	4.81	4.73	4.65	4.48	4.40	4.31
10	0.100	2.32	2.30	2.28	2.24	2.20	2.18	2.16	2.11	2.08	2.06
	0.050	2.98	2.94	2.91	2.85	2.77	2.74	2.70	2.62	2.58	2.54
	0.025	3.72	3.66	3.62	3.52	3.42	3.37	3.31	3.20	3.14	3.08
	0.010	4.85	4.77	4.71	4.56	4.41	4.33	4.25	4.08	4.00	3.91
11	0.100	2.25	2.23	2.21	2.17	2.12	2.10	2.08	2.03	2.00	1.97
	0.050	2.85	2.82	2.79	2.72	2.65	2.61	2.57	2.49	2.45	2.41
	0.025	3.53	3.47	3.43	3.33	3.23	3.17	3.12	3.00	2.94	2.88
	0.010	4.54	4.46	4.40	4.25	4.10	4.02	3.94	3.78	3.69	3.60
12	0.100	2.19	2.17	2.15	2.10	2.06	2.04	2.01	1.96	1.93	1.90
	0.050	2.75	2.72	2.69	2.62	2.54	2.51	2.47	2.38	2.34	2.30
	0.025	3.37	3.32	3.28	3.18	3.07	3.02	2.96	2.85	2.79	2.73
	0.010	4.30	4.22	4.16	4.01	3.86	3.78	3.70	3.54	3.45	3.36
13	0.100	2.14	2.12	2.10	2.05	2.01	1.98	1.96	1.90	1.88	1.85
	0.050	2.67	2.63	2.60	2.53	2.46	2.42	2.38	2.30	2.25	2.21
	0.025	3.25	3.20	3.15	3.05	2.95	2.89	2.84	2.72	2.66	2.60
	0.010	4.10	4.02	3.96	3.82	3.66	3.59	3.51	3.34	3.25	3.17

Appendix D (Continued).

		ϕ_1									
ϕ_2	α	10	11	12	15	20	24	30	60	120	∞
14	0.100	2.10	2.07	2.05	2.01	1.96	1.94	1.91	1.86	1.83	1.80
	0.050	2.60	2.57	2.53	2.46	2.39	2.35	2.31	2.22	2.18	2.13
	0.025	3.15	3.09	3.05	2.95	2.84	2.79	2.73	2.61	2.55	2.49
	0.010	3.94	3.86	3.80	3.66	3.51	3.43	3.35	3.18	3.09	3.01
15	0.100	2.06	2.04	2.02	1.97	1.92	1.90	1.87	1.82	1.79	1.76
	0.050	2.54	2.51	2.48	2.40	2.33	2.29	2.25	2.16	2.11	2.07
	0.025	3.06	3.01	2.96	2.86	2.76	2.70	2.64	2.52	2.46	2.40
	0.010	3.80	3.73	3.67	3.52	3.37	3.29	3.21	3.05	2.96	2.87
16	0.100	2.03	2.01	1.99	1.94	1.89	1.87	1.84	1.78	1.75	1.72
	0.050	2.49	2.46	2.42	2.35	2.28	2.24	2.19	2.11	2.06	2.01
	0.025	2.99	2.93	2.89	2.79	2.68	2.63	2.57	2.45	2.38	2.32
	0.010	3.69	3.62	3.55	3.41	3.26	3.18	3.10	2.93	2.84	2.75
17	0.100	2.00	1.98	1.96	1.91	1.86	1.84	1.81	1.75	1.72	1.69
	0.050	2.45	2.41	2.38	2.31	2.23	2.19	2.15	2.06	2.01	1.96
	0.025	2.92	2.87	2.82	2.72	2.62	2.56	2.50	2.38	2.32	2.25
	0.010	3.59	3.52	3.46	3.31	3.16	3.08	3.00	2.83	2.75	2.65
18	0.100	1.98	1.95	1.93	1.89	1.84	1.81	1.78	1.72	1.69	1.66
	0.050	2.41	2.37	2.34	2.27	2.19	2.15	2.11	2.02	1.97	1.92
	0.025	2.87	2.81	2.77	2.67	2.56	2.50	2.44	2.32	2.26	2.19
	0.010	3.51	3.43	3.37	3.23	3.08	3.00	2.92	2.75	2.66	2.57
19	0.100	1.96	1.93	1.91	1.86	1.81	1.79	1.76	1.70	1.67	1.63
	0.050	2.38	2.34	2.31	2.23	2.16	2.11	2.07	1.98	1.93	1.88
	0.025	2.82	2.76	2.72	2.62	2.51	2.45	2.39	2.27	2.20	2.13
	0.010	3.43	3.36	3.30	3.15	3.00	2.92	2.84	2.67	2.58	2.49
20	0.100	1.94	1.91	1.89	1.84	1.79	1.77	1.74	1.68	1.64	1.61
	0.050	2.35	2.31	2.28	2.20	2.12	2.08	2.04	1.95	1.90	1.84
	0.025	2.77	2.72	2.68	2.57	2.46	2.41	2.35	2.22	2.16	2.09
	0.010	3.37	3.29	3.23	3.09	2.94	2.86	2.78	2.61	2.52	2.42
24	0.100	1.88	1.85	1.83	1.78	1.73	1.70	1.67	1.61	1.57	1.53
	0.050	2.25	2.22	2.18	2.11	2.03	1.98	1.94	1.84	1.79	1.73
	0.025	2.64	2.59	2.54	2.44	2.33	2.27	2.21	2.08	2.01	1.94
	0.010	3.17	3.09	3.03	2.89	2.74	2.66	2.58	2.40	2.31	2.21
30	0.100	1.82	1.79	1.77	1.72	1.67	1.64	1.61	1.54	1.50	1.46
	0.050	2.16	2.13	2.09	2.01	1.93	1.89	1.84	1.74	1.68	1.62
	0.025	2.51	2.46	2.41	2.31	2.20	2.14	2.07	1.94	1.87	1.79
	0.010	2.98	2.91	2.84	2.70	2.55	2.47	2.39	2.21	2.11	2.01
60	0.100	1.71	1.68	1.66	1.60	1.54	1.51	1.48	1.40	1.35	1.29
	0.050	1.99	1.95	1.92	1.84	1.75	1.70	1.65	1.53	1.47	1.39
	0.025	2.27	2.22	2.17	2.06	1.94	1.88	1.82	1.67	1.58	1.48
	0.010	2.63	2.56	2.50	2.35	2.20	2.12	2.03	1.84	1.73	1.60
120	0.100	1.65	1.63	1.60	1.55	1.48	1.45	1.41	1.32	1.26	1.19
	0.050	1.91	1.87	1.83	1.75	1.66	1.61	1.55	1.43	1.35	1.26
	0.025	2.16	2.10	2.05	1.94	1.82	1.76	1.69	1.53	1.43	1.31
	0.010	2.47	2.40	2.34	2.19	2.03	1.95	1.86	1.66	1.53	1.38
∞	0.100	1.60	1.57	1.55	1.49	1.42	1.38	1.34	1.24	1.17	1.00
	0.050	1.83	1.79	1.75	1.67	1.57	1.52	1.46	1.32	1.22	1.00
	0.025	2.05	1.99	1.95	1.83	1.71	1.64	1.57	1.39	1.27	1.00
	0.010	2.32	2.25	2.19	2.04	1.88	1.79	1.70	1.48	1.33	1.00

Appendix E Omega transformation table

$$dB = -10\log(\frac{1}{p} - 1)$$

p(%)	dB	p(%)	dB	p(%)	dB	p(%)	dB	p(%)	dB	p(%)	dB
0.0	∞*	3.5	-14.405	7.0	-11.234	10.5	-9.306	14.0	-7.884	17.5	-6.734
0.1	-29.996	3.6	-14.278	7.1	-11.168	10.6	-9.260	14.1	-7.848	17.6	-6.704
0.2	-26.981	3.7	-14.154	7.2	-11.102	10.7	-9.215	14.2	-7.812	17.7	-6.674
0.3	-25.216	3.8	-14.034	7.3	-11.038	10.8	-9.169	14.3	-7.776	17.8	-6.645
0.4	-23.962	3.9	-13.917	7.4	-10.974	10.9	-9.125	14.4	-7.741	17.9	-6.615
0.5	-22.989	4.0	-13.802	7.5	-10.911	11.0	-9.080	14.5	-7.706	18.0	-6.585
0.6	-22.192	4.1	-13.690	7.6	-10.849	11.1	-9.036	14.6	-7.671	18.1	-6.556
0.7	-21.519	4.2	-13.581	7.7	-10.787	11.2	-8.992	14.7	-7.636	18.2	-6.527
0.8	-20.934	4.3	-13.474	7.8	-10.726	11.3	-8.948	14.8	-7.602	18.3	-6.498
0.9	-20.418	4.4	-13.370	7.9	-10.666	11.4	-8.905	14.9	-7.567	18.4	-6.469
1.0	-19.956	4.5	-13.268	8.0	-10.607	11.5	-8.862	15.0	-7.533	18.5	-6.440
1.1	-19.538	4.6	-13.168	8.1	-10.548	11.6	-8.820	15.1	-7.499	18.6	-6.411
1.2	-19.156	4.7	-13.070	8.2	-10.490	11.7	-8.778	15.2	-7.466	18.7	-6.382
1.3	-18.804	4.8	-12.974	8.3	-10.433	11.8	-8.736	15.3	-7.432	18.8	-6.354
1.4	-18.477	4.9	-12.880	8.4	-10.376	11.9	-8.694	15.4	-7.398	18.9	-6.326
1.5	-18.173	5.0	-12.788	8.5	-10.320	12.0	-8.653	15.5	-7.365	19.0	-6.297
1.6	-17.889	5.1	-12.697	8.6	-10.264	12.1	-8.612	15.6	-7.332	19.1	-6.269
1.7	-17.621	5.2	-12.608	8.7	-10.210	12.2	-8.571	15.7	-7.299	19.2	-6.241
1.8	-17.368	5.3	-12.521	8.8	-10.155	12.3	-8.531	15.8	-7.267	19.3	-6.213
1.9	-17.129	5.4	-12.435	8.9	-10.101	12.4	-8.491	15.9	-7.234	19.4	-6.185
2.0	-16.902	5.5	-12.351	9.0	-10.048	12.5	-8.451	16.0	-7.202	19.5	-6.158
2.1	-16.686	5.6	-12.268	9.1	-9.995	12.6	-8.411	16.1	-7.169	19.6	-6.130
2.2	-16.479	5.7	-12.186	9.2	-9.943	12.7	-8.372	16.2	-7.137	19.7	-6.102
2.3	-16.282	5.8	-12.106	9.3	-9.891	12.8	-8.333	16.3	-7.105	19.8	-6.075
2.4	-16.092	5.9	-12.027	9.4	-9.840	12.9	-8.294	16.4	-7.074	19.9	-6.048
2.5	-15.911	6.0	-11.950	9.5	-9.789	13.0	-8.256	16.5	-7.042	20.0	-6.021
2.6	-15.736	6.1	-11.873	9.6	-9.739	13.1	-8.217	16.6	-7.011	20.1	-5.994
2.7	-15.567	6.2	-11.798	9.7	-9.689	13.2	-8.179	16.7	-6.979	20.2	-5.967
2.8	-15.405	6.3	-11.724	9.8	-9.640	13.3	-8.142	16.8	-6.948	20.3	-5.940
2.9	-15.248	6.4	-11.651	9.9	-9.591	13.4	-8.104	16.9	-6.917	20.4	-5.913
3.0	-15.097	6.5	-11.579	10.0	-9.542	13.5	-8.067	17.0	-6.886	20.5	-5.886
3.1	-14.950	6.6	-11.508	10.1	-9.494	13.6	-8.030	17.1	-6.856	20.6	-5.860
3.2	-14.807	6.7	-11.438	10.2	-9.447	13.7	-7.993	17.2	-6.825	20.7	-5.833
3.3	-14.669	6.8	-11.369	10.3	-9.400	13.8	-7.956	17.3	-6.795	20.8	-5.807
3.4	-14.535	6.9	-11.301	10.4	-9.353	13.9	-7.920	17.4	-6.764	20.9	-5.780
21.0	-5.754	26.0	-4.543	31.0	-3.475	36.0	-2.499	41.0	-1.581	46.0	-0.696
21.1	-5.728	26.1	-4.520	31.1	-3.455	36.1	-2.480	41.1	-1.563	46.1	-0.679
21.2	-5.702	26.2	-4.498	31.2	-3.434	36.2	-2.461	41.2	-1.545	46.2	-0.661
21.3	-5.676	26.3	-4.475	31.3	-3.414	36.3	-2.442	41.3	-1.527	46.3	-0.644

Appendix E (Continued).

p(%)	dB	p(%)	dB	p(%)	dB	p(%)	dB	p(%)	dB	p(%)	dB
21.4	-5.650	26.4	-4.453	31.4	-3.394	36.4	-2.424	41.4	-1.509	46.4	-0.626
21.5	-5.624	26.5	-4.430	31.5	-3.374	36.5	-2.405	41.5	-1.491	46.5	-0.609
21.6	-5.599	26.6	-4.408	31.6	-3.354	36.6	-2.386	41.6	-1.473	46.6	-0.592
21.7	-5.573	26.7	-4.386	31.7	-3.334	36.7	-2.367	41.7	-1.455	46.7	-0.574
21.8	-5.548	26.8	-4.364	31.8	-3.314	36.8	-2.349	41.8	-1.437	46.8	-0.557
21.9	-5.522	26.9	-4.342	31.9	-3.294	36.9	-2.330	41.9	-1.420	46.9	-0.539
22.0	-5.497	27.0	-4.320	32.0	-3.274	37.0	-2.311	42.0	-1.402	47.0	-0.522
22.1	-5.471	27.1	-4.298	32.1	-3.254	37.1	-2.293	42.1	-1.384	47.1	-0.504
22.2	-5.446	27.2	-4.276	32.2	-3.234	37.2	-2.274	42.2	-1.366	47.2	-0.487
22.3	-5.421	27.3	-4.254	32.3	-3.214	37.3	-2.256	42.3	-1.348	47.3	-0.469
22.4	-5.396	27.4	-4.232	32.4	-3.194	37.4	-2.237	42.4	-1.331	47.4	-0.452
22.5	-5.371	27.5	-4.210	32.5	-3.174	37.5	-2.218	42.5	-1.313	47.5	-0.435
22.6	-5.346	27.6	-4.188	32.6	-3.154	37.6	-2.200	42.6	-1.295	47.6	-0.417
22.7	-5.322	27.7	-4.167	32.7	-3.135	37.7	-2.181	42.7	-1.277	47.7	-0.400
22.8	-5.297	27.8	-4.145	32.8	-3.115	37.8	-2.163	42.8	-1.260	47.8	-0.382
22.9	-5.272	27.9	-4.123	32.9	-3.095	37.9	-2.145	42.9	-1.242	47.9	-0.365
23.0	-5.248	28.0	-4.102	33.0	-3.076	38.0	-2.126	43.0	-1.224	48.0	-0.348
23.1	-5.223	28.1	-4.080	33.1	-3.056	38.1	-2.108	43.1	-1.206	48.1	-0.330
23.2	-5.199	28.2	-4.059	33.2	-3.036	38.2	-2.089	43.2	-1.189	48.2	-0.313
23.3	-5.174	28.3	-4.037	33.3	-3.017	38.3	-2.071	43.3	-1.171	48.3	-0.295
23.4	-5.150	28.4	-4.016	33.4	-2.997	38.4	-2.052	43.4	-1.153	48.4	-0.278
23.5	-5.126	28.5	-3.995	33.5	-2.978	38.5	-2.034	43.5	-1.136	48.5	-0.261
23.6	-5.102	28.6	-3.973	33.6	-2.958	38.6	-2.016	43.6	-1.118	48.6	-0.243
23.7	-5.078	28.7	-3.952	33.7	-2.939	38.7	-1.997	43.7	-1.100	48.7	-0.226
23.8	-5.054	28.8	-3.931	33.8	-2.919	38.8	-1.979	43.8	-1.083	48.8	-0.209
23.9	-5.030	28.9	-3.910	33.9	-2.900	38.9	-1.961	43.9	-1.065	48.9	-0.191
24.0	-5.006	29.0	-3.889	34.0	-2.881	39.0	-1.943	44.0	-1.047	49.0	-0.174
24.1	-4.982	29.1	-3.868	34.1	-2.861	39.1	-1.924	44.1	-1.030	49.1	-0.156
24.2	-4.959	29.2	-3.847	34.2	-2.842	39.2	-1.906	44.2	-1.012	49.2	-0.139
24.3	-4.935	29.3	-3.826	34.3	-2.823	39.3	-1.888	44.3	-0.995	49.3	-0.122
24.4	-4.911	29.4	-3.805	34.4	-2.803	39.4	-1.870	44.4	-0.977	49.4	-0.104
24.5	-4.888	29.5	-3.784	34.5	-2.784	39.5	-1.852	44.5	-0.959	49.5	-0.087
24.6	-4.864	29.6	-3.763	34.6	-2.765	39.6	-1.833	44.6	-0.942	49.6	-0.069
24.7	-4.841	29.7	-3.742	34.7	-2.746	39.7	-1.815	44.7	-0.924	49.7	-0.052
24.8	-4.818	29.8	-3.721	34.8	-2.727	39.8	-1.797	44.8	-0.907	49.8	-0.035
24.9	-4.794	29.9	-3.700	34.9	-2.708	39.9	-1.779	44.9	-0.889	49.9	-0.017
25.0	-4.771	30.0	-3.680	35.0	-2.688	40.0	-1.761	45.0	-0.872	50.0	0.000
25.1	-4.748	30.1	-3.659	35.1	-2.669	40.1	-1.743	45.1	-0.854	50.1	0.017
25.2	-4.725	30.2	-3.638	35.2	-2.650	40.2	-1.725	45.2	-0.836	50.2	0.035
25.3	-4.702	30.3	-3.618	35.3	-2.631	40.3	-1.707	45.3	-0.819	50.3	0.052
25.4	-4.679	30.4	-3.597	35.4	-2.612	40.4	-1.689	45.4	-0.801	50.4	0.069
25.5	-4.656	30.5	-3.577	35.5	-2.593	40.5	-1.671	45.5	-0.784	50.5	0.087
25.6	-4.633	30.6	-3.556	35.6	-2.574	40.6	-1.653	45.6	-0.766	50.6	0.104
25.7	-4.611	30.7	-3.536	35.7	-2.555	40.7	-1.635	45.7	-0.749	50.7	0.122
25.8	-4.588	30.8	-3.516	35.8	-2.537	40.8	-1.617	45.8	-0.731	50.8	0.139

Appendix E (Continued).

p(%)	dB	p(%)	dB	p(%)	dB	p(%)	dB	p(%)	dB	p(%)	dB
25.9	-4.565	30.9	-3.495	35.9	-2.518	40.9	-1.599	45.9	-0.714	50.9	0.156
51.0	0.174	56.0	1.047	61.0	1.943	66.0	2.881	71.0	3.889	76.0	5.006
51.1	0.191	56.1	1.065	61.1	1.961	66.1	2.900	71.1	3.910	76.1	5.030
51.2	0.209	56.2	1.083	61.2	1.979	66.2	2.919	71.2	3.931	76.2	5.054
51.3	0.226	56.3	1.100	61.3	1.997	66.3	2.939	71.3	3.952	76.3	5.078
51.4	0.243	56.4	1.118	61.4	2.016	66.4	2.958	71.4	3.973	76.4	5.102
51.5	0.261	56.5	1.136	61.5	2.034	66.5	2.978	71.5	3.995	76.5	5.126
51.6	0.278	56.6	1.153	61.6	2.052	66.6	2.997	71.6	4.016	76.6	5.150
51.7	0.295	56.7	1.171	61.7	2.071	66.7	3.017	71.7	4.037	76.7	5.174
51.8	0.313	56.8	1.189	61.8	2.089	66.8	3.036	71.8	4.059	76.8	5.199
51.9	0.330	56.9	1.206	61.9	2.108	66.9	3.056	71.9	4.080	76.9	5.223
52.0	0.348	57.0	1.224	62.0	2.126	67.0	3.076	72.0	4.102	77.0	5.248
52.1	0.365	57.1	1.242	62.1	2.145	67.1	3.095	72.1	4.123	77.1	5.272
52.2	0.382	57.2	1.260	62.2	2.163	67.2	3.115	72.2	4.145	77.2	5.297
52.3	0.400	57.3	1.277	62.3	2.181	67.3	3.135	72.3	4.167	77.3	5.322
52.4	0.417	57.4	1.295	62.4	2.200	67.4	3.154	72.4	4.188	77.4	5.346
52.5	0.435	57.5	1.313	62.5	2.218	67.5	3.174	72.5	4.210	77.5	5.371
52.6	0.452	57.6	1.331	62.6	2.237	67.6	3.194	72.6	4.232	77.6	5.396
52.7	0.469	57.7	1.348	62.7	2.256	67.7	3.214	72.7	4.254	77.7	5.421
52.8	0.487	57.8	1.366	62.8	2.274	67.8	3.234	72.8	4.276	77.8	5.446
52.9	0.504	57.9	1.384	62.9	2.293	67.9	3.254	72.9	4.298	77.9	5.471
53.0	0.522	58.0	1.402	63.0	2.311	68.0	3.274	73.0	4.320	78.0	5.497
53.1	0.539	58.1	1.420	63.1	2.330	68.1	3.294	73.1	4.342	78.1	5.522
53.2	0.557	58.2	1.437	63.2	2.349	68.2	3.314	73.2	4.364	78.2	5.548
53.3	0.574	58.3	1.455	63.3	2.367	68.3	3.334	73.3	4.386	78.3	5.573
53.4	0.592	58.4	1.473	63.4	2.386	68.4	3.354	73.4	4.408	78.4	5.599
53.5	0.609	58.5	1.491	63.5	2.405	68.5	3.374	73.5	4.430	78.5	5.624
53.6	0.626	58.6	1.509	63.6	2.424	68.6	3.394	73.6	4.453	78.6	5.650
53.7	0.644	58.7	1.527	63.7	2.442	68.7	3.414	73.7	4.475	78.7	5.676
53.8	0.661	58.8	1.545	63.8	2.461	68.8	3.434	73.8	4.498	78.8	5.702
53.9	0.679	58.9	1.563	63.9	2.480	68.9	3.455	73.9	4.520	78.9	5.728
54.0	0.696	59.0	1.581	64.0	2.499	69.0	3.475	74.0	4.543	79.0	5.754
54.1	0.714	59.1	1.599	64.1	2.518	69.1	3.495	74.1	4.565	79.1	5.780
54.2	0.731	59.2	1.617	64.2	2.537	69.2	3.516	74.2	4.588	79.2	5.807
54.3	0.749	59.3	1.635	64.3	2.555	69.3	3.536	74.3	4.611	79.3	5.833
54.4	0.766	59.4	1.653	64.4	2.574	69.4	3.556	74.4	4.633	79.4	5.860
54.5	0.784	59.5	1.671	64.5	2.593	69.5	3.577	74.5	4.656	79.5	5.886
54.6	0.801	59.6	1.689	64.6	2.612	69.6	3.597	74.6	4.679	79.6	5.913
54.7	0.819	59.7	1.707	64.7	2.631	69.7	3.618	74.7	4.702	79.7	5.940
54.8	0.836	59.8	1.725	64.8	2.650	69.8	3.638	74.8	4.725	79.8	5.967
54.9	0.854	59.9	1.743	64.9	2.669	69.9	3.659	74.9	4.748	79.9	5.994
55.0	0.872	60.0	1.761	65.0	2.688	70.0	3.680	75.0	4.771	80.0	6.021
55.1	0.889	60.1	1.779	65.1	2.708	70.1	3.700	75.1	4.794	80.1	6.048
55.2	0.907	60.2	1.797	65.2	2.727	70.2	3.721	75.2	4.818	80.2	6.075
55.3	0.924	60.3	1.815	65.3	2.746	70.3	3.742	75.3	4.841	80.3	6.102

Appendix E (Continued).

p(%)	dB	p(%)	dB	p(%)	dB	p(%)	dB	p(%)	dB	p(%)	dB
55.4	0.942	60.4	1.833	65.4	2.765	70.4	3.763	75.4	4.864	80.4	6.130
55.5	0.959	60.5	1.852	65.5	2.784	70.5	3.784	75.5	4.888	80.5	6.158
55.6	0.977	60.6	1.870	65.6	2.803	70.6	3.805	75.6	4.911	80.6	6.185
55.7	0.995	60.7	1.888	65.7	2.823	70.7	3.826	75.7	4.935	80.7	6.213
55.8	1.012	60.8	1.906	65.8	2.842	70.8	3.847	75.8	4.959	80.8	6.241
55.9	1.030	60.9	1.924	65.9	2.861	70.9	3.868	75.9	4.982	80.9	6.269
81.0	6.297	84.5	7.365	88.0	8.653	91.5	10.320	95.0	12.788	98.5	18.173
81.1	6.326	84.6	7.398	88.1	8.694	91.6	10.376	95.1	12.880	98.6	18.477
81.2	6.354	84.7	7.432	88.2	8.736	91.7	10.433	95.2	12.974	98.7	18.804
81.3	6.382	84.8	7.466	88.3	8.778	91.8	10.490	95.3	13.070	98.8	19.156
81.4	6.411	84.9	7.499	88.4	8.820	91.9	10.548	95.4	13.168	98.9	19.538
81.5	6.440	85.0	7.533	88.5	8.862	92.0	10.607	95.5	13.268	99.0	19.956
81.6	6.469	85.1	7.567	88.6	8.905	92.1	10.666	95.6	13.370	99.1	20.418
81.7	6.498	85.2	7.602	88.7	8.948	92.2	10.726	95.7	13.474	99.2	20.934
81.8	6.527	85.3	7.636	88.8	8.992	92.3	10.787	95.8	13.581	99.3	21.519
81.9	6.556	85.4	7.671	88.9	9.036	92.4	10.849	95.9	13.690	99.4	22.192
82.0	6.585	85.5	7.706	89.0	9.080	92.5	10.911	96.0	13.802	99.5	22.989
82.1	6.615	85.6	7.741	89.1	9.125	92.6	10.974	96.1	13.917	99.6	23.962
82.2	6.645	85.7	7.776	89.2	9.169	92.7	11.038	96.2	14.034	99.7	25.216
82.3	6.674	85.8	7.812	89.3	9.215	92.8	11.102	96.3	14.154	99.8	26.981
82.4	6.704	85.9	7.848	89.4	9.260	92.9	11.168	96.4	14.278	99.9	29.996
82.5	6.734	86.0	7.884	89.5	9.306	93.0	11.234	96.5	14.405		
82.6	6.764	86.1	7.920	89.6	9.353	93.1	11.301	96.6	14.535		
82.7	6.795	86.2	7.956	89.7	9.400	93.2	11.369	96.7	14.669		
82.8	6.825	86.3	7.993	89.8	9.447	93.3	11.438	96.8	14.807		
82.9	6.856	86.4	8.030	89.9	9.494	93.4	11.508	96.9	14.950		
83.0	6.886	86.5	8.067	90.0	9.542	93.5	11.579	97.0	15.097		
83.1	6.917	86.6	8.104	90.1	9.591	93.6	11.651	97.1	15.248		
83.2	6.948	86.7	8.142	90.2	9.640	93.7	11.724	97.2	15.405		
83.3	6.979	86.8	8.179	90.3	9.689	93.8	11.798	97.3	15.567		
83.4	7.011	86.9	8.217	90.4	9.739	93.9	11.873	97.4	15.736		
83.5	7.042	87.0	8.256	90.5	9.789	94.0	11.950	97.5	15.911		
83.6	7.074	87.1	8.294	90.6	9.840	94.1	12.027	97.6	16.092		
83.7	7.105	87.2	8.333	90.7	9.891	94.2	12.106	97.7	16.282		
83.8	7.137	87.3	8.372	90.8	9.943	94.3	12.186	97.8	16.479		
83.9	7.169	87.4	8.411	90.9	9.995	94.4	12.268	97.9	16.686		
84.0	7.202	87.5	8.451	91.0	10.048	94.5	12.351	98.0	16.902		
84.1	7.234	87.6	8.491	91.1	10.101	94.6	12.435	98.1	17.129		
84.2	7.267	87.7	8.531	91.2	10.155	94.7	12.521	98.2	17.368		
84.3	7.299	87.8	8.571	91.3	10.210	94.8	12.608	98.3	17.621		
84.4	7.332	87.9	8.612	91.4	10.264	94.9	12.697	98.4	17.889		

* When p is less than 0.1% or greater than 99.9%, convert ρ to a ratio, set $\eta = p/(1-p)$ and find the decibel value using Appendix F.

Appendix F Devibel table

$$dB = 10\log\eta$$

η	Decibels	η	Decibels	η	Decibels	η	Decibels	η	Decibels	η	Decibels
1.00	0.00	1.30	1.14	1.60	2.04	1.90	2.79	2.20	3.42	2.50	3.98
1.01	0.04	1.31	1.17	1.61	2.07	1.91	2.81	2.21	3.44	2.51	4.00
1.02	0.09	1.32	1.21	1.62	2.10	1.92	2.83	2.22	3.46	2.52	4.01
1.03	0.13	1.33	1.24	1.63	2.12	1.93	2.86	2.23	3.48	2.53	4.03
1.04	0.17	1.34	1.27	1.64	2.15	1.94	2.88	2.24	3.50	2.54	4.05
1.05	0.21	1.35	1.30	1.65	2.17	1.95	2.90	2.25	3.52	2.55	4.07
1.06	0.25	1.36	1.34	1.66	2.20	1.96	2.92	2.26	3.54	2.56	4.08
1.07	0.29	1.37	1.37	1.67	2.23	1.97	2.94	2.27	3.56	2.57	4.10
1.08	0.33	1.38	1.40	1.68	2.25	1.98	2.97	2.28	3.58	2.58	4.12
1.09	0.37	1.39	1.43	1.69	2.28	1.99	2.99	2.29	3.60	2.59	4.13
1.10	0.41	1.40	1.46	1.70	2.30	2.00	3.01	2.30	3.62	2.60	4.15
1.11	0.45	1.41	1.49	1.71	2.33	2.01	3.03	2.31	3.64	2.61	4.17
1.12	0.49	1.42	1.52	1.72	2.36	2.02	3.05	2.32	3.65	2.62	4.18
1.13	0.53	1.43	1.55	1.73	2.38	2.03	3.07	2.33	3.67	2.63	4.20
1.14	0.57	1.44	1.58	1.74	2.41	2.04	3.10	2.34	3.69	2.64	4.22
1.15	0.61	1.45	1.61	1.75	2.43	2.05	3.12	2.35	3.71	2.65	4.23
1.16	0.64	1.46	1.64	1.76	2.46	2.06	3.14	2.36	3.73	2.66	4.25
1.17	0.68	1.47	1.67	1.77	2.48	2.07	3.16	2.37	3.75	2.67	4.27
1.18	0.72	1.48	1.70	1.78	2.50	2.08	3.18	2.38	3.77	2.68	4.28
1.19	0.76	1.49	1.73	1.79	2.53	2.09	3.20	2.39	3.78	2.69	4.30
1.20	0.79	1.50	1.76	1.80	2.55	2.10	3.22	2.40	3.80	2.70	4.31
1.21	0.83	1.51	1.79	1.81	2.58	2.11	3.24	2.41	3.82	2.71	4.33
1.22	0.86	1.52	1.82	1.82	2.60	2.12	3.26	2.42	3.84	2.72	4.35
1.23	0.90	1.53	1.85	1.83	2.62	2.13	3.28	2.43	3.86	2.73	4.36
1.24	0.93	1.54	1.88	1.84	2.65	2.14	3.30	2.44	3.87	2.74	4.38
1.25	0.97	1.55	1.90	1.85	2.67	2.15	3.32	2.45	3.89	2.75	4.39
1.26	1.00	1.56	1.93	1.86	2.70	2.16	3.34	2.46	3.91	2.76	4.41
1.27	1.04	1.57	1.96	1.87	2.72	2.17	3.36	2.47	3.93	2.77	4.42
1.28	1.07	1.58	1.99	1.88	2.74	2.18	3.38	2.48	3.94	2.78	4.44
1.29	1.11	1.59	2.01	1.89	2.76	2.19	3.40	2.49	3.96	2.79	4.46
2.80	4.47	3.25	5.12	3.70	5.68	4.15	6.18	4.60	6.63	5.05	7.03
2.81	4.49	3.26	5.13	3.71	5.69	4.16	6.19	4.61	6.64	5.06	7.04
2.82	4.50	3.27	5.15	3.72	5.71	4.17	6.20	4.62	6.65	5.07	7.05
2.83	4.52	3.28	5.16	3.73	5.72	4.18	6.21	4.63	6.66	5.08	7.06
2.84	4.53	3.29	5.17	3.74	5.73	4.19	6.22	4.64	6.67	5.09	7.07
2.85	4.55	3.30	5.19	3.75	5.74	4.20	6.23	4.65	6.67	5.10	7.08
2.86	4.56	3.31	5.20	3.76	5.75	4.21	6.24	4.66	6.68	5.11	7.08
2.87	4.58	3.32	5.21	3.77	5.76	4.22	6.25	4.67	6.69	5.12	7.09
2.88	4.59	3.33	5.22	3.78	5.77	4.23	6.26	4.68	6.70	5.13	7.10

Appendix F (Continued).

η	Decibels	η	Decibels	η	Decibels	η	Decibels	η	Decibels	η	Decibels
2.90	4.62	3.35	5.25	3.80	5.80	4.25	6.28	4.70	6.72	5.15	7.12
2.91	4.64	3.36	5.26	3.81	5.81	4.26	6.29	4.71	6.73	5.16	7.13
2.92	4.65	3.37	5.28	3.82	5.82	4.27	6.30	4.72	6.74	5.17	7.13
2.93	4.67	3.38	5.29	3.83	5.83	4.28	6.31	4.73	6.75	5.18	7.14
2.94	4.68	3.39	5.30	3.84	5.84	4.29	6.32	4.74	6.76	5.19	7.15
2.95	4.70	3.40	5.31	3.85	5.85	4.30	6.33	4.75	6.77	5.20	7.16
2.96	4.71	3.41	5.33	3.86	5.87	4.31	6.34	4.76	6.78	5.21	7.17
2.97	4.73	3.42	5.34	3.87	5.88	4.32	6.35	4.77	6.79	5.22	7.18
2.98	4.74	3.43	5.35	3.88	5.89	4.33	6.36	4.78	6.79	5.23	7.19
2.99	4.76	3.44	5.37	3.89	5.90	4.34	6.37	4.79	6.80	5.24	7.19
3.00	4.77	3.45	5.38	3.90	5.91	4.35	6.38	4.80	6.81	5.25	7.20
3.01	4.79	3.46	5.39	3.91	5.92	4.36	6.39	4.81	6.82	5.26	7.21
3.02	4.80	3.47	5.40	3.92	5.93	4.37	6.40	4.82	6.83	5.27	7.22
3.03	4.81	3.48	5.42	3.93	5.94	4.38	6.41	4.83	6.84	5.28	7.23
3.04	4.83	3.49	5.43	3.94	5.95	4.39	6.42	4.84	6.85	5.29	7.23
3.05	4.84	3.50	5.44	3.95	5.97	4.40	6.43	4.85	6.86	5.30	7.24
3.06	4.86	3.51	5.45	3.96	5.98	4.41	6.44	4.86	6.87	5.31	7.25
3.07	4.87	3.52	5.47	3.97	5.99	4.42	6.45	4.87	6.88	5.32	7.26
3.08	4.89	3.53	5.48	3.98	6.00	4.43	6.46	4.88	6.88	5.33	7.27
3.09	4.90	3.54	5.49	3.99	6.01	4.44	6.47	4.89	6.89	5.34	7.28
3.10	4.91	3.55	5.50	4.00	6.02	4.45	6.48	4.90	6.90	5.35	7.28
3.11	4.93	3.56	5.51	4.01	6.03	4.46	6.49	4.91	6.91	5.36	7.29
3.12	4.94	3.57	5.53	4.02	6.04	4.47	6.50	4.92	6.92	5.37	7.30
3.13	4.96	3.58	5.54	4.03	6.05	4.48	6.51	4.93	6.93	5.38	7.31
3.14	4.97	3.59	5.55	4.04	6.06	4.49	6.52	4.94	6.94	5.39	7.32
3.15	4.98	3.60	5.56	4.05	6.07	4.50	6.53	4.95	6.95	5.40	7.32
3.16	5.00	3.61	5.58	4.06	6.09	4.51	6.54	4.96	6.95	5.41	7.33
3.17	5.01	3.62	5.59	4.07	6.10	4.52	6.55	4.97	6.96	5.42	7.34
3.18	5.02	3.63	5.60	4.08	6.11	4.53	6.56	4.98	6.97	5.43	7.35
3.19	5.04	3.64	5.61	4.09	6.12	4.54	6.57	4.99	6.98	5.44	7.36
3.20	5.05	3.65	5.62	4.10	6.13	4.55	6.58	5.00	6.99	5.45	7.36
3.21	5.07	3.66	5.63	4.11	6.14	4.56	6.59	5.01	7.00	5.46	7.37
3.22	5.08	3.67	5.65	4.12	6.15	4.57	6.60	5.02	7.01	5.47	7.38
3.23	5.09	3.68	5.66	4.13	6.16	4.58	6.61	5.03	7.02	5.48	7.39
3.24	5.11	3.69	5.67	4.14	6.17	4.59	6.62	5.04	7.02	5.49	7.40
5.50	7.40	5.95	7.75	6.40	8.06	6.85	8.36	7.30	8.63	7.75	8.89
5.51	7.41	5.96	7.75	6.41	8.07	6.86	8.36	7.31	8.64	7.76	8.90
5.52	7.42	5.97	7.76	6.42	8.08	6.87	8.37	7.32	8.65	7.77	8.90
5.53	7.43	5.98	7.77	6.43	8.08	6.88	8.38	7.33	8.65	7.78	8.91
5.54	7.44	5.99	7.77	6.44	8.09	6.89	8.38	7.34	8.66	7.79	8.92
5.55	7.44	6,00	7.78	6.45	8.10	6.90	8.39	7.35	8.66	7.80	8.92
5.56	7.45	6.01	7.79	6.46	8.10	6.91	8.39	7.36	8.67	7.81	8.93
5.57	7.46	6.02	7.80	6.47	8.11	6.92	8.40	7.37	8.67	7.82	8.93
5.58	7.47	6.03	7.80	6.48	8.12	6.93	8.41	7.38	8.68	7.83	8.94
5.59	7.47	6.04	7.81	6.49	8.12	6.94	8.41	7.39	8.69	7.84	8.94

Appendix F (Continued).

η	Decibels	η	Decibels	η	Decibels	η	Decibels	η	Decibels	η	Decibels
5.60	7.48	6.05	7.82	6.50	8.13	6.95	8.42	7.40	8.69	7.85	8.95
5.61	7.49	6.06	7.82	6.51	8.14	6.96	8.43	7.41	8.70	7.86	8.95
5.62	7.50	6.07	7.83	6.52	8.14	6.97	8.43	7.42	8.70	7.87	8.96
5.63	7.51	6.08	7.84	6.53	8.15	6.98	8.44	7.43	8.71	7.88	8.97
5.64	7.51	6.09	7.85	6.54	8.16	6.99	8.44	7.44	8.72	7.89	8.97
5.65	7.52	6.10	7.85	6.55	8.16	7.00	8.45	7.45	8.72	7.90	8.98
5.66	7.53	6.11	7.86	6.56	8.17	7.01	8.46	7.46	8.73	7.91	8.98
5.67	7.54	6.12	7.87	6.57	8.18	7.02	8.46	7.47	8.73	7.92	8.99
5.68	7.54	6.13	7.87	6.58	8.18	7.03	8.47	7.48	8.74	7.93	8.99
5.69	7.55	6.14	7.88	6.59	8.19	7.04	8.48	7.49	8.74	7.94	9.00
5.70	7.56	6.15	7.89	6.60	8.20	7.05	8.48	7.50	8.75	7.95	9.00
5.71	7.57	6.16	7.90	6.61	8.20	7.06	8.49	7.51	8.76	7.96	9.01
5.72	7.57	6.17	7.90	6.62	8.21	7.07	8.49	7.52	8.76	7.97	9.01
5.73	7.58	6.18	7.91	6.63	8.22	7.08	8.50	7.53	8.77	7.98	9.02
5.74	7.59	6.19	7.92	6.64	8.22	7.09	8.51	7.54	8.77	7.99	9.03
5.75	7.60	6.20	7.92	6.65	8.23	7.10	8.51	7.55	8.78	8.00	9.03
5.76	7.60	6.21	7.93	6.66	8.23	7.11	8.52	7.56	8.79	8.01	9.04
5.77	7.61	6.22	7.94	6.67	8.24	7.12	8.52	7.57	8.79	8.02	9.04
5.78	7.62	6.23	7.94	6.68	8.25	7.13	8.53	7.58	8.80	8.03	9.05
5.79	7.63	6.24	7.95	6.69	8.25	7.14	8.54	7.59	8.80	8.04	9.05
5.80	7.63	6.25	7.96	6.70	8.26	7.15	8.54	7.60	8.81	8.05	9.06
5.81	7.64	6.26	7.97	6.71	8.27	7.16	8.55	7.61	8.81	8.06	9.06
5.82	7.65	6.27	7.97	6.72	8.27	7.17	8.56	7.62	8.82	8.07	9.07
5.83	7.66	6.28	7.98	6.73	8.28	7.18	8.56	7.63	8.83	8.08	9.07
5.84	7.66	6.29	7.99	6.74	8.29	7.19	8.57	7.64	8.83	8.09	9.08
5.85	7.67	6.30	7.99	6.75	8.29	7.20	8.57	7.65	8.84	8.10	9.08
5.86	7.68	6.31	8.00	6.76	8.30	7.21	8.58	7.66	8.84	8.11	9.09
5.87	7.69	6.32	8.01	6.77	8.31	7.22	8.59	7.67	8.85	8.12	9.10
5.88	7.69	6.33	8.01	6.78	8.31	7.23	8.59	7.68	8.85	8.13	9.10
5.89	7.70	6.34	8.02	6.79	8.32	7.24	8.60	7.69	8.86	8.14	9.11
5.90	7.71	6.35	8.03	6.80	8.33	7.25	8.60	7.70	8.86	8.15	9.11
5.91	7.72	6.36	8.03	6.81	8.33	7.26	8.61	7.71	8.87	8.16	9.12
5.92	7.72	6.37	8.04	6.82	8.34	7.27	8.62	7.72	8.88	8.17	9.12
5.93	7.73	6.38	8.05	6.83	8.34	7.28	8.62	7.73	8.88	8.18	9.13
5.94	7.74	6.39	8.06	6.84	8.35	7.29	8.63	7.74	8.89	8.19	9.13
8.20	9.14	8.50	9.29	8.80	9.44	9.10	9.59	9.40	9.73	9.70	9.87
8.21	9.14	8.51	9.30	8.81	9.45	9.11	9.60	9.41	9.74	9.71	9.87
8.22	9.15	8.52	9.30	8.82	9.45	9.12	9.60	9.42	9.74	9.72	9.88
8.23	9.15	8.53	9.31	8.83	9.46	9.13	9.60	9.43	9.75	9.73	9.88
8.24	9.16	8.54	9.31	8.84	9.46	9.14	9.61	9.44	9.75	9.74	9.89
8.25	9.16	8.55	9.32	8.85	9.47	9.15	9.61	9.45	9.75	9.75	9.89
8.26	9.17	8.56	9.32	8.86	9.47	9.16	9.62	9.46	9.76	9.76	9.89
8.27	9.18	8.57	9.33	8.87	9.48	9.17	9.62	9.47	9.76	9.77	9.90
8.28	9.18	8.58	9.33	8.88	9.48	9.18	9.63	9.48	9.77	9.78	9.90
8.29	9.19	8.59	9.34	8.89	9.49	9.19	9.63	9.49	9.77	9.79	9.91

Appendix F (Continued).

η	Decibels	η	Decibels	η	Decibels	η	Decibels	η	Decibels	η	Decibels
8.30	9.19	8.60	9.34	8.90	9.49	9.20	9.64	9.50	9.78	9.80	9.91
8.31	9.20	8.61	9.35	8.91	9.50	9.21	9.64	9.51	9.78	9.81	9.92
8.32	9.20	8.62	9.36	8.92	9.50	9.22	9.65	9.52	9.79	9.82	9.92
8.33	9.21	8.63	9.36	8.93	9.51	9.23	9.65	9.53	9.79	9.83	9.93
8.34	9.21	8.64	9.37	8.94	9.51	9.24	9.66	9.54	9.80	9.84	9.93
8.35	9.22	8.65	9.37	8.95	9.52	9.25	9.66	9.55	9.80	9.85	9.93
8.36	9.22	8.66	9.38	8.96	9.52	9.26	9.67	9.56	9.80	9.86	9.94
8.37	9.23	8.67	9.38	8.97	9.53	9.27	9.67	9.57	9.81	9.87	9.94
8.38	9.23	8.68	9.39	8.98	9.53	9.28	9.68	9.58	9.81	9.88	9.95
8.39	9.24	8.69	9.39	8.99	9.54	9.29	9.68	9.59	9.82	9.89	9.95
8.40	9.24	8.70	9.40	9.00	9.54	9.30	9.68	9.60	9.82	9.90	9.96
8.41	9.25	8.71	9.40	9.01	9.55	9.31	9.69	9.61	9.83	9.91	9.96
8.42	9.25	8.72	9.41	9.02	9.55	9.32	9.69	9.62	9.83	9.92	9.97
8.43	9.26	8.73	9.41	9.03	9.56	9.33	9.70	9.63	9.84	9.93	9.97
8.44	9.26	8.74	9.42	9.04	9.56	9.34	9.70	9.64	9.84	9.94	9.97
8.45	9.27	8.75	9.42	9.05	9.57	9.35	9.71	9.65	9.85	9.95	9.98
8.46	9.27	8.76	9.43	9.06	9.57	9.36	9.71	9.66	9.85	9.96	9.98
8.47	9.28	8.77	9.43	9.07	9.58	9.37	9.72	9.67	9.85	9.97	9.99
8.48	9.28	8.78	9.43	9.08	9.58	9.38	9.72	9.68	9.86	9.98	9.99
8.49	9.29	8.79	9.44	9.09	9.59	9.39	9.73	9.69	9.86	9.99	10.00
										10.00	10.00

Appendix G Othogonal arrays and linear graphs

Note:

(i) Most of these orthogonal arrays and linear graphs are taken from Taguchi (1987).

(ii) The classification by groups of the columns of the orthogonal array is indicated by the following symbols in the linear graphs shown in this Appendix.

$L_{32}(2^{31})$		$L_{64}(2^{31})$		Other arrays	
Symbol	Group	Symbol	Group	Symbol	Group
○	Gp1 and Gp2	○	Gp1, Gp2, and Gp3	○	Group 1
◎	Group 3	◎	Group 4	◎	Group 2
⊙	Group 4	⊙	Group 5	⊙	Group 3
·	Group 5	·	Group 6	·	Group 4

G.1 $L_{12}(2^{11})$

Experiment number	Column number										
	1	2	3	4	5	6	7	8	9	10	11
1	0	0	0	0	0	0	0	0	0	0	0
2	0	0	0	0	0	1	1	1	1	1	1
3	0	0	1	1	1	0	0	0	1	1	1
4	0	1	0	1	1	0	1	1	0	0	1
5	0	1	1	0	1	1	0	1	0	1	0
6	0	1	1	1	0	1	1	0	1	0	0
7	1	0	1	1	0	0	1	1	0	1	0
8	1	0	1	0	1	1	1	0	0	0	1
9	1	0	0	1	1	1	0	1	1	0	0
10	1	1	1	0	0	0	0	1	1	0	1
11	1	1	0	1	0	1	0	0	0	1	1
12	1	1	0	0	1	0	1	0	1	1	0
Group	1	2									

Note: The components of interaction of any two columns become confounded a little with each of the remaining nine columns. Therefore, do not use this array for experiments where interactions are necessary.

G.2 $L_{32}(2^{31})$

Experiment number	Column number																														
	1	2	3	4	5	6	7	8	9	10	11	12	13	14	15	16	17	18	19	20	21	22	23	24	25	26	27	28	29	30	31
1	0	0	0	0	0	0	0	0	0	0	0	0	0	0	0	0	0	0	0	0	0	0	0	0	0	0	0	0	0	0	0
2	0	0	0	0	0	0	0	0	0	0	0	0	0	0	0	1	1	1	1	1	1	1	1	1	1	1	1	1	1	1	1
3	0	0	0	0	0	0	0	1	1	1	1	1	1	1	1	0	0	0	0	0	0	0	0	1	1	1	1	1	1	1	1
4	0	0	0	0	0	0	0	1	1	1	1	1	1	1	1	1	1	1	1	1	1	1	1	0	0	0	0	0	0	0	0
5	0	0	0	1	1	1	1	0	0	0	0	1	1	1	1	0	0	0	0	1	10	1	1	0	0	0	0	1	1	1	1
6	0	0	0	1	1	1	1	0	0	0	0	1	1	1	1	1	1	1	1	0	0	0	0	1	1	1	1	0	0	0	0
7	0	0	0	1	1	1	1	1	1	1	1	0	0	0	0	0	0	0	0	1	1	1	1	1	1	1	1	0	0	0	0
8	0	0	0	1	1	1	1	1	1	1	1	0	0	0	0	1	1	1	1	0	0	0	0	0	0	0	0	1	1	1	1
9	0	1	1	0	0	1	1	0	0	1	1	0	0	1	1	0	0	1	1	0	0	1	1	0	0	1	1	0	0	1	1
10	0	1	1	0	0	1	1	0	0	1	1	0	0	1	1	1	1	0	0	1	1	0	0	1	1	0	0	1	1	0	0
11	0	1	1	0	0	1	1	1	1	0	0	1	1	0	0	0	0	1	1	0	0	1	1	1	1	0	0	1	1	0	0
12	0	1	1	0	0	1	1	1	1	0	0	1	1	0	0	1	1	0	0	1	1	0	0	0	0	1	1	0	0	1	1
13	0	1	1	1	1	0	0	0	0	1	1	1	1	0	0	0	0	1	1	1	1	0	0	0	0	1	1	1	1	0	0
14	0	1	1	1	1	0	0	0	0	1	1	1	1	0	0	1	1	0	0	0	0	1	1	1	1	0	0	0	0	1	1
15	0	1	1	1	1	0	0	1	1	0	0	0	0	1	1	0	0	1	1	1	10	0	0	1	1	0	0	0	0	1	1
16	0	1	1	1	1	0	0	1	1	0	0	0	0	1	1	1	1	0	0	0	0	1	1	0	0	1	1	1	1	0	0
17	1	0	1	0	1	0	1	0	1	0	1	0	1	0	1	0	1	0	1	0	1	01	1	0	10	01	1	0	1	0	1
18	1	0	1	0	1	0	1	0	1	0	1	0	1	0	1	1	0	1	0	1	0	1	0	1	0	1	0	1	0	1	0
19	1	0	1	0	1	0	1	1	0	1	0	1	0	1	0	0	1	0	1	0	1	0	1	0	1	0	1	0	1	0	1
20	1	0	1	0	1	0	1	1	0	1	0	1	0	1	0	1	0	1	0	1	0	1	0	1	0	1	0	1	0	1	0
21	1	0	1	1	0	1	0	0	1	0	1	1	0	1	0	0	1	0	1	1	0	1	0	0	1	0	1	1	0	1	0
22	1	0	1	1	0	1	0	0	1	0	1	1	0	1	0	1	0	1	0	0	1	0	1	1	0	1	0	0	1	0	1
23	1	0	1	1	0	1	0	1	0	1	0	0	1	0	1	0	1	0	1	1	0	1	0	1	0	1	0	0	10	0	1
24	1	0	1	1	0	1	0	1	0	1	0	0	1	0	1	1	0	1	0	0	1	0	1	0	1	0	1	1	0	1	0
25	1	1	0	0	1	1	0	0	1	1	0	0	1	1	0	0	1	1	0	0	1	1	0	0	1	1	0	0	1	1	0
26	1	1	0	0	1	1	0	0	1	1	0	0	1	1	0	1	0	0	1	1	0	0	1	1	0	0	1	1	0	0	1
27	1	1	0	0	1	1	0	1	0	0	1	1	0	0	1	0	1	1	0	0	1	1	0	1	0	01	1	1	0	0	1
28	1	1	0	0	1	1	0	1	0	0	1	1	0	0	1	1	0	0	1	1	0	0	1	0	1	1	0	0	1	1	0
29	1	1	0	1	0	0	1	0	1	1	0	1	0	0	1	0	1	1	0	10	0	0	1	0	1	1	1	1	0	0	1
30	1	1	0	1	0	0	1	0	1	1	0	1	0	0	1	1	0	0	1	0	1	1	0	1	0	0	1	0	1	1	0
31	1	1	0	1	0	0	1	1	0	0	1	0	1	1	0	0	1	1	0	1	0	0	1	1	0	01	0	0	1	1	0
32	1	1	0	1	0	0	1	1	0	0	1	0	1	1	0	1	0	0	1	0	1	1	0	0	1	1	0	1	0	0	1
Basic mark	a	b	a b	c	a c	b c	a b c	d	a d	b d	a b d	c d	a c d	B c d	a b c d	e	a e	b e	a b e	c e	a c e	b c e	a b c e	d e	a d e	b d e	a b d e	c d e	a c d e	b c d e	a b c d e
Group	1	2		3				4								5															

LINEAR GRAPHS OF $L_{32}(2^{31})$

There are altogether 13 different types of linear graph in $L_{32}(2^{31})$. Of these, two typical types of linear graph are shown below.

(1) a

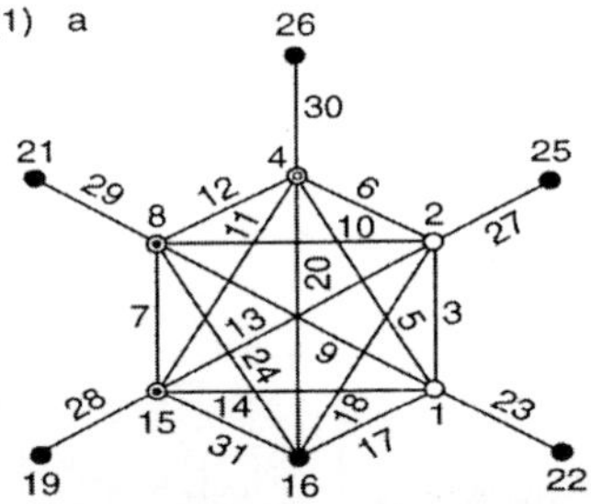

b

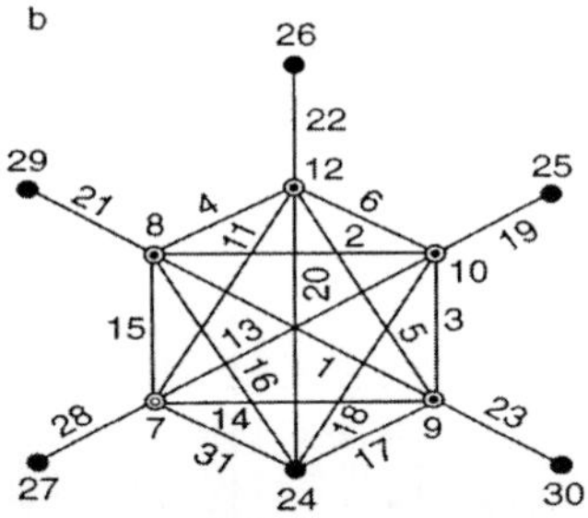

c

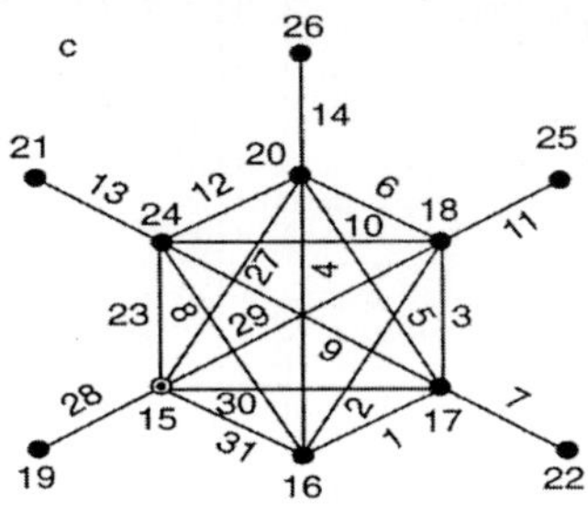

(2) a

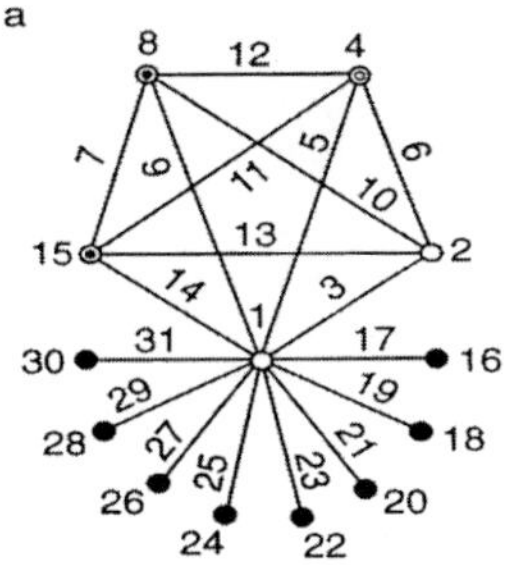

b

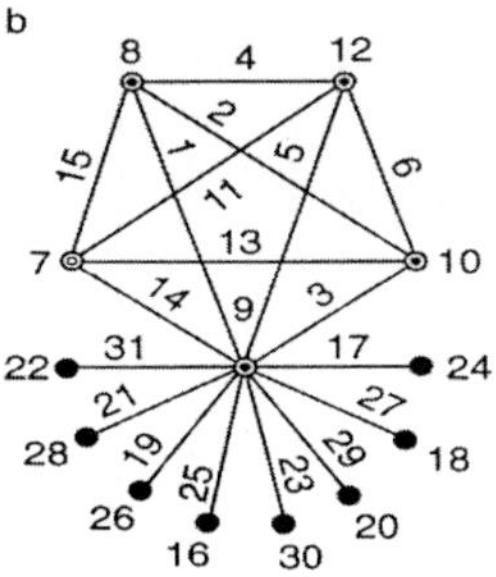

c

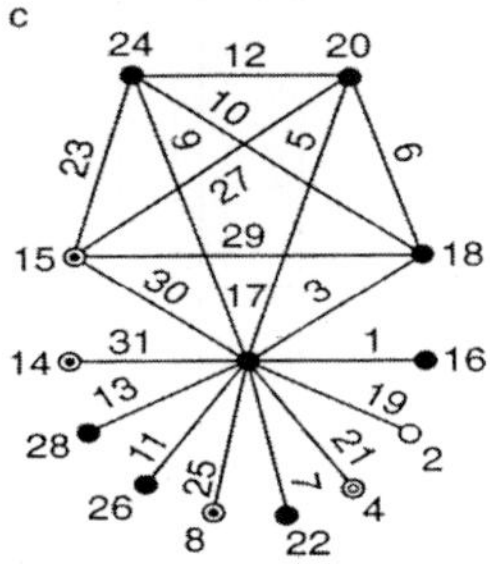

G.3 $L_{64}(2^{63})$

Experiment number	Column number																													
	1	2	3	4	5	6	7	8	9	10	11	12	13	14	15	16	17	18	19	20	21	22	23	24	25	26	27	28	29	30	31
1	0	0	0	0	0	0	0	0	0	0	0	0	0	0	0	0	0	0	0	0	0	0	0	0	0	0	0	0	0	0	0
2	0	0	0	0	0	0	0	0	0	0	0	0	0	0	0	0	0	0	0	0	0	0	0	0	0	0	0	0	0	0	0
3	0	0	0	0	0	0	0	0	0	0	0	0	0	0	0	1	1	1	1	1	1	1	1	1	1	1	1	1	1	1	1
4	0	0	0	0	0	0	0	0	0	0	0	0	0	0	0	1	1	1	1	1	1	1	1	1	1	1	1	1	1	1	1
5	0	0	0	0	0	0	0	1	1	1	1	1	1	1	1	0	0	0	0	0	0	0	0	1	1	1	1	1	1	1	1
6	0	0	0	0	0	0	0	1	1	1	1	1	1	1	1	0	0	0	0	0	0	0	0	1	1	1	1	1	1	1	1
7	0	0	0	0	0	0	0	1	1	1	1	1	1	1	1	1	1	1	1	1	1	1	1	0	0	0	0	0	0	0	0
8	0	0	0	0	0	0	0	1	1	1	1	1	1	1	1	1	1	1	1	1	1	1	1	0	0	0	0	0	0	0	0
9	0	0	0	1	1	1	1	0	0	0	0	1	1	1	1	0	0	0	0	1	1	1	1	0	0	0	0	1	1	1	1
10	0	0	0	1	1	1	1	0	0	0	0	1	1	1	1	0	0	0	0	1	1	1	1	0	0	0	0	1	1	1	1
11	0	0	0	1	1	1	1	0	0	0	0	1	1	1	1	1	1	1	1	0	0	0	0	1	1	1	1	0	0	0	0
12	0	0	0	1	1	1	1	0	0	0	0	1	1	1	1	1	1	1	1	0	0	0	0	1	1	1	1	0	0	0	0
13	0	0	0	1	1	1	1	1	1	1	1	0	0	0	0	0	0	0	0	1	1	1	1	1	1	1	1	0	0	0	0
14	0	0	0	1	1	1	1	1	1	1	1	0	0	0	0	0	0	0	0	1	1	1	1	1	1	1	1	0	0	0	0
15	0	0	0	1	1	1	1	1	1	1	1	0	0	0	0	1	1	1	1	0	0	0	0	0	0	0	0	1	1	1	1
16	0	0	0	1	1	1	1	1	1	1	1	0	0	0	0	1	1	1	1	0	0	0	0	0	0	0	0	1	1	1	1
17	0	1	1	0	0	1	1	0	0	1	1	0	0	1	1	0	0	1	1	0	0	1	1	0	0	1	1	0	0	1	1
18	0	1	1	0	0	1	1	0	0	1	1	0	0	1	1	0	0	1	1	0	0	1	1	0	0	1	1	0	0	1	1
19	0	1	1	0	0	1	1	0	0	1	1	0	0	1	1	1	1	0	0	1	1	0	0	1	1	0	0	1	1	0	0
20	0	1	1	0	0	1	1	0	0	1	1	0	0	1	1	1	1	0	0	1	1	0	0	1	1	0	0	1	1	0	0
21	0	1	1	0	0	1	1	1	1	0	0	1	1	0	0	0	0	1	1	0	0	1	1	1	1	0	0	1	1	0	0
22	0	1	1	0	0	1	1	1	1	0	0	1	1	0	0	0	0	1	1	0	0	1	1	1	1	0	0	1	1	0	0
23	0	1	1	0	0	1	1	1	1	0	0	1	1	0	0	1	1	0	0	1	1	0	0	0	0	1	1	0	0	1	1
24	0	1	1	0	0	1	1	1	1	0	0	1	1	0	0	1	1	0	0	1	1	0	0	0	0	1	1	0	0	1	1
25	0	1	1	1	1	0	0	0	0	1	1	1	1	0	0	0	0	1	1	1	1	0	0	0	0	1	1	1	1	0	0
26	0	1	1	1	1	0	0	0	0	1	1	1	1	0	0	0	0	1	1	1	1	0	0	0	0	1	1	1	1	0	0
27	0	1	1	1	1	0	0	0	0	1	1	1	1	0	0	1	1	0	0	0	0	1	1	1	1	0	0	0	0	1	1
28	0	1	1	1	1	0	0	0	0	1	1	1	1	0	0	1	1	0	0	0	0	1	1	1	1	0	0	0	0	1	1
29	0	1	1	1	1	0	0	1	1	0	0	0	0	1	1	0	0	1	1	1	1	0	0	1	1	0	0	0	0	1	1
30	0	1	1	1	1	0	0	1	1	0	0	0	0	1	1	0	0	1	1	1	1	0	0	1	1	0	0	0	0	1	1
31	0	1	1	1	1	0	0	1	1	0	0	0	0	1	1	1	1	0	0	0	0	1	1	0	0	1	1	1	1	0	0
32	0	1	1	1	1	0	0	1	1	0	0	0	0	1	1	1	1	0	0	0	0	1	1	0	0	1	1	1	1	0	0
33	1	0	1	0	1	0	1	0	1	0	1	0	1	0	1	0	1	0	1	0	1	0	1	0	1	0	1	0	1	0	1
34	1	0	1	0	1	0	1	0	1	0	1	0	1	0	1	0	1	0	1	0	1	0	1	0	1	0	1	0	1	0	1
35	1	0	1	0	1	0	1	0	1	0	1	0	1	0	1	1	0	1	0	1	0	1	0	1	0	1	0	1	0	1	0
36	1	0	1	0	1	0	1	0	1	0	1	0	1	0	1	1	0	1	0	1	0	1	0	1	0	1	0	1	0	1	0
37	1	0	1	0	1	0	1	1	0	1	0	1	0	1	0	0	1	0	1	0	1	0	1	1	0	1	0	1	0	1	0
38	1	0	1	0	1	0	1	1	0	1	0	1	0	1	0	0	1	0	1	0	1	0	1	1	0	1	0	1	0	1	0
39	1	0	1	0	1	0	1	1	0	1	0	1	0	1	0	1	0	1	0	1	0	1	0	0	1	0	1	0	1	0	1
40	1	0	1	0	1	0	1	1	0	1	0	1	0	1	0	1	0	1	0	1	0	1	0	0	1	0	1	0	1	0	1
41	1	0	1	1	0	1	0	0	1	0	1	1	0	1	0	0	1	0	1	1	0	1	0	0	1	0	1	1	0	1	0
42	1	0	1	1	0	1	0	0	1	0	1	1	0	1	0	0	1	0	1	1	0	1	0	0	1	0	1	1	0	1	0
43	1	0	1	1	0	1	0	0	1	0	1	1	0	1	0	1	0	1	0	0	1	0	1	1	0	1	0	0	1	0	1
44	1	0	1	1	0	1	0	0	1	0	1	1	0	1	0	1	0	1	0	0	1	0	1	1	0	1	0	0	1	0	1
45	1	0	1	1	0	1	0	1	0	1	0	0	1	0	1	0	1	0	1	1	0	1	0	1	0	1	0	0	1	0	1
46	1	0	1	1	0	1	0	1	0	1	0	0	1	0	1	0	1	0	1	1	0	1	0	1	0	1	0	0	1	0	1
47	1	0	1	1	0	1	0	1	0	1	0	0	1	0	1	1	0	1	0	0	1	0	1	0	1	0	1	1	0	1	0
48	1	0	1	1	0	1	0	1	0	1	0	0	1	0	1	1	0	1	0	0	1	0	1	0	1	0	1	1	0	1	0
49	1	1	0	0	1	1	0	0	1	1	0	0	1	1	0	0	1	1	0	0	1	1	0	0	1	1	0	0	1	1	0
50	1	1	0	0	1	1	0	0	1	1	0	0	1	1	0	0	1	1	0	0	1	1	0	0	1	1	0	0	1	1	0
51	1	1	0	0	1	1	0	0	1	1	0	0	1	1	0	1	0	0	1	1	0	0	1	1	0	0	1	1	0	0	1
52	1	1	0	0	1	1	0	0	1	1	0	0	1	1	0	1	0	0	1	1	0	0	1	1	0	0	1	1	0	0	1
53	1	1	0	0	1	1	0	1	0	0	1	1	0	0	1	0	1	1	0	0	1	1	0	1	0	0	1	1	0	0	1
54	1	1	0	0	1	1	0	1	0	0	1	1	0	0	1	0	1	1	0	0	1	1	0	1	0	0	1	1	0	0	1
55	1	1	0	0	1	1	0	1	0	0	1	1	0	0	1	1	0	0	1	1	0	0	1	0	1	1	0	0	1	1	0
56	1	1	0	0	1	1	0	1	0	0	1	1	0	0	1	1	0	0	1	1	0	0	1	0	1	1	0	0	1	1	0
57	1	1	0	1	0	0	1	0	1	1	0	1	0	0	1	0	1	1	0	1	0	0	1	0	1	1	0	1	0	0	1
58	1	1	0	1	0	0	1	0	1	1	0	1	0	0	1	0	1	1	0	1	0	0	1	0	1	1	0	1	0	0	1
59	1	1	0	1	0	0	1	0	1	1	0	1	0	0	1	1	0	0	1	0	1	1	0	1	0	0	1	0	1	1	0
60	1	1	0	1	0	0	1	0	1	1	0	1	0	0	1	1	0	0	1	0	1	1	0	1	0	0	1	0	1	1	0
61	1	1	0	1	0	0	1	1	0	0	1	0	1	1	0	0	1	1	0	1	0	0	1	1	0	0	1	0	1	1	0
62	1	1	0	1	0	0	1	1	0	0	1	0	1	1	0	0	1	1	0	1	0	0	1	1	0	0	1	0	1	1	0
63	1	1	0	1	0	0	1	1	0	0	1	0	1	1	0	1	0	0	1	0	1	1	0	0	1	1	0	1	0	0	1
64	1	1	0	1	0	0	1	1	0	0	1	0	1	1	0	1	0	0	1	0	1	1	0	0	1	1	0	1	0	0	1
Basic mark	a	b	a b	c	a c	b c	a b c	d	a d	b d	a b d	c d	a c d	b c d	a b c d	e	a e	b e	a b e	c e	a c e	b c e	a b c e	d e	a d e	b d e	a b d e	c d e	a c d e	b c d e	a b c d e
Group	1	2		3				4								5															

G.3 $L_{64}(2^{63})$ (Continued).

Experiment number	32	33	34	35	36	37	38	39	40	41	42	43	44	45	46	47	48	49	50	51	52	53	54	55	56	57	58	59	60	61	62	63
1	0	0	0	0	0	0	0	0	0	0	0	0	0	0	0	0	0	0	0	0	0	0	0	0	0	0	0	0	0	0	0	0
2	1	1	1	1	1	1	1	1	1	1	1	1	1	1	1	1	1	1	1	1	1	1	1	1	1	1	1	1	1	1	1	1
3	0	0	0	0	0	0	0	0	0	0	0	0	0	0	0	0	1	1	1	1	1	1	1	1	1	1	1	1	1	1	1	1
4	1	1	1	1	1	1	1	1	1	1	1	1	1	1	1	1	0	0	0	0	0	0	0	0	0	0	0	0	0	0	0	0
5	0	0	0	0	0	0	0	0	1	1	1	1	1	1	1	1	0	0	0	0	0	0	0	0	1	1	1	1	1	1	1	1
6	1	1	1	1	1	1	1	1	0	0	0	0	0	0	0	0	1	1	1	1	1	1	1	1	0	0	0	0	0	0	0	0
7	0	0	0	0	0	0	0	0	1	1	1	1	1	1	1	1	1	1	1	1	1	1	1	1	0	0	0	0	0	0	0	0
8	1	1	1	1	1	1	1	1	0	0	0	0	0	0	0	0	0	0	0	0	0	0	0	0	1	1	1	1	1	1	1	1
9	0	0	0	0	1	1	1	1	0	0	0	0	1	1	1	1	0	0	0	0	1	1	1	1	0	0	0	0	1	1	1	1
10	1	1	1	1	0	0	0	0	1	1	1	1	0	0	0	0	1	1	1	1	0	0	0	0	1	1	1	1	0	0	0	0
11	0	0	0	0	1	1	1	1	0	0	0	0	1	1	1	1	1	1	1	1	0	0	0	0	1	1	1	1	0	0	0	0
12	1	1	1	1	0	0	0	0	1	1	1	1	0	0	0	0	0	0	0	0	1	1	1	1	0	0	0	0	1	1	1	1
13	0	0	0	0	1	1	1	1	1	1	1	1	0	0	0	0	0	0	0	0	1	1	1	1	1	1	1	1	0	0	0	0
14	1	1	1	1	0	0	0	0	0	0	0	0	1	1	1	1	1	1	1	1	0	0	0	0	0	0	0	0	1	1	1	1
15	0	0	0	0	1	1	1	1	1	1	1	1	0	0	0	0	1	1	1	1	0	0	0	0	0	0	0	0	1	1	1	1
16	1	1	1	1	0	0	0	0	0	0	0	0	1	1	1	1	0	0	0	0	1	1	1	1	1	1	1	1	0	0	0	0
17	0	0	1	1	0	0	1	1	0	0	1	1	0	0	1	1	0	0	1	1	0	0	1	1	0	0	1	1	0	0	1	1
18	1	1	0	0	1	1	0	0	1	1	0	0	1	1	0	0	1	1	0	0	1	1	0	0	1	1	0	0	1	1	0	0
19	0	0	1	1	0	0	1	1	0	0	1	1	0	0	1	1	1	1	0	0	1	1	0	0	1	1	0	0	1	1	0	0
20	1	1	0	0	1	1	0	0	1	1	0	0	1	1	0	0	0	0	1	1	0	0	1	1	0	0	1	1	0	0	1	1
21	0	0	1	1	0	0	1	1	1	1	0	0	1	1	0	0	0	0	1	1	0	0	1	1	1	1	0	0	1	1	0	0
22	1	1	0	0	1	1	0	0	0	0	1	1	0	0	1	1	1	1	0	0	1	1	0	0	0	0	1	1	0	0	1	1
23	0	0	1	1	0	0	1	1	1	1	0	0	1	1	0	0	1	1	0	0	1	1	0	0	0	0	1	1	0	0	1	1
24	1	1	0	0	1	1	0	0	0	0	1	1	0	0	1	1	0	0	1	1	0	0	1	1	1	1	0	0	1	1	0	0
25	0	0	1	1	1	1	0	0	0	0	1	1	1	1	0	0	0	0	1	1	1	1	0	0	0	0	1	1	1	1	0	0
26	1	1	0	0	0	0	1	1	1	1	0	0	0	0	1	1	1	1	0	0	0	0	1	1	1	1	0	0	0	0	1	1
27	0	0	1	1	1	1	0	0	0	0	1	1	1	1	0	0	1	1	0	0	0	0	1	1	1	1	0	0	0	0	1	1
28	1	1	0	0	0	0	1	1	1	1	0	0	0	0	1	1	0	0	1	1	1	1	0	0	0	0	1	1	1	1	0	0
29	0	0	1	1	1	1	0	0	1	1	0	0	0	0	1	1	0	0	1	1	1	1	0	0	1	1	0	0	0	0	1	1
30	1	1	0	0	0	0	1	1	0	0	1	1	1	1	0	0	1	1	0	0	0	0	1	1	0	0	1	1	1	1	0	0
31	0	0	1	1	1	1	0	0	1	1	0	0	0	0	1	1	1	1	0	0	0	0	1	1	0	0	1	1	1	1	0	0
32	1	1	0	0	0	0	1	1	0	0	1	1	1	1	0	0	0	0	1	1	1	1	0	0	1	1	0	0	0	0	1	1
33	0	1	0	1	0	1	0	1	0	1	0	1	0	1	0	1	0	1	0	1	0	1	0	1	0	1	0	1	0	1	0	1
34	1	0	1	0	1	0	1	0	1	0	1	0	1	0	1	0	1	0	1	0	1	0	1	0	1	0	1	0	1	0	1	0
35	0	1	0	1	0	1	0	1	0	1	0	1	0	1	0	1	1	0	1	0	1	0	1	0	1	0	1	0	1	0	1	0
36	1	0	1	0	1	0	1	0	1	0	1	0	1	0	1	0	0	1	0	1	0	1	0	1	0	1	0	1	0	1	0	1
37	0	1	0	1	0	1	0	1	1	0	1	0	1	0	1	0	0	1	0	1	0	1	0	1	1	0	1	0	1	0	1	0
38	1	0	1	0	1	0	1	0	0	1	0	1	0	1	0	1	1	0	1	0	1	0	1	0	0	1	0	1	0	1	0	1
39	0	1	0	1	0	1	0	1	1	0	1	0	1	0	1	0	1	0	1	0	1	0	1	0	0	1	0	1	0	1	0	1
40	1	0	1	0	1	0	1	0	0	1	0	1	0	1	0	1	0	1	0	1	0	1	0	1	1	0	1	0	1	0	1	0
41	0	1	0	1	1	0	1	0	0	1	0	1	1	0	1	0	0	1	0	1	1	0	1	0	0	1	0	1	1	0	1	0
42	1	0	1	0	0	1	0	1	1	0	1	0	0	1	0	1	1	0	1	0	0	1	0	1	1	0	1	0	0	1	0	1
43	0	1	0	1	1	0	1	0	0	1	0	1	1	0	1	0	1	0	1	0	0	1	0	1	1	0	1	0	0	1	0	1
44	1	0	1	0	0	1	0	1	1	0	1	0	0	1	0	1	0	1	0	1	1	0	1	0	0	1	0	1	1	0	1	0
45	0	1	0	1	1	0	1	0	1	0	1	0	0	1	0	1	0	1	0	1	1	0	1	0	1	0	1	0	0	1	0	1
46	1	0	1	0	0	1	0	1	0	1	0	1	1	0	1	0	1	0	1	0	0	1	0	1	0	1	0	1	1	0	1	0
47	0	1	0	1	1	0	1	0	1	0	1	0	0	1	0	1	1	0	1	0	0	1	0	1	0	1	0	1	1	0	1	0
48	1	0	1	0	0	1	0	1	0	1	0	1	1	0	1	0	0	1	0	1	1	0	1	0	1	0	1	0	0	1	0	1
49	0	1	1	0	0	1	1	0	0	1	1	0	0	1	1	0	0	1	1	0	0	1	1	0	0	1	1	0	0	1	1	0
50	1	0	0	1	1	0	0	1	1	0	0	1	1	0	0	1	1	0	0	1	1	0	0	1	1	0	0	1	1	0	0	1
51	0	1	1	0	0	1	1	0	0	1	1	0	0	1	1	0	1	0	0	1	1	0	0	1	1	0	0	1	1	0	0	1
52	1	0	0	1	1	0	0	1	1	0	0	1	1	0	0	1	0	1	1	0	0	1	1	0	0	1	1	0	0	1	1	0
53	0	1	1	0	0	1	1	0	1	0	0	1	1	0	0	1	0	1	1	0	0	1	1	0	1	0	0	1	1	0	0	1
54	1	0	0	1	1	0	0	1	0	1	1	0	0	1	1	0	1	0	0	1	1	0	0	1	0	1	1	0	0	1	1	0
55	0	1	1	0	0	1	1	0	1	0	0	1	1	0	0	1	1	0	0	1	1	0	0	1	0	1	1	0	0	1	1	0
56	1	0	0	1	1	0	0	1	0	1	1	0	0	1	1	0	0	1	1	0	0	1	1	0	1	0	0	1	1	0	0	1
57	0	1	1	0	1	0	0	1	0	1	1	0	1	0	0	1	0	1	1	0	1	0	0	1	0	1	1	0	1	0	0	1
58	1	0	0	1	0	1	1	0	1	0	0	1	0	1	1	0	1	0	0	1	0	1	1	0	1	0	0	1	0	1	1	0
59	0	1	1	0	1	0	0	1	0	1	1	0	1	0	0	1	1	0	0	1	0	1	1	0	1	0	0	1	0	1	1	0
60	1	0	0	1	0	1	1	0	1	0	0	1	0	1	1	0	0	1	1	0	1	0	0	1	0	1	1	0	1	0	0	1
61	0	1	1	0	1	0	0	1	1	0	0	1	0	1	1	0	0	1	1	0	1	0	0	1	1	0	0	1	0	1	1	0
62	1	0	0	1	0	1	1	0	0	1	1	0	1	0	0	1	1	0	0	1	0	1	1	0	0	1	1	0	1	0	0	1
63	0	1	1	0	1	0	0	1	1	0	0	1	0	1	1	0	1	0	0	1	0	1	1	0	0	1	1	0	1	0	0	1
64	1	0	0	1	0	1	1	0	0	1	1	0	1	0	0	1	0	1	1	0	1	0	0	1	1	0	0	1	0	1	1	0
Basic mark	f	a f	b f	a b f	c f	a c f	b c f	a b c f	d f	a d f	b d f	a b d f	c d f	a c d f	b c d f	a b c d f	e f	a e f	b e f	a b e f	c e f	a c e f	B c e f	a b c e f	d e f	a d e f	b d e f	a b d e f	c d e f	a c d e f	b c d e f	a b c d e f
Group	6																															

LINEAR GRAPHS OF $L_{64}(2^{63})$

There are altogether ten different types of linear graph in $L_{64}(2^{63})$. Of these, two typical types of linear graph are shown below.

(1)

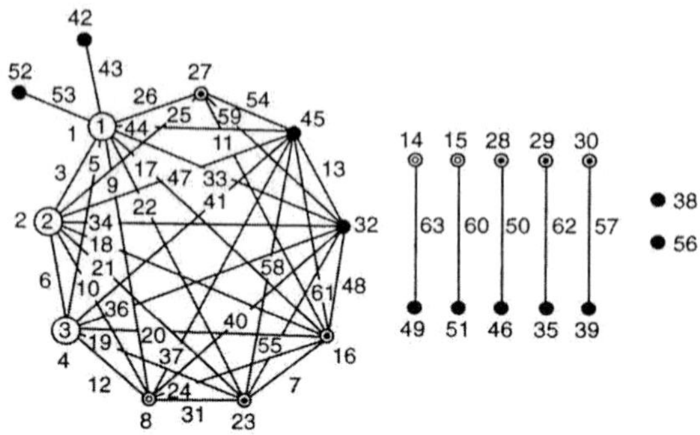

a

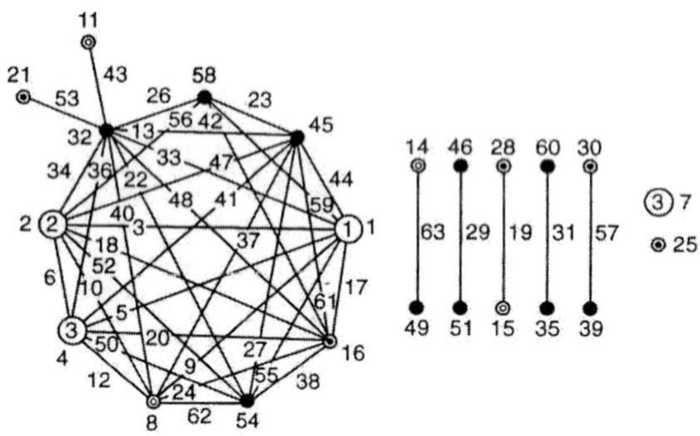

b

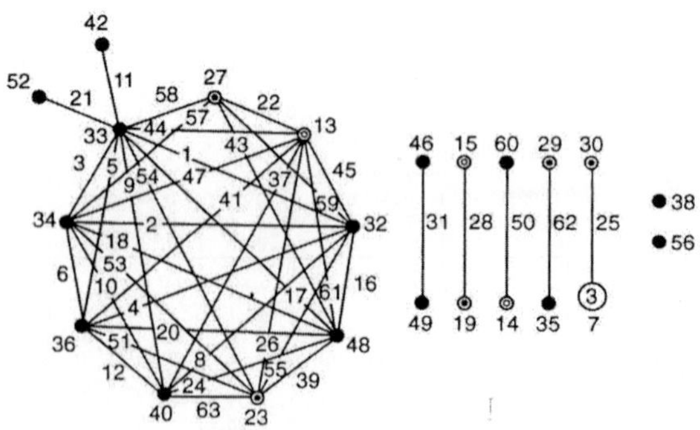

c

(2)

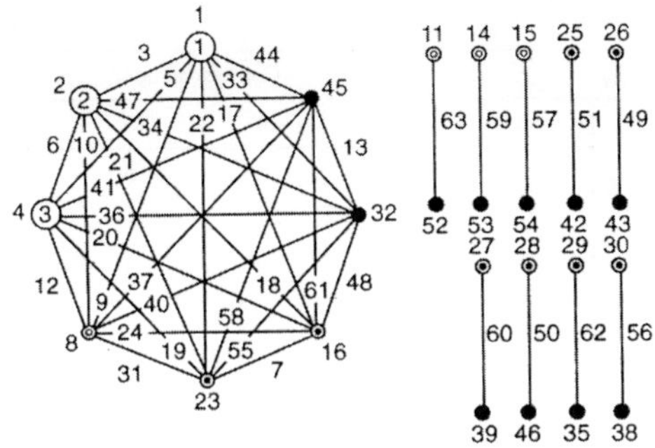

a

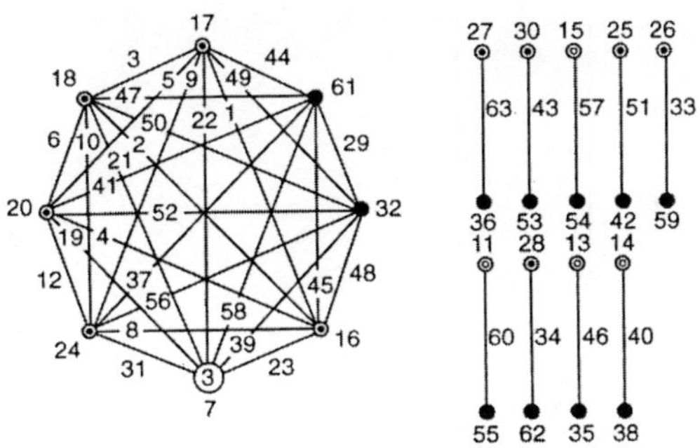

b

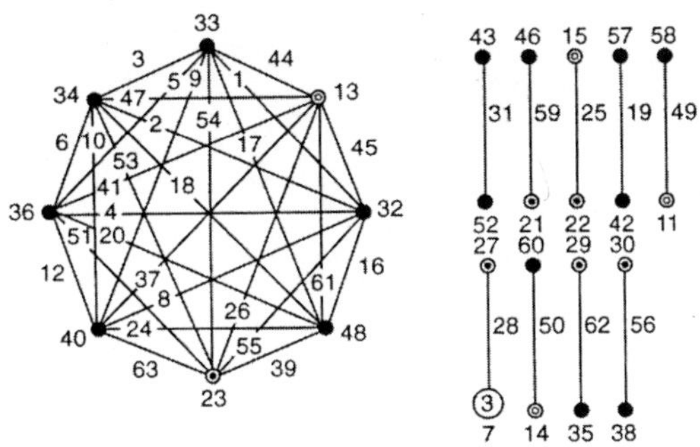

c

G.4 $L_{18}(2^1 \times 3^7)$

Experiment number	Column number							
	1	2	3	4	5	6	7	8
1	0	0	0	0	0	0	0	0
2	0	0	1	1	1	1	1	1
3	0	0	2	2	2	2	2	2
4	0	1	0	0	1	1	2	2
5	0	1	1	1	2	2	0	0
6	0	1	2	2	0	0	1	1
7	0	2	0	1	0	2	1	2
8	0	2	1	2	1	0	2	0
9	0	2	2	0	2	1	0	1
10	1	0	0	2	2	1	1	0
11	1	0	1	0	0	2	2	1
12	1	0	2	1	1	0	0	2
13	1	1	0	1	2	0	2	1
14	1	1	1	2	0	1	0	2
15	1	1	2	0	1	2	1	0
16	1	2	0	2	1	2	0	1
17	1	2	1	0	2	0	1	2
18	1	2	2	1	0	1	2	0
Group	1	2	3					

Note: Interactions among three levels are partially confounded a little in each of the columns of the remaining three levels. Therefore, do not use this array for experiments where interactions are necessary.

G.5 $L_{54}(2^1 \times 3^{25})$

Experiment number	Column number																								
	1	2	3	4	5	6	7	8	9	10	11	12	13	14	15	16	17	18	19	20	21	22	23	24	25	26
1	0	0	0	0	0	0	0	0	0	0	0	0	0	0	0	0	0	0	0	0	0	0	0	0	0	0
2	0	0	0	0	0	0	0	0	1	1	1	1	1	1	1	1	1	1	1	1	1	1	1	1	1	1
3	0	0	0	0	0	0	0	0	2	2	2	2	2	2	2	2	2	2	2	2	2	2	2	2	2	2
4	0	0	1	1	1	1	1	1	0	0	0	0	0	0	1	2	1	2	1	2	1	2	1	2	1	2
5	0	0	1	1	1	1	1	1	1	1	1	1	1	1	2	0	2	0	2	0	2	0	2	0	2	0
6	0	0	1	1	1	1	1	1	2	2	2	2	2	2	0	1	0	1	0	1	0	1	0	1	0	1
7	0	0	2	2	2	2	2	2	0	0	0	0	0	0	2	1	2	1	2	1	2	1	2	1	2	1
8	0	0	2	2	2	2	2	2	1	1	1	1	1	1	0	2	0	2	0	2	0	2	0	2	0	2
9	0	0	2	2	2	2	2	2	2	2	2	2	2	2	1	0	1	0	1	0	1	0	1	0	1	0
10	0	1	0	0	1	1	2	2	0	0	1	1	2	2	0	0	0	0	1	2	1	2	2	1	2	1
11	0	1	0	0	1	1	2	2	1	1	2	2	0	0	1	1	1	1	2	0	2	0	0	2	0	2
12	0	1	0	0	1	1	2	2	2	2	0	0	1	1	2	2	2	2	0	1	0	1	1	0	1	0
13	0	1	1	1	2	2	0	0	0	0	1	1	2	2	1	2	1	2	2	1	2	1	0	0	0	0
14	0	1	1	1	2	2	0	0	1	1	2	2	0	0	2	0	2	0	0	2	0	2	1	1	1	1
15	0	1	1	1	2	2	0	0	2	2	0	0	1	1	0	1	0	1	1	0	1	0	2	2	2	2
16	0	1	2	2	0	0	1	1	0	0	1	1	2	2	2	1	2	1	0	0	0	0	1	2	1	2
17	0	1	2	2	0	0	1	1	1	1	2	2	0	0	0	2	0	2	1	1	1	1	2	0	2	0
18	0	1	2	2	0	0	1	1	2	2	0	0	1	1	1	0	1	0	2	2	2	2	0	1	0	1
19	0	2	0	1	0	2	1	2	0	1	0	2	1	2	0	0	1	2	0	0	2	1	1	2	2	1
20	0	2	0	1	0	2	1	2	1	2	1	0	2	0	1	1	2	0	1	1	0	2	2	0	0	2
21	0	2	0	1	0	2	1	2	2	0	2	1	0	1	2	2	0	1	2	2	1	0	0	1	1	0
22	0	2	1	2	1	0	2	0	0	1	0	2	1	2	1	2	2	1	1	2	0	0	2	1	0	0
23	0	2	1	2	1	0	2	0	1	2	1	0	2	0	2	0	0	2	2	0	1	1	0	2	1	1
24	0	2	1	2	1	0	2	0	2	0	2	1	0	1	0	1	1	0	0	1	2	2	1	0	2	2
25	0	2	2	0	2	1	0	1	0	1	0	2	1	2	2	1	0	0	2	1	1	2	0	0	1	2
26	0	2	2	0	2	1	0	1	1	2	1	0	2	0	0	2	1	1	0	2	2	0	1	1	2	0
27	0	2	2	0	2	1	0	1	2	0	2	1	0	1	1	0	2	2	1	0	0	1	2	2	0	1
28	1	0	0	2	2	1	1	0	0	2	2	1	1	0	0	0	2	1	2	1	1	2	1	2	0	0
29	1	0	0	2	2	1	1	0	1	0	0	2	2	1	1	1	0	2	0	2	2	0	2	0	1	1
30	1	0	0	2	2	1	1	0	2	1	1	0	0	2	2	2	1	0	1	0	0	1	0	1	2	2
31	1	0	1	0	0	2	2	1	0	2	2	1	1	0	1	2	0	0	0	0	2	1	2	1	1	2
32	1	0	1	0	0	2	2	1	1	0	0	2	2	1	2	0	1	1	1	1	0	2	0	2	2	0
33	1	0	1	0	0	2	2	1	2	1	1	0	0	2	0	1	2	2	2	2	1	0	1	0	0	1
34	1	0	2	1	1	0	0	2	0	2	2	1	1	0	2	1	1	2	1	2	0	0	0	0	2	1
35	1	0	2	1	1	0	0	2	1	0	0	2	2	1	0	2	2	0	2	0	1	1	1	1	0	2
36	1	0	2	1	1	0	0	2	2	1	1	0	0	2	1	0	0	1	0	1	2	2	2	2	1	0
37	1	1	0	1	2	0	2	1	0	1	2	0	2	1	0	0	1	2	2	1	0	0	2	1	1	2
38	1	1	0	1	2	0	2	1	1	2	0	1	0	2	1	1	2	0	0	2	1	1	0	2	2	0
39	1	1	0	1	2	0	2	1	2	0	1	2	1	0	2	2	0	1	1	0	2	2	1	0	0	1
40	1	1	1	2	0	1	0	2	0	1	2	0	2	1	1	2	2	1	0	0	1	2	0	0	2	1
41	1	1	1	2	0	1	0	2	1	2	0	1	0	2	2	0	0	2	1	1	2	0	1	1	0	2
42	1	1	1	2	0	1	0	2	2	0	1	2	1	0	0	1	1	0	2	2	0	1	2	2	1	0
43	1	1	2	0	1	2	1	0	0	1	2	0	2	1	2	1	0	0	1	2	2	1	1	2	0	0
44	1	1	2	0	1	2	1	0	1	2	0	1	0	2	0	2	1	1	2	0	0	2	2	0	1	1
45	1	1	2	0	1	2	1	0	2	0	1	2	1	0	1	0	2	2	0	1	1	0	0	1	2	2
46	1	2	0	2	1	2	0	1	0	2	1	2	0	1	0	0	2	1	1	2	2	1	0	0	1	2
47	1	2	0	2	1	2	0	1	1	0	2	0	1	2	1	1	0	2	2	0	0	2	1	1	2	0
48	1	2	0	2	1	2	0	1	2	1	0	1	2	0	2	2	1	0	0	1	1	0	2	2	0	1
49	1	2	1	0	2	0	1	2	0	2	1	2	0	1	1	2	0	0	2	1	0	0	1	2	2	1
50	1	2	1	0	2	0	1	2	1	0	2	0	1	2	2	0	1	1	0	2	1	1	2	0	0	2
51	1	2	1	0	2	0	1	2	2	1	0	1	2	0	0	1	2	2	1	0	2	2	0	1	1	0
52	1	2	2	1	0	1	2	0	0	2	1	2	0	1	2	1	1	2	0	0	1	2	2	1	0	0
53	1	2	2	1	0	1	2	0	1	0	2	0	1	2	0	2	2	0	1	1	2	0	0	2	1	1
54	1	2	2	1	0	1	2	0	2	1	0	1	2	0	1	0	0	1	2	2	0	1	1	0	2	2
Group	1	2	3						4																	

LINEAR GRAPH OF $L_{54}(2^1 \times 3^{25})$

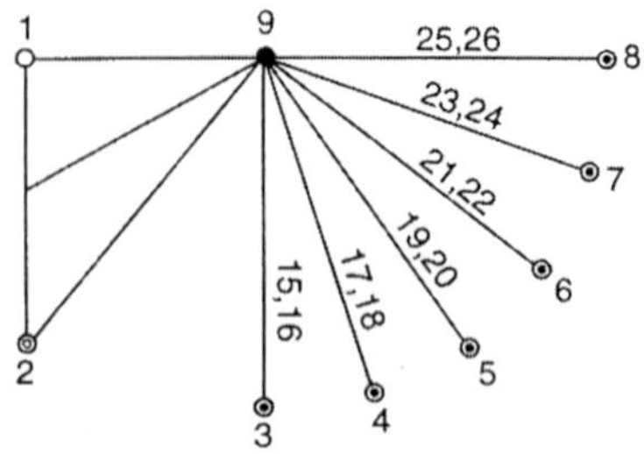

Note:

(i) Column 1 is at two levels.

(ii) Interaction of column 1 and column 2 can be found without sacrifice of other columns. Find this interaction from the two-way array of column 1 and column 2.

(iii) The three interactions of column 1 × column 9, column 2 × column 9, and column 1 × column 2 × column 9 appear comprehensively in the five columns 10, 11, 12, 13 and 14. The numbers of degrees of freedom are 2, 4 and 4, making a total of 10. The total of the numbers of degrees of freedom of columns 10–14 is also 10. Therefore, of two-level, three-level, and three-level factors are made to correspond to columns 1, 2, and 9, it is possible to find three two-factor interactions and one three-factor interaction from the three-way array.

(iv) By letting six levels correspond to the combination of column 1 and column 2, it is also possible to find interactions with column 9.

G.6(a) $L_{81}(3^{40})$

Experiment number	Column number																																						
	1	2	3	4	5	6	7	8	9	10	11	12	13	14	15	16	17	18	19	20	21	22	23	24	25	26	27	28	29	30	31	32	33	34	35	36	37	38	39	40
1	0	0	0	0	0	0	0	0	0	0	0	0	0	0	0	0	0	0	0	0	0	0	0	0	0	0	0	0	0	0	0	0	0	0	0	0	0	0	0	0
2	0	0	0	0	0	0	0	0	0	0	0	0	0	1	1	2	1	1	2	2	2	2	1	1	2	1	1	2	2	2	2	2	2	2	2	2	2	2	2	2
3	0	0	0	0	0	0	0	0	0	0	0	0	0	2	2	1	2	2	1	1	1	1	2	2	1	2	2	1	1	1	1	1	1	1	1	1	1	1	1	1
4	0	0	0	0	1	1	2	1	1	2	2	2	2	0	0	0	0	0	0	0	0	0	1	1	2	1	1	2	2	2	2	1	1	1	1	1	1	1	1	1
5	0	0	0	0	1	1	2	1	1	2	2	2	2	1	1	2	1	1	2	2	2	2	2	2	1	2	2	1	1	1	1	0	0	0	0	0	0	0	0	0
6	0	0	0	0	1	1	2	1	1	2	2	2	2	2	2	1	2	2	1	1	1	1	0	0	0	0	0	0	0	0	0	2	2	2	2	2	2	2	2	2
7	0	0	0	0	2	2	1	2	2	1	1	1	1	0	0	0	0	0	0	0	0	0	2	2	1	2	2	1	1	1	1	2	2	2	2	2	2	2	2	2
8	0	0	0	0	2	2	1	2	2	1	1	1	1	1	1	2	1	1	2	2	2	2	0	0	0	0	0	0	0	0	0	1	1	1	1	1	1	1	1	1
9	0	0	0	0	2	2	1	2	2	1	1	1	1	2	2	1	2	2	1	1	1	1	1	1	2	1	1	2	2	2	2	0	0	0	0	0	0	0	0	0
10	0	1	1	2	0	0	0	1	1	2	1	1	1	0	0	0	1	1	2	1	1	1	0	0	0	1	1	2	1	1	1	0	0	0	2	2	2	1	1	1
11	0	1	1	2	0	0	0	1	1	2	1	1	1	1	1	2	2	2	1	0	0	0	1	1	2	2	2	1	0	0	0	2	2	2	1	1	1	0	0	0
12	0	1	1	2	0	0	0	1	1	2	1	1	1	2	2	1	0	0	0	2	2	2	2	2	1	0	0	0	2	2	2	1	1	1	0	0	0	2	2	2
13	0	1	1	2	1	1	2	2	2	1	0	0	0	0	0	0	1	1	2	1	1	1	1	1	2	2	2	1	0	0	0	1	1	1	0	0	0	2	2	2
14	0	1	1	2	1	1	2	2	2	1	0	0	0	1	1	2	2	2	1	0	0	0	2	2	1	0	0	0	2	2	2	0	0	0	2	2	2	1	1	1
15	0	1	1	2	1	1	2	2	2	1	0	0	0	2	2	1	0	0	0	2	2	2	0	0	0	1	1	2	1	1	1	2	2	2	1	1	1	0	0	0
16	0	1	1	2	2	2	1	0	0	0	2	2	2	0	0	0	1	1	2	1	1	1	2	2	1	0	0	0	2	2	2	2	2	2	1	1	1	0	0	0
17	0	1	1	2	2	2	1	0	0	0	2	2	2	1	1	2	2	2	1	0	0	0	0	0	0	1	1	2	1	1	1	1	1	1	0	0	0	2	2	2
18	0	1	1	2	2	2	1	0	0	0	2	2	2	2	2	1	0	0	0	2	2	2	1	1	2	2	2	1	0	0	0	0	0	0	2	2	2	1	1	1
19	0	2	2	1	0	0	0	2	2	1	2	2	2	0	0	0	2	2	1	2	2	2	0	0	0	2	2	1	2	2	2	0	0	0	1	1	1	2	2	2
20	0	2	2	1	0	0	0	2	2	1	2	2	2	1	1	2	0	0	0	1	1	1	1	1	2	0	0	0	1	1	1	2	2	2	0	0	0	1	1	1
21	0	2	2	1	0	0	0	2	2	1	2	2	2	2	2	1	1	1	2	0	0	0	2	2	1	1	1	2	0	0	0	1	1	1	2	2	2	0	0	0
22	0	2	2	1	1	1	2	0	0	0	1	1	1	0	0	0	2	2	1	2	2	2	1	1	2	0	0	0	1	1	1	1	1	1	2	2	2	0	0	0
23	0	2	2	1	1	1	2	0	0	0	1	1	1	1	1	2	0	0	0	1	1	1	2	2	1	1	1	2	0	0	0	0	0	0	1	1	1	2	2	2
24	0	2	2	1	1	1	2	0	0	0	1	1	1	2	2	1	1	1	2	0	0	0	0	0	0	2	2	1	2	2	2	2	2	2	0	0	0	1	1	1
25	0	2	2	1	2	2	1	1	1	2	0	0	0	0	0	0	2	2	1	2	2	2	2	2	1	1	1	2	0	0	0	2	2	2	0	0	0	1	1	1
26	0	2	2	1	2	2	1	1	1	2	0	0	0	1	1	2	0	0	0	1	1	1	0	0	0	2	2	1	2	2	2	1	1	1	2	2	2	0	0	0
27	0	2	2	1	2	2	1	1	1	2	0	0	0	2	2	1	1	1	2	0	0	0	1	1	2	0	0	0	1	1	1	0	0	0	1	1	1	2	2	2
28	1	0	1	1	0	1	1	0	1	1	0	2	1	0	1	1	0	1	1	0	2	1	0	1	1	0	1	1	0	2	1	0	2	1	0	1	2	0	2	1
29	1	0	1	1	0	1	1	0	1	1	0	2	1	1	2	0	1	2	0	2	1	0	1	2	0	1	2	0	2	1	0	2	1	0	2	0	1	2	1	0
30	1	0	1	1	0	1	1	0	1	1	0	2	1	2	0	2	2	0	2	1	0	2	2	0	2	2	0	2	1	0	2	1	0	2	1	2	0	1	0	2
31	1	0	1	1	1	2	0	1	2	0	2	1	0	0	1	1	0	1	1	0	2	1	1	2	0	1	2	0	2	1	0	1	0	2	1	2	0	1	0	2
32	1	0	1	1	1	2	0	1	2	0	2	1	0	1	2	0	1	2	0	2	1	0	2	0	2	2	0	2	1	0	2	0	2	1	0	1	2	0	2	1
33	1	0	1	1	1	2	0	1	2	0	2	1	0	2	0	2	2	0	2	1	0	2	0	1	1	0	1	1	0	2	1	2	1	0	2	0	1	2	1	0
34	1	0	1	1	2	0	2	2	0	2	1	0	2	0	1	1	0	1	1	0	2	1	2	0	2	2	0	2	1	0	2	2	1	0	2	0	1	2	1	0
35	1	0	1	1	2	0	2	2	0	2	1	0	2	1	2	0	1	2	0	2	1	0	0	1	1	0	1	1	0	2	1	1	0	2	1	2	0	1	0	2
36	1	0	1	1	2	0	2	2	0	2	1	0	2	2	0	2	2	0	2	1	0	2	1	2	0	1	2	0	2	1	0	0	2	1	0	1	2	0	2	1
37	1	1	2	0	0	1	1	1	2	0	1	0	2	0	1	1	1	2	0	1	0	2	0	1	1	1	2	0	1	0	2	0	2	1	2	0	1	1	0	2
38	1	1	2	0	0	1	1	1	2	0	1	0	2	1	2	0	2	0	2	0	2	1	1	2	0	2	0	2	0	2	1	2	1	0	1	2	0	0	2	1

G.6(a) $L_{81}(3^{40})$ (Continued).

Experiment number	Column number																																						
	1	2	3	4	5	6	7	8	9	10	11	12	13	14	15	16	17	18	19	20	21	22	23	24	25	26	27	28	29	30	31	32	33	34	35	36	37	38	39	40
39	1	1	2	0	0	1	1	1	2	0	1	0	2	2	0	2	0	1	1	2	1	0	2	0	2	0	1	1	2	1	0	1	0	2	0	1	2	2	1	0
40	1	1	2	0	1	2	0	2	0	2	0	2	1	0	1	1	1	2	0	1	0	2	1	2	0	2	0	2	0	2	1	1	0	2	0	1	2	2	1	0
41	1	1	2	0	1	2	0	2	0	2	0	2	1	1	2	0	2	0	2	0	2	1	2	0	2	0	1	1	2	1	0	0	2	1	2	0	1	1	0	2
42	1	1	2	0	1	2	0	2	0	2	0	2	1	2	0	2	0	1	1	2	1	0	0	1	1	1	2	0	1	0	2	2	1	0	1	2	0	0	2	1
43	1	1	2	0	2	0	2	0	1	1	2	1	0	0	1	1	1	2	0	1	0	2	2	0	2	0	1	1	2	1	0	2	1	0	1	2	0	0	2	1
44	1	1	2	0	2	0	2	0	1	1	2	1	0	1	2	0	2	0	2	0	2	1	0	1	1	1	2	0	1	0	2	1	0	2	0	1	2	2	1	0
45	1	1	2	0	2	0	2	0	1	1	2	1	0	2	0	2	0	1	1	2	1	0	1	2	0	2	0	2	0	2	1	0	2	1	2	0	1	1	0	2
46	1	2	0	2	0	1	1	2	0	2	2	1	0	0	1	1	2	0	2	2	1	0	0	1	1	2	0	2	2	1	0	0	2	1	1	2	0	2	1	0
47	1	2	0	2	0	1	1	2	0	2	2	1	0	1	2	0	0	1	1	1	0	2	1	2	0	0	1	1	1	0	2	2	1	0	0	1	2	1	0	2
48	1	2	0	2	0	1	1	2	0	2	2	1	0	2	0	2	1	2	0	0	2	1	2	0	2	1	2	0	0	2	1	1	0	2	2	0	1	0	2	1
49	1	2	0	2	1	2	0	0	1	1	1	0	2	0	1	1	2	0	2	2	1	0	1	2	0	0	1	1	1	0	2	1	0	2	2	0	1	0	2	1
50	1	2	0	2	1	2	0	0	1	1	1	0	2	1	2	0	0	1	1	1	0	2	2	0	2	1	2	0	0	2	1	0	2	1	1	2	0	2	1	0
51	1	2	0	2	1	2	0	0	1	1	1	0	2	2	0	2	1	2	0	0	2	1	0	1	1	2	0	2	2	1	0	2	1	0	0	1	2	1	0	2
52	1	2	0	2	2	0	2	1	2	0	0	2	1	0	1	1	2	0	2	2	1	0	2	0	2	1	2	0	0	2	1	2	1	0	0	1	2	1	0	2
53	1	2	0	2	2	0	2	1	2	0	0	2	1	1	2	0	0	1	1	1	0	2	0	1	1	2	0	2	2	1	0	1	0	2	2	0	1	0	2	1
54	1	2	0	2	2	0	2	1	2	0	0	2	1	2	0	2	1	2	0	0	2	1	1	2	0	0	1	1	1	0	2	0	2	1	1	2	0	2	1	0
55	2	0	2	2	0	2	2	0	2	2	0	1	2	0	2	2	0	2	2	0	1	2	0	2	2	0	2	2	0	1	2	0	1	2	0	2	1	0	1	2
56	2	0	2	2	0	2	2	0	2	2	0	1	2	1	0	1	1	0	1	2	0	1	1	0	1	1	0	1	2	0	1	2	0	1	2	1	0	2	0	1
57	2	0	2	2	0	2	2	0	2	2	0	1	2	2	1	0	2	1	0	1	2	0	2	1	0	2	1	0	1	2	0	1	2	0	1	0	2	1	2	0
58	2	0	2	2	1	0	1	1	0	1	2	0	1	0	2	2	0	2	2	0	1	2	1	0	1	1	0	1	2	0	1	1	2	0	1	0	2	1	2	0
59	2	0	2	2	1	0	1	1	0	1	2	0	1	1	0	1	1	0	1	2	0	1	2	1	0	2	1	0	1	2	0	0	1	2	0	2	1	0	1	2
60	2	0	2	2	1	0	1	1	0	1	2	0	1	2	1	0	2	1	0	1	2	0	0	2	2	0	2	2	0	1	2	2	0	1	2	1	0	2	0	1
61	2	0	2	2	2	1	0	2	1	0	1	2	0	0	2	2	0	2	2	0	1	2	2	1	0	2	1	0	1	2	0	2	0	1	2	1	0	2	0	1
62	2	0	2	2	2	1	0	2	1	0	1	2	0	1	0	1	1	0	1	2	0	1	0	2	2	0	2	2	0	1	2	1	2	0	1	0	2	1	2	0
63	2	0	2	2	2	1	0	2	1	0	1	2	0	2	1	0	2	1	0	1	2	0	1	0	1	1	0	1	2	0	1	0	1	2	0	2	1	0	1	2
64	2	1	0	1	0	2	2	1	0	1	1	2	0	0	2	2	1	0	1	1	2	0	0	2	2	1	0	1	1	2	0	0	1	2	2	1	0	1	2	0
65	2	1	0	1	0	2	2	1	0	1	1	2	0	1	0	1	2	1	0	0	1	2	1	0	1	2	1	0	0	1	2	2	0	1	1	0	2	0	1	2
66	2	1	0	1	0	2	2	1	0	1	1	2	0	2	1	0	0	2	2	2	0	1	2	1	0	0	2	2	2	0	1	1	2	0	0	2	1	2	0	1
67	2	1	0	1	1	0	1	2	1	0	0	1	2	0	2	2	1	0	1	1	2	0	1	0	1	2	1	0	0	1	2	1	2	0	0	2	1	2	0	1
68	2	1	0	1	1	0	1	2	1	0	0	1	2	1	0	1	2	1	0	0	1	2	2	1	0	0	2	2	2	0	1	0	1	2	2	1	0	1	2	0
69	2	1	0	1	1	0	1	2	1	0	0	1	2	2	1	0	0	2	2	2	0	1	0	2	2	1	0	1	1	2	0	2	0	1	1	0	2	0	1	2
70	2	1	0	1	2	1	0	0	2	2	2	0	1	0	2	2	1	0	1	1	2	0	2	1	0	0	2	2	2	0	1	2	0	1	1	0	2	0	1	2
71	2	1	0	1	2	1	0	0	2	2	2	0	1	1	0	1	2	1	0	0	1	2	0	2	2	1	0	1	1	2	0	1	2	0	0	2	1	2	0	1
72	2	1	0	1	2	1	0	0	2	2	2	0	1	2	1	0	0	2	2	2	0	1	1	0	1	2	1	0	0	1	2	0	1	2	2	1	0	1	2	0
73	2	2	1	0	0	2	2	2	1	0	2	0	1	0	2	2	2	1	0	2	0	1	0	2	2	2	1	0	2	0	1	0	1	2	1	0	2	2	0	1

G.6(a) $L_{81}(3^{40})$ (Continued).

Experiment number	Column number																																						
	1	2	3	4	5	6	7	8	9	10	11	12	13	14	15	16	17	18	19	20	21	22	23	24	25	26	27	28	29	30	31	32	33	34	35	36	37	38	39	40
74	2	2	1	0	0	2	2	2	1	0	2	0	1	1	0	1	0	2	2	1	2	0	1	0	1	0	2	2	1	2	0	2	0	1	0	2	1	1	2	0
75	2	2	1	0	0	2	2	2	1	0	2	0	1	2	1	0	1	0	1	0	1	2	2	1	0	1	0	1	0	1	2	1	2	0	2	1	0	0	1	2
76	2	2	1	0	1	0	1	0	2	2	1	2	0	0	2	2	2	1	0	2	0	1	1	0	1	0	2	2	1	2	0	1	2	0	2	1	0	0	1	2
77	2	2	1	0	1	0	1	0	2	2	1	2	0	1	0	1	0	2	2	1	2	0	2	1	0	1	0	1	0	1	2	0	1	2	1	0	2	2	0	1
78	2	2	1	0	1	0	1	0	2	2	1	2	0	2	1	0	1	0	1	0	1	2	0	2	2	2	1	0	2	0	1	2	0	1	0	2	1	1	2	0
79	2	2	1	0	2	1	0	1	0	1	0	1	2	0	2	2	2	1	0	2	0	1	2	1	0	1	0	1	0	1	2	2	0	1	0	2	1	1	2	0
80	2	2	1	0	2	1	0	1	0	1	0	1	2	1	0	1	0	2	2	1	2	0	0	2	2	2	1	0	2	0	1	1	2	0	2	1	0	0	1	2
81	2	2	1	0	2	1	0	1	0	1	0	1	2	2	1	0	1	0	1	0	1	2	1	0	1	0	2	2	1	2	0	0	1	2	1	0	2	2	0	1
Basic mark	a	b	ab	ab^2	c	ac	aC^2	bc	abc	ab^2c^2	bc^2	ab^2c	abc^2	d	ad	ad^2	bd	abd	ab^2d^2	bd^2	ab^2d	abd^2	cd	acd	ac^2d^2	bcd	$abcd$	$ab^2c^2d^2$	bc^2d^2	ab^2cd	abc^2d^2	cd^2	ac^2d	acd^2	bc^2d	abc^2d	ab^2cd^2	bcd^2	ab^2c^2d	$abcd^2$
Group	1	2			3									4																										

Interactions between two columns in $L_{81}(3^{40})$

Column number	1	2	3	4	5	6	7	8	9	10	11	12	13	14	15	16	17	18	19	20	21	22	23	24	25	26	27	28	29	30	31	32	33	34	35	36	37	38	39	40
	(1)	3	2	2	6	5	5	9	8	8	12	11	11	15	14	14	18	17	17	21	20	20	24	23	23	27	26	26	30	29	29	33	32	32	36	35	35	39	38	38
		4	4	3	7	7	6	10	10	9	13	13	12	16	16	15	19	19	18	22	22	21	25	25	24	28	28	27	31	31	30	34	34	33	37	37	36	40	40	39
		(2)	1	1	8	9	10	5	6	7	5	6	7	17	18	19	14	15	16	14	15	16	26	27	28	23	24	25	23	24	25	35	36	37	32	33	34	32	33	34
			4	3	11	12	13	11	12	13	8	9	10	20	21	22	20	21	22	17	18	19	29	30	31	29	30	31	26	27	28	38	39	40	38	39	40	35	36	37
			(3)	1	9	10	8	7	5	6	6	7	5	18	19	17	16	14	15	15	16	14	27	28	26	25	23	24	24	25	23	36	37	35	34	32	33	33	34	32
				2	13	11	12	12	13	11	10	8	9	22	20	21	21	22	20	19	17	18	31	29	30	30	31	29	28	26	27	40	38	39	39	40	38	37	35	36
				(4)	10	8	9	6	7	5	7	5	6	19	17	18	15	16	14	16	14	15	28	26	27	24	25	23	25	23	24	37	35	36	33	34	32	34	32	33
					12	13	11	13	11	12	9	10	8	21	22	20	22	20	21	18	19	17	30	31	29	31	29	30	27	28	26	39	40	38	40	38	39	36	37	35
					(5)	1	1	2	3	4	2	4	3	23	24	25	26	27	28	29	30	31	14	15	16	17	18	19	20	21	22	14	15	16	17	18	19	20	21	22
						7	6	11	13	12	8	10	9	32	33	34	35	36	37	38	39	40	32	33	34	35	36	37	38	39	40	23	24	25	26	27	28	29	30	31
						(6)	1	4	2	3	3	2	4	24	25	23	27	28	26	30	31	29	16	14	15	19	17	18	22	20	21	15	16	14	18	19	17	21	22	20
							5	13	12	11	10	9	8	34	32	33	37	35	36	40	38	39	33	34	32	36	37	35	39	40	38	25	23	24	28	26	27	31	29	30
							(7)	3	4	2	4	3	2	25	23	24	28	26	27	31	29	30	15	16	14	18	19	17	21	22	20	16	14	15	19	17	18	22	20	21
								12	11	13	9	8	10	33	34	32	36	37	35	39	40	38	34	32	33	37	35	36	40	38	39	24	25	23	27	28	26	30	31	29
								(8)	1	1	2	3	4	26	27	28	29	30	31	23	24	25	20	21	22	14	15	16	17	18	19	17	18	19	20	21	22	14	15	16
									10	9	5	7	6	38	39	40	32	33	34	35	36	37	35	36	37	38	39	40	32	33	34	29	30	31	23	24	25	26	27	18
									(9)	1	4	2	3	27	28	26	30	31	29	24	25	23	22	20	21	16	14	15	19	17	18	18	19	17	21	22	20	15	16	14
										8	7	6	5	40	38	39	34	32	33	37	35	36	36	37	35	39	40	38	33	34	32	31	29	30	25	23	24	28	26	27
										(10)	3	4	2	28	26	27	31	29	30	25	23	24	21	22	20	15	16	14	18	19	17	19	17	18	22	20	21	16	14	15
											6	5	7	39	40	38	33	34	32	36	37	35	27	35	36	40	38	39	34	32	33	30	31	29	24	25	23	27	28	26
											(11)	1	1	29	30	31	23	24	25	26	27	28	17	18	19	20	21	22	14	15	16	20	21	22	14	15	16	17	18	19
												13	12	35	36	37	38	39	40	32	33	34	38	39	40	32	33	34	35	36	37	26	27	28	29	30	31	23	24	25
												(12)	1	30	31	29	24	25	23	27	28	26	19	17	18	22	20	21	16	14	15	21	22	20	15	16	14	18	19	17
													11	37	35	36	40	38	39	34	32	33	39	40	38	33	34	32	36	37	35	28	26	27	31	29	30	25	23	24
													(13)	31	29	30	25	23	24	28	26	27	18	19	17	21	22	20	15	16	14	22	20	21	16	14	15	19	17	18
														36	37	35	39	40	38	33	34	32	40	38	39	34	32	33	37	35	36	27	28	26	30	31	29	24	25	23
														(14)	1	1	2	3	4	2	4	3	5	6	7	8	9	10	11	12	13	5	7	6	11	13	11	8	10	9
															16	15	20	22	21	17	19	18	32	34	33	38	40	39	35	37	36	23	25	24	29	31	31	26	28	27
															(15)	11	4	2	3	3	2	4	7	5	6	10	8	9	13	11	12	6	5	7	12	11	6	9	8	10
																14	22	21	20	19	18	17	34	33	32	40	39	38	37	36	35	25	24	23	31	30	27	28	27	26
																(16)	3	4	2	4	3	2	6	7	5	9	10	8	12	13	11	7	6	5	13	12	7	10	9	28
																	21	20	22	18	17	19	33	32	34	39	38	40	36	35	37	24	23	25	30	29	26	27	26	28
																	(17)	1	1	2	3	4	11	12	13	5	6	7	8	9	10	8	10	9	5	7	6	11	13	12
																		19	18	14	16	15	38	40	39	35	37	36	32	34	33	29	31	30	26	28	27	23	25	24
																		(18)	1	4	2	3	13	11	12	7	5	6	10	8	9	9	8	10	6	5	7	12	11	13
																			17	16	15	14	40	39	38	37	36	35	34	33	33	31	30	29	28	27	26	25	24	24
																			(19)	3	4	2	12	13	11	6	7	5	9	10	8	10	9	8	7	6	5	13	12	11
																				15	14	16	39	38	40	36	35	37	33	32	34	30	29	31	27	26	28	24	23	25

Interactions between two columns in $L_{81}(3^{40})$ (Continued).

Column number	1	2	3	4	5	6	7	8	9	10	11	12	13	14	15	16	17	18	19	20	21	22	23	24	25	26	27	28	29	30	31	32	33	34	35	36	37	38	39	40
																				(20)	1	1	8	9	10	11	12	13	5	6	7	11	13	12	8	10	9	5	7	6
																					22	21	35	37	36	32	34	33	38	40	39	26	28	27	23	25	24	29	31	30
																					(21)	1	10	8	9	13	11	12	7	5	6	12	11	13	9	8	10	6	5	7
																						20	37	36	35	34	33	32	40	39	38	28	27	26	25	24	23	31	30	29
																						(22)	9	10	8	12	13	11	6	7	5	13	12	11	10	9	8	7	6	5
																							36	35	37	33	32	34	39	38	40	27	26	28	24	23	25	30	29	31
																							(23)	1	1	2	3	4	2	4	3	5	6	7	8	9	10	11	12	13
																								25	24	29	31	30	26	28	27	14	16	15	20	22	21	17	19	18
																								(24)	1	4	2	3	3	2	4	7	5	6	10	8	9	13	11	12
																									23	31	30	29	28	27	26	16	15	14	22	21	20	19	18	17
																									(25)	3	4	2	4	3	2	6	7	5	9	10	8	12	13	11
																										30	29	31	27	26	28	15	14	16	21	20	22	18	17	19
																										(26)	1	1	2	3	4	11	12	13	5	6	7	8	9	10
																											28	27	23	25	24	20	22	21	17	19	18	14	16	15
																											(27)	1	4	2	3	13	11	12	7	5	6	10	8	9
																												26	25	24	23	22	21	20	19	18	17	16	15	14
																												(28)	3	4	2	12	13	11	6	7	5	9	10	8
																													24	23	25	21	20	22	18	17	19	15	14	16
																													(29)	1	1	8	9	10	11	12	13	5	6	7
																														31	30	17	19	18	14	16	15	20	22	21
																														(30)	1	10	8	9	13	11	12	7	5	6
																															29	19	18	17	16	15	14	22	21	20
																															(31)	9	10	8	12	13	11	6	7	5
																																18	17	19	15	14	16	21	20	22
																																(32)	1	1	2	3	4	2	4	3
																																	34	33	38	40	39	35	37	36
																																	(33)	1	4	2	3	3	2	4
																																		32	40	39	38	37	36	35
																																		(34)	3	4	2	4	3	2
																																			39	38	40	36	35	37
																																			(35)	1	1	2	3	4
																																				37	36	32	34	33
																																				(36)	1	4	2	3
																																					35	34	33	32
																																					(37)	3	4	2
																																						33	32	34
																																						(38)	1	1
																																							40	39
																																							(39)	1
																																								38

LINEAR GRAPHS OF $L_{81}(3^{40})$

There are altogether 14 different types of linear graphs in $L_{81}(3^{40})$. Of these two typical different type of linear graphs are shown below.

(1)

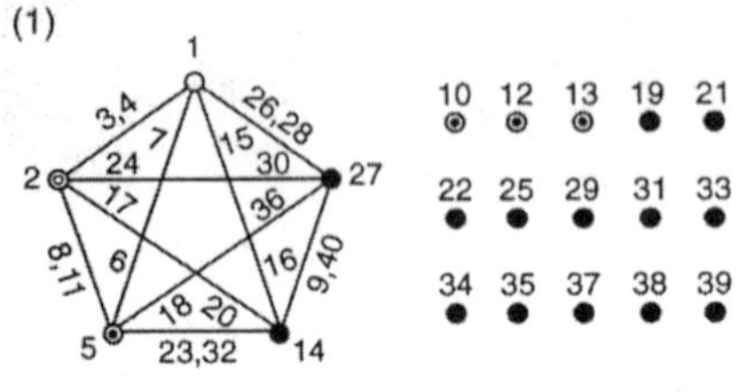

a

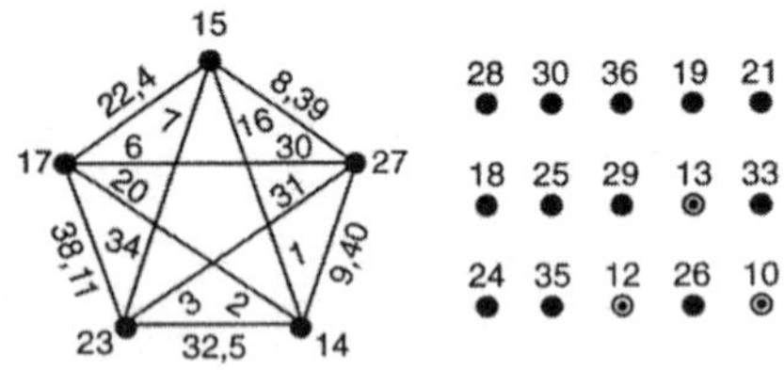

b

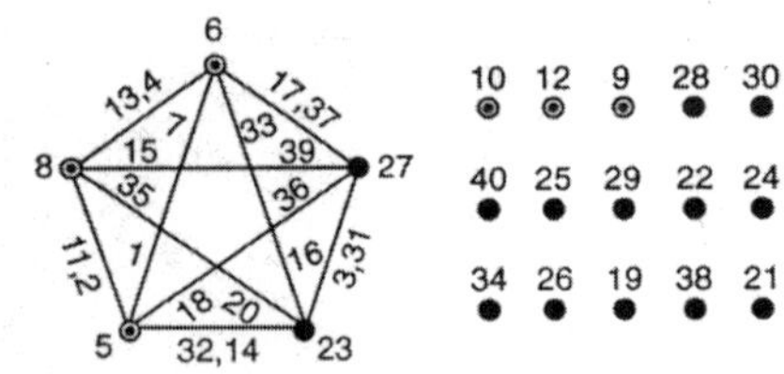

c

(2)

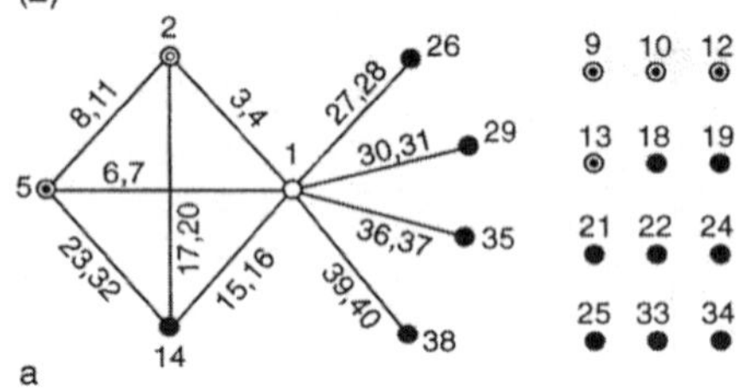

a

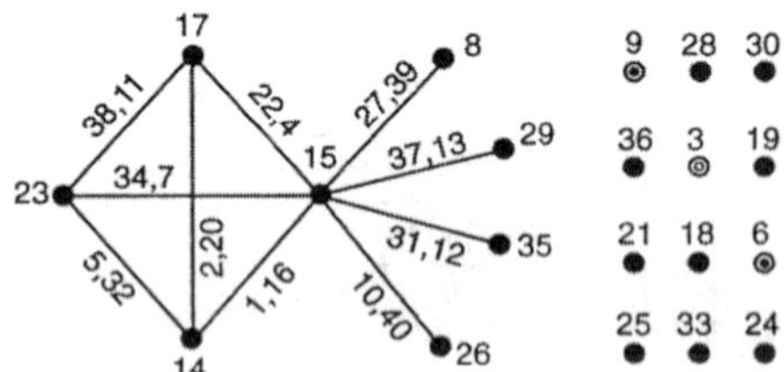

b

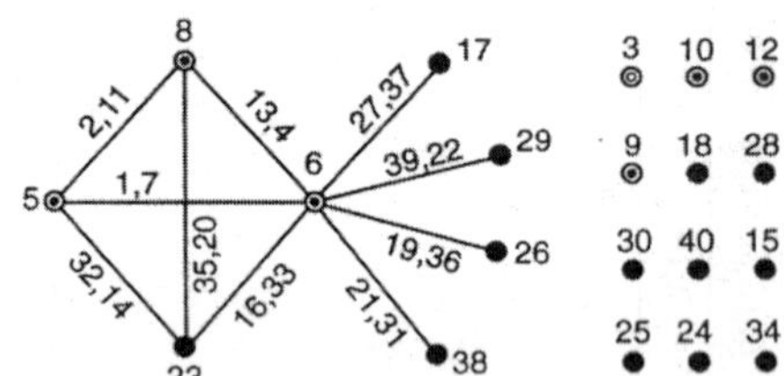

c

G.7 $L_{16}(4^5)$

Experiment number	Column number				
	1	2	3	4	5
1	0	0	0	0	0
2	0	1	1	1	1
3	0	2	2	2	2
4	0	3	3	3	3
5	1	0	1	2	3
6	1	1	0	3	2
7	1	2	3	0	1
8	1	3	2	1	0
9	2	0	2	3	1
10	2	1	3	2	0
11	2	2	0	1	3
12	2	3	1	0	2
13	3	0	3	1	2
14	3	1	2	0	3
15	3	2	1	3	0
16	3	3	0	2	1
Group	1	2			

LINEAR GRAPHS OF $L_{16}(4^5)$

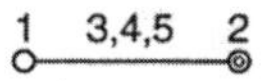

G.8 $L_{32}(2^1 \times 4^9)$

Experiment number	Column number									
	1	2	3	4	5	6	7	8	9	10
1	0	0	0	0	0	0	0	0	0	0
2	0	0	1	1	1	1	1	1	1	1
3	0	0	2	2	2	2	2	2	2	2
4	0	0	3	3	3	3	3	3	3	3
5	0	1	0	0	1	1	2	2	3	3
6	0	1	1	1	0	0	3	3	2	2
7	0	1	2	2	3	3	0	0	1	1
8	0	1	3	3	2	2	1	1	0	0
9	0	2	0	1	2	3	0	1	2	3
10	0	2	1	0	3	2	1	0	3	2
11	0	2	2	3	0	1	2	3	0	1
12	0	2	3	2	1	0	3	2	1	0
13	0	3	0	1	3	2	2	3	1	0
14	0	3	1	0	2	3	3	2	0	1
15	0	3	2	3	1	0	0	1	3	2
16	0	3	3	2	0	1	1	0	2	3
17	1	0	0	3	0	3	1	2	1	2
18	1	0	1	2	1	2	0	3	0	3
19	1	0	2	1	2	1	3	0	3	0
20	1	0	3	0	3	0	2	1	2	1
21	1	1	0	3	1	2	3	0	2	1
22	1	1	1	2	0	3	2	1	3	0
23	1	1	2	1	3	0	1	2	0	3
24	1	1	3	0	2	1	0	3	1	2
25	1	2	0	2	2	0	1	3	3	1
26	1	2	1	3	3	1	0	2	2	0
27	1	2	2	0	0	2	3	1	1	3
28	1	2	3	1	1	3	2	0	0	2
29	1	3	0	2	3	1	3	1	0	2
30	1	3	1	3	2	0	2	0	1	3
31	1	3	2	0	1	3	1	3	2	0
32	1	3	3	1	0	2	0	2	3	1
Group	1	2	3							

Note:

(i) Interactions between column 1 (two levels) and column 2 (four levels) can be found without sacrifice of other columns.

(ii) Interactions between two columns. It is necessary to use the sequential correction method if one wishes to find the interactions. It is therefore better to use L_{84} if one wishes to find interactions.

G.9 $L_{64}(4^{21})$

Experiment number	Column number																				
	1	2	3	4	5	6	7	8	9	10	11	12	13	14	15	16	17	18	19	20	21
1	0	0	0	0	0	0	0	0	0	0	0	0	0	0	0	0	0	0	0	0	0
2	0	0	0	0	0	1	1	1	1	1	1	1	1	1	1	1	1	1	1	1	1
3	0	0	0	0	0	2	2	2	2	2	2	2	2	2	2	2	2	2	2	2	2
4	0	0	0	0	0	3	3	3	3	3	3	3	3	3	3	3	3	3	3	3	3
5	0	1	1	1	1	0	0	0	0	1	1	1	1	2	2	2	2	3	3	3	3
6	0	1	1	1	1	1	1	1	1	0	0	0	0	3	3	3	3	2	2	2	2
7	0	1	1	1	1	2	2	2	2	3	3	3	3	0	0	0	0	1	1	1	1
8	0	1	1	1	1	3	3	3	3	2	2	2	2	1	1	1	1	0	0	0	0
9	0	2	2	2	2	0	0	0	0	2	2	2	2	3	3	3	3	1	1	1	1
10	0	2	2	2	2	1	1	1	1	3	3	3	3	2	2	2	2	0	0	0	0
11	0	2	2	2	2	2	2	2	2	0	0	0	0	1	1	1	1	3	3	3	3
12	0	2	2	2	2	3	3	3	3	1	1	1	1	0	0	0	0	2	2	2	2
13	0	3	3	3	3	0	0	0	0	3	3	3	3	1	1	1	1	2	2	2	2
14	0	3	3	3	3	1	1	1	1	2	2	2	2	0	0	0	0	3	3	3	3
15	0	3	3	3	3	2	2	2	2	1	1	1	1	3	3	3	3	0	0	0	0
16	0	3	3	3	3	3	3	3	3	0	0	0	0	2	2	2	2	1	1	1	1
17	1	0	1	2	3	0	1	2	3	0	1	2	3	0	1	2	3	0	1	2	3
18	1	0	1	2	3	1	0	3	2	1	0	3	2	1	0	3	2	1	0	3	2
19	1	0	1	2	3	2	3	0	1	2	3	0	1	2	3	0	1	2	3	0	1
20	1	0	1	2	3	3	2	1	0	3	2	1	0	3	2	1	0	3	2	1	0
21	1	1	0	3	2	0	1	2	3	1	0	3	2	2	3	0	1	3	2	1	0
22	1	1	0	3	2	1	0	3	2	0	1	2	3	3	2	1	0	2	3	0	1
23	1	1	0	3	2	2	3	0	1	3	2	1	0	0	1	2	3	1	0	3	2
24	1	1	0	3	2	3	2	1	0	2	3	0	1	1	0	3	2	0	1	2	3
25	1	2	3	0	1	0	1	2	3	2	3	0	1	3	2	1	0	1	0	3	2
26	1	2	3	0	1	1	0	3	2	3	2	1	0	2	3	0	1	0	1	2	3
27	1	2	3	0	1	2	3	0	1	0	1	2	3	1	0	3	2	3	2	1	0
28	1	2	3	0	1	3	2	1	0	1	0	3	2	0	1	2	3	2	3	0	1
29	1	3	2	1	0	0	1	2	3	3	2	1	0	1	0	3	2	2	3	0	1
30	1	3	2	1	0	1	0	3	2	2	3	0	1	0	1	2	3	3	2	1	0
31	1	3	2	1	0	2	3	0	1	1	0	3	2	3	2	1	0	0	1	2	3
32	1	3	2	1	0	3	2	1	0	0	1	2	3	2	3	0	1	1	0	3	2
33	2	0	2	3	1	0	2	3	1	0	2	3	1	0	2	3	1	0	2	3	1
34	2	0	2	3	1	1	3	2	0	1	3	2	0	1	3	2	0	1	3	2	0
35	2	0	2	3	1	2	0	1	3	2	0	1	3	2	0	1	3	2	0	1	3
36	2	0	2	3	1	3	1	0	2	3	1	0	2	3	1	0	2	3	1	0	2
37	2	1	3	2	0	0	2	3	1	1	3	2	0	2	0	1	3	3	1	0	2
38	2	1	3	2	0	1	3	2	0	0	2	3	1	3	1	0	2	2	0	1	3
39	2	1	3	2	0	2	0	1	3	3	0	0	2	0	2	3	1	1	3	2	0
40	2	1	3	2	0	3	1	0	2	2	1	1	3	1	3	2	0	0	2	3	1
41	2	2	0	1	3	0	2	3	1	2	0	1	3	3	1	0	2	1	3	2	0
42	2	2	0	1	3	1	3	2	0	3	1	0	2	2	0	1	3	0	2	3	1
43	2	2	0	1	3	2	0	1	3	0	2	3	1	1	3	2	0	3	1	0	2
44	2	2	0	1	3	3	1	0	2	1	3	2	0	0	2	3	1	2	0	1	3
45	2	3	1	0	2	0	2	3	1	3	1	0	2	1	3	2	0	2	0	1	3
46	2	3	1	0	2	1	3	2	0	2	0	1	3	0	2	3	1	3	1	0	2
47	2	3	1	0	2	2	0	1	3	1	3	2	0	3	1	0	2	0	2	3	1
48	2	3	1	0	2	3	1	0	2	0	2	3	1	2	0	1	3	1	3	2	0
49	3	0	3	1	2	0	3	1	2	0	3	1	2	0	3	1	2	0	3	1	2
50	3	0	3	1	2	1	2	0	3	1	2	0	3	1	2	0	3	1	2	0	3
51	3	0	3	1	2	2	1	3	0	2	1	3	0	2	1	3	0	2	1	3	0
52	3	0	3	1	2	3	0	2	1	3	0	2	1	3	0	2	1	3	0	2	1
53	3	1	2	0	3	0	3	1	2	1	2	0	3	2	1	3	0	3	0	2	1
54	3	1	2	0	3	1	2	0	3	0	3	1	2	3	0	2	1	2	1	3	0
55	3	1	2	0	3	2	1	3	0	3	0	2	1	0	3	1	2	1	2	0	3
56	3	1	2	0	3	3	0	2	1	2	1	3	0	1	2	0	3	0	3	1	2
57	3	2	1	3	0	0	3	1	2	2	1	3	0	3	0	2	1	1	2	0	3
58	3	2	1	3	0	1	2	0	3	3	0	2	1	2	1	3	0	0	3	1	2
59	3	2	1	3	0	2	1	3	0	0	3	1	2	1	2	0	3	3	0	2	1
60	3	2	1	3	0	3	0	2	1	1	2	0	3	0	3	1	2	2	1	3	0
61	3	3	0	2	1	0	3	1	2	3	0	2	1	1	2	0	3	2	1	3	0
62	3	3	0	2	1	1	2	0	3	2	1	3	0	0	3	1	2	3	0	2	1
63	3	3	0	2	1	2	1	3	0	1	2	0	3	3	0	2	1	0	3	1	2
64	3	3	0	2	1	3	0	2	1	0	3	1	2	2	1	3	0	1	2	0	3
Group	1	2				3															

Interactions between two columns in $L_{64}(4^{21})$

Column number	1	2	3	4	5	6	7	8	9	10	11	12	13	14	15	16	17	18	19	20	21
		3	2	2	2	7	6	6	6	11	10	10	10	15	14	14	14	19	18	18	18
	(1)	4	4	3	3	8	8	7	7	12	12	11	11	16	16	15	15	20	20	19	19
		5	5	5	4	9	9	9	8	13	13	13	12	17	17	17	16	21	21	21	21
			1	1	1	10	11	12	13	6	7	8	9	6	7	8	9	6	7	8	9
		(2)	4	3	3	14	15	16	17	14	15	16	17	10	11	12	13	10	11	12	13
			5	5	4	18	19	20	21	18	19	20	21	18	19	20	21	14	15	16	17
				1	1	11	10	13	12	7	6	9	8	8	9	6	7	9	8	7	6
			(3)	2	2	16	17	14	15	17	16	15	14	13	12	11	10	12	13	10	11
				5	4	21	20	19	18	29	21	18	19	19	18	21	20	15	14	17	16
					1	12	13	10	11	8	9	6	7	9	8	7	6	7	6	9	8
				(4)	2	17	16	15	14	15	14	17	16	11	10	13	12	13	12	11	10
					3	19	18	21	20	21	20	19	18	20	21	18	19	16	17	14	15
						13	12	11	10	9	8	7	6	7	6	9	8	8	9	6	7
					(5)	15	14	17	16	16	17	14	15	12	13	10	11	11	10	13	12
						20	21	18	19	19	18	21	20	21	20	19	18	17	16	15	14
							1	1	1	2	3	4	5	2	5	3	4	2	4	5	3
						(6)	8	7	7	14	16	17	15	10	13	11	12	10	12	13	11
							9	9	8	18	21	19	20	18	20	21	19	14	17	15	16
								1	1	3	2	5	4	5	2	4	3	4	2	3	5
							(7)	6	6	17	15	14	16	12	11	13	10	13	11	10	12
								9	8	20	19	21	18	21	19	18	20	16	15	17	14
									1	4	5	2	3	3	4	2	5	5	3	2	4
								(8)	6	15	17	16	14	13	10	12	11	11	13	12	10
									7	11	18	20	19	19	21	20	18	17	14	16	15
										5	4	3	2	4	3	5	2	3	5	4	2
									(9)	16	14	15	17	11	12	10	13	12	10	11	13
										19	20	18	21	20	18	19	21	15	16	14	17
											1	1	1	2	4	5	3	2	5	3	4
										(10)	12	11	11	6	8	9	7	6	9	7	4
											13	13	12	18	21	19	20	14	16	17	15
												1	1	4	2	3	5	5	2	4	3
											(11)	10	10	9	8	9	7	6	9	7	8
												13	12	20	19	21	18	17	15	14	16
													1	5	3	2	4	3	4	2	5
												(12)	10	7	9	8	6	9	6	8	7
													11	21	18	20	19	15	17	16	14
														3	5	4	2	4	3	5	2
													(13)	8	6	7	9	7	8	6	9
														19	20	18	21	16	14	15	17
															1	1	1	2	3	4	5
														(14)	16	15	15	6	8	9	7
															17	17	16	10	13	11	12
																1	1	3	2	5	4
															(15)	14	14	9	7	6	8
																17	16	12	11	13	10
																	1	4	5	2	3
																(16)	14	7	9	8	6
																	15	13	10	12	11
																		5	4	3	2
																	(17)	8	6	7	9
																		11	12	10	13
																			1	1	1
																		(18)	20	19	19
																			21	21	20
																				1	1
																			(19)	18	18
																				21	20
																					1
																				(20)	18
																					19

LINEAR GRAPHS OF $L_{64}(4^{21})$

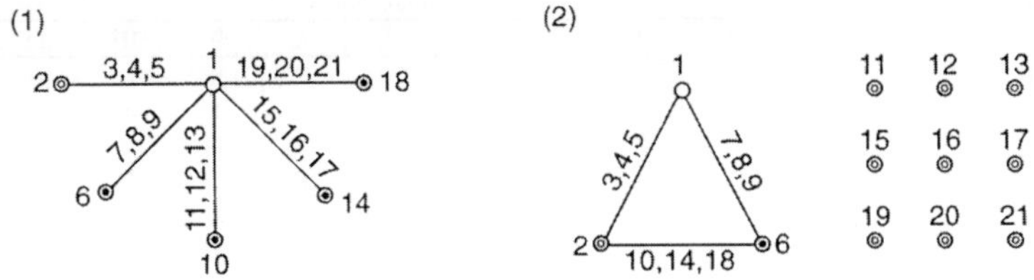

G.10 $L_{25}(5^6)$

Experiment number	Column number 1	2	3	4	5	6
1	0	0	0	0	0	0
2	0	1	1	1	1	1
3	0	2	2	2	2	2
4	0	3	3	3	3	3
5	0	4	4	4	4	4
6	1	0	1	2	3	4
7	1	1	2	3	4	0
8	1	2	3	4	0	1
9	1	3	4	0	1	2
10	1	4	0	1	2	3
11	2	0	2	4	1	3
12	2	1	3	0	2	4
13	2	2	4	1	3	0
14	2	3	0	2	4	1
15	2	4	1	3	0	2
16	3	0	3	1	4	2
17	3	1	4	2	0	3
18	3	2	0	3	1	4
19	3	3	1	4	2	0
20	3	4	2	0	3	1
21	4	0	4	3	2	1
22	4	1	0	4	3	2
23	4	2	1	0	4	3
24	4	3	2	1	0	4
25	4	4	3	2	1	0
Group	1			2		

LINEAR GRAPH OF $L_{25}(5^6)$

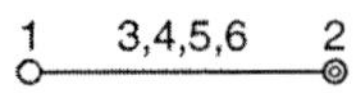

G.11 $L_{50}(2^1 \times 5^{11})$

Experiment number	Column number											
	1	2	3	4	5	6	7	8	9	10	11	12
1	0	0	0	0	0	0	0	0	0	0	0	0
2	0	0	1	1	1	1	1	1	1	1	1	1
3	0	0	2	2	2	2	2	2	2	2	2	2
4	0	0	3	3	3	3	3	3	3	3	3	3
5	0	0	4	4	4	4	4	4	4	4	4	4
6	0	1	0	1	2	3	4	0	1	2	3	4
7	0	1	1	2	3	4	0	1	2	3	4	0
8	0	1	2	3	4	0	1	2	3	4	0	1
9	0	1	3	4	0	1	2	3	4	0	1	2
10	0	1	4	0	1	2	3	4	0	1	2	3
11	0	2	0	2	4	1	3	3	0	2	4	1
12	0	2	1	3	0	2	4	4	1	3	0	2
13	0	2	2	4	1	3	0	0	2	4	1	3
14	0	2	3	0	2	4	1	1	3	0	2	4
15	0	2	4	1	3	0	2	2	4	1	3	0
16	0	3	0	3	1	4	2	4	2	0	3	1
17	0	3	1	4	2	0	3	0	3	1	4	2
18	0	3	2	0	3	1	4	1	4	2	0	3
19	0	3	3	1	4	2	0	2	0	3	1	4
20	0	3	4	2	0	3	1	3	1	4	2	0
21	0	4	0	4	3	2	1	3	2	1	0	4
22	0	4	1	0	4	3	2	4	3	2	1	0
23	0	4	2	1	0	4	3	0	4	3	2	1
24	0	4	3	2	1	0	4	1	0	4	3	2
25	0	4	4	3	2	1	0	2	1	0	4	3
26	1	0	0	0	3	4	3	2	1	4	1	2
27	1	0	1	1	4	0	4	3	2	0	2	3
28	1	0	2	2	0	1	0	4	3	1	3	4
29	1	0	3	3	1	2	1	0	4	2	4	0
30	1	0	4	4	2	3	2	1	0	3	0	1
31	1	1	0	1	0	2	2	1	3	4	4	3
32	1	1	1	2	1	3	3	2	4	0	0	4
33	1	1	2	3	2	4	4	3	0	1	1	0
34	1	1	3	4	3	0	0	4	1	2	2	1
35	1	1	4	0	4	1	1	0	2	3	3	2
36	1	2	0	2	2	0	1	4	4	3	1	3
37	1	2	1	3	3	1	2	0	0	4	2	4
38	1	2	2	4	4	2	3	1	1	0	3	0
39	1	2	3	0	0	3	4	2	2	1	4	1
40	1	2	4	1	1	4	0	3	3	2	0	2
41	1	3	0	3	4	3	0	1	4	1	2	2
42	1	3	1	4	0	4	1	2	0	2	3	3
43	1	3	2	0	1	0	2	3	1	3	4	4
44	1	3	3	1	2	1	3	4	2	4	0	0
45	1	3	4	2	3	2	4	0	3	0	1	1
46	1	4	0	4	1	1	4	2	3	3	2	0
47	1	4	1	0	2	2	0	3	4	4	3	1
48	1	4	2	1	3	3	1	4	0	0	4	2
49	1	4	3	2	4	4	2	0	1	1	0	3
50	1	4	4	3	0	0	3	1	2	2	1	4
Group	1	2	3									

G.12 $L_{36}(2^{11} \times 3^{12})$, $L_{36}(2^{3} \times 3^{13})$ and $L_{36}(3^{13})$

Experiment number	Column number																									
	1	2	3	4	5	6	7	8	9	10	11	12	13	14	15	16	17	18	19	20	21	22	23	1'	2'	3'	4'
1	0	0	0	0	0	0	0	0	0	0	0	0	0	0	0	0	0	0	0	0	0	0	0	0	0	0	0
2	0	0	0	0	0	0	0	0	0	0	0	1	1	1	1	1	1	1	1	1	1	1	1	0	0	0	0
3	0	0	0	0	0	0	0	0	0	0	0	2	2	2	2	2	2	2	2	2	2	2	2	0	0	0	0
4	0	0	0	0	0	1	1	1	1	1	1	0	0	0	0	1	1	1	1	2	2	2	2	0	1	1	0
5	0	0	0	0	0	1	1	1	1	1	1	1	1	1	1	2	2	2	2	0	0	0	0	0	1	1	0
6	0	0	0	0	0	1	1	1	1	1	1	2	2	2	2	0	0	0	0	1	1	1	1	0	1	1	0
7	0	0	1	1	1	0	0	0	1	1	1	0	0	1	2	0	1	2	2	0	1	1	2	1	0	1	0
8	0	0	1	1	1	0	0	0	1	1	1	1	1	2	0	1	2	0	0	1	2	2	0	1	0	1	0
9	0	0	1	1	1	0	0	0	1	1	1	2	2	0	1	2	0	1	1	2	0	0	1	1	0	1	0
10	0	1	0	1	1	0	1	1	0	0	1	0	0	2	1	0	2	1	2	1	0	2	1	1	1	0	0
11	0	1	0	1	1	0	1	1	0	0	1	1	1	0	2	1	0	2	0	2	1	0	2	1	1	0	0
12	0	1	0	1	1	0	1	1	0	0	1	2	2	1	0	2	1	0	1	0	2	1	0	1	1	0	0
13	0	1	1	0	1	1	0	1	0	1	0	0	1	2	0	2	1	0	2	2	1	0	1	0	0	0	1
14	0	1	1	0	1	1	0	1	0	1	0	1	2	0	1	0	2	1	0	0	2	1	2	0	0	0	1
15	0	1	1	0	1	1	0	1	0	1	0	2	0	1	2	1	0	2	1	1	0	2	0	0	0	0	1
16	0	1	1	1	0	1	1	0	1	0	0	0	1	2	1	0	0	2	1	2	2	1	0	0	1	1	1
17	0	1	1	1	0	1	1	0	1	0	0	1	2	0	2	1	1	0	2	0	0	2	1	0	1	1	1
18	0	1	1	1	0	1	1	0	1	0	0	2	0	1	0	2	2	1	0	1	1	0	2	0	1	1	1
19	1	0	1	1	0	0	1	1	0	1	0	0	1	0	2	2	2	0	1	1	0	1	2	1	0	1	1
20	1	0	1	1	0	0	1	1	0	1	0	1	2	1	0	0	0	1	2	2	1	2	0	1	0	1	1
21	1	0	1	1	0	0	1	1	0	1	0	2	0	2	1	1	1	2	0	0	2	0	1	1	0	1	1
22	1	0	1	0	1	1	1	0	0	0	1	0	1	1	2	2	0	1	0	0	2	2	1	1	1	0	1
23	1	0	1	0	1	1	1	0	0	0	1	1	2	2	0	0	1	2	1	1	0	0	2	1	1	0	1
24	1	0	1	0	1	1	1	0	0	0	1	2	0	0	1	1	2	0	2	2	1	1	0	1	1	0	1
25	1	0	0	1	1	1	0	1	1	0	0	0	2	1	0	1	2	2	0	2	0	1	1	0	0	0	2
26	1	0	0	1	1	1	0	1	1	0	0	1	0	2	1	2	0	0	1	0	1	2	2	0	0	0	2
27	1	0	0	1	1	1	0	1	1	0	0	2	1	0	2	0	1	1	2	1	2	0	0	0	0	0	2
28	1	1	1	0	0	0	0	1	1	0	1	0	2	1	1	1	0	0	2	1	2	0	2	0	1	1	2
29	1	1	1	0	0	0	0	1	1	0	1	1	0	2	2	2	1	1	0	2	0	1	0	0	1	1	2
30	1	1	1	0	0	0	0	1	1	0	1	2	1	0	0	0	2	2	1	0	1	2	1	0	1	1	2
31	1	1	0	1	0	1	0	0	0	1	1	0	2	2	2	1	2	1	1	0	1	0	0	1	0	1	2
32	1	1	0	1	0	1	0	0	0	1	1	1	0	0	0	2	0	2	2	1	2	1	1	1	0	1	2
33	1	1	0	1	0	1	0	0	0	1	1	2	1	1	1	0	1	0	0	2	0	2	2	1	0	1	2
34	1	1	0	0	1	0	1	0	1	1	0	0	2	0	1	2	1	2	0	1	1	2	0	1	1	0	2
35	1	1	0	0	1	0	1	0	1	1	0	1	0	1	2	0	2	0	1	2	2	0	1	1	1	0	2
36	1	1	0	0	1	0	1	0	1	1	0	2	1	2	0	1	0	1	2	0	0	1	2	1	1	0	2
Group	1	2										3															

Notes:

(i) By introducing columns 1',2',3' and 4' in place of columns 1,2,⋯,11, one obtains $L_{36}(2^{3} \times 3^{13})$.

(ii) Since interactions are not orthogonal to other columns in the case of $L_{36}(2^{11} \times 3^{12})$, it is best to avoid assignments to find such interactions.

G.13 $L'_{27}(3^{22})$

Experiment number	Column number																				
	1	2	3	4	5	6	7	8	9	10	11	12	13	14	15	16	17	18	19	20	21	22
1	0	0	0	0	0	0	0	0	0	0	0	0	0	2	2	2	1	1	1	0	0	0
2	0	0	0	0	1	1	1	1	1	1	1	1	1	2	2	2	2	2	2	1	1	1
3	0	0	0	0	2	2	2	2	2	2	2	2	2	0	0	0	1	1	1	1	1	1
4	0	1	1	1	0	0	0	1	1	1	2	2	2	1	1	1	2	2	2	0	0	0
5	0	1	1	1	1	1	1	2	2	2	0	0	0	0	0	0	2	2	2	2	2	2
6	0	1	1	1	2	2	2	0	0	0	1	1	1	0	0	0	0	0	0	0	0	0
7	0	2	2	2	0	0	0	2	2	2	1	1	1	1	1	1	1	1	1	2	2	2
8	0	2	2	2	1	1	1	0	0	0	2	2	2	2	2	2	0	0	0	2	2	2
9	0	2	2	2	2	2	2	1	1	1	0	0	0	1	1	1	0	0	0	1	1	1
10	1	0	1	2	0	1	2	0	1	2	0	1	2	0	1	2	2	0	1	2	0	1
11	1	0	1	2	1	2	0	1	2	0	1	2	0	2	0	1	2	0	1	1	2	0
12	1	0	1	2	2	0	1	2	0	1	2	0	1	2	0	1	0	1	2	2	0	1
13	1	1	2	0	0	1	2	1	2	0	2	0	1	0	1	2	1	2	0	1	2	0
14	1	1	2	0	1	2	0	2	0	1	0	1	2	1	2	0	0	1	2	1	2	0
15	1	1	2	0	2	0	1	0	1	2	1	2	0	0	1	2	0	1	2	0	1	2
16	1	2	0	1	0	1	2	2	0	1	1	2	0	1	2	0	1	2	0	2	0	1
17	1	2	0	1	1	2	0	0	1	2	2	0	1	1	2	0	2	0	1	0	1	2
18	1	2	0	1	2	0	1	1	2	0	0	1	2	2	0	1	1	2	0	0	1	2
19	2	0	2	1	0	2	1	0	2	1	0	2	1	1	0	2	0	2	1	1	0	2
20	2	0	2	1	1	0	2	1	0	2	1	0	2	2	1	0	2	1	0	1	0	2
21	2	0	2	1	2	1	0	2	1	0	2	1	0	1	0	2	2	1	0	0	2	1
22	2	1	0	2	0	0	2	1	0	2	2	1	0	2	1	0	0	2	1	2	1	0
23	2	1	0	2	1	1	0	2	1	0	0	2	1	2	1	0	1	0	2	0	2	1
24	2	1	0	2	2	2	1	0	2	1	1	0	2	0	2	1	0	2	1	0	2	1
25	2	2	1	0	0	2	1	2	1	0	1	0	2	1	0	2	1	0	2	2	1	0
26	2	2	1	0	1	0	2	0	2	1	2	1	0	0	2	1	2	0	2	1	0	2
27	2	2	1	0	2	1	0	1	0	2	0	2	1	0	2	1	2	1	0	2	1	0
Group	1	2			3									Mixed								

Note:

(i) The four columns 1,11,12 and 13 are mutually orthogonal with any of the columns. The nine columns 2,3,4,5,6,7,8,9 and 10 are mutually orthogonal and are also orthogonal to 1, 11, 12 and 13, but are not orthogonal to the nine columns 14–22. The nine columns 14,15,16,17,18,19,20,21 and 22 are mutually orthogonal and are also orthogonal to 1, 11, 12 and 13.

(ii) Columns 14–22 are all partially group 1, partially group 2, and partially group 3. Classed-by-group expression is , therefore, not used in the following linear graph.

LINEAR GRAPHS OF $L'_{27}(3^{22})$

(1)

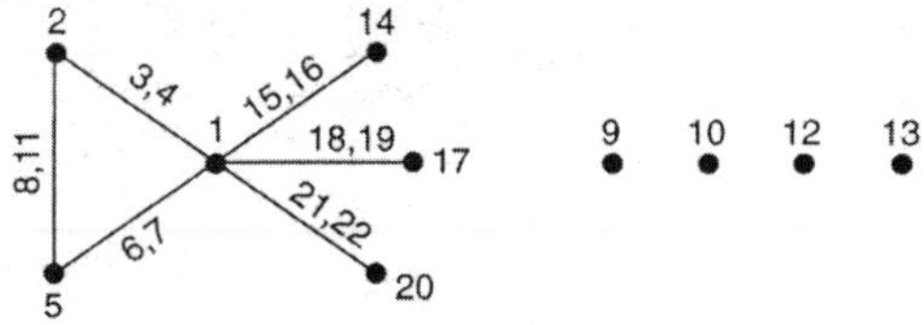

a

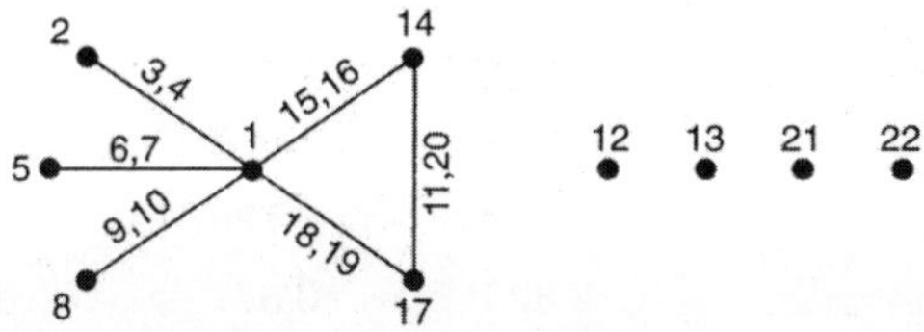

b

(2)

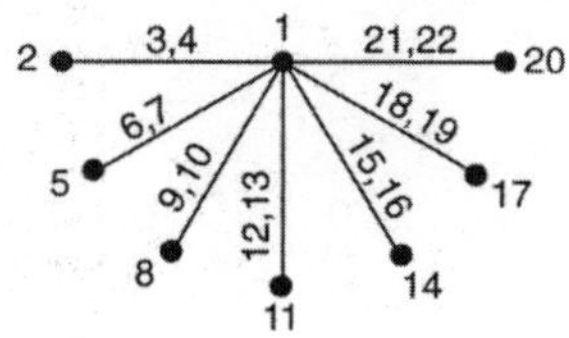

Appendix H Constants for $\bar{X} - R$ control chart

Number of observations in subgroup/sample	d_2	A_2	D_3	D_4
2	1.128	1.880	0	3.267
3	1.693	1.023	0	2.575
4	2.059	0.729	0	2.282
5	2.326	0.577	0	2.115
6	2.534	0.483	0	2.004
7	2.704	0.419	0.076	1.924
8	2.847	0.373	0.136	1.864
9	2.970	0.337	0.184	1.816
10	3.078	0.308	0.223	1.777
15	3.472	0.223	0.348	1.652
20	3.735	0.180	0.414	1.586

Appendix I GE Quality 2000: A dream with a great plan

John F. Welch, Jr., was chairman and CEO of GE Company. His speech was presented at the GE 1996 Annual Meeting in Charlottesville, Virginia, on April 24, 1996. The speech was published in the August/September issue of Executive Speeches, 1996. His speech is regarded as a milestone of Six Sigma history in the world. The part of his speech which is related to quality and Six Sigma is cited here.

"...The business performance of 222,000 employees worldwide has made us very proud as well. 1995 was another outstanding year for the company by any measure: a 17% growth in revenues to $70 billion, 11% earnings growth to $6.6 billion, and earnings per share up 13%. Our shareowners had a 45% return on their investment in 1995. GE, whose market capitalization already was the highest in the U.S., achieved that status globally in 1995, and is now the world's most valuable company.

Self-confidence and stretch thinking were two of the key factors that encouraged us to launch, in 1995, the most challenging stretch goal of all the biggest opportunity for growth, increased profitability and individual employee satisfaction in the history of our company. We have set for ourselves the goal of becoming, by the year 2000, a Six Sigma quality company, which means a company that produces virtually defect free products, services and transactions. Six sigma is a level of quality that to date has been approached by only a handful of companies, among them several in Japan, with Motorola being the acknowledged leader in this country.

GE today is a quality company. It has always been a quality company. Quality improvement at GE has never taken a back seat. We have operated under the theory that if we improved our speed, our productivity, our employee and supplier involvement, and pursued other business and cultural initiatives, quality would be a natural by product. And it has been. It's gotten better with each succeeding generation of product and service. But it has not improved enough to get us to the quality levels of that small circle of excellent global companies that had survived the intense competitive assault by themselves, achieving new levels of quality.

This Six Sigma journey will change the paradigm from fixing products so that they are perfect to fixing processes so that they produce nothing but perfection, or close to it. Typical processes at GE generate about 35,000 defects per million, which sounds like a lot, and is a lot, but it is consistent with the defect levels of most successful companies. The number of defects per million is referred to in the very precise jargon of statistics as about three and one-half sigma. For those of you who flew to Charlottesville, you are sitting here in your seats today because the airlines' record in getting passengers safely from one place to another is even better than six sigma, with less than one-half failure per million. However, if your bags did not arrive with you, it's because airline baggage operations are in the 35,000 to 50,000 defect range, which is typical of manufacturing and service operations, as well as other human activities such as writing up restaurant bills, payroll processing, and prescription writing by doctors.

The experience of others indicates that the cost of this three to four sigma quality is typically 10%-15% of revenues. In GE's case, with over $70 billion in revenues, that amounts to some $7-10 billion annually, mostly in scrap, reworking of parts and rectifying mistakes in transactions. So the financial rationale for embarking on this quality journey is clear. But beyond the pure financials, there are even more important rewards that will come with dramatically improved quality. Among them: the unlimited growth from selling products and services universally recognized by customers as being on a completely different plane of quality than those of our competitors; and the resulting pride, job satisfaction and job security from this volume growth for GE employees.

Six Sigma will be an exciting journey and the most difficult and invigorating stretch goal we have ever undertaken. The magnitude of the challenge of going from 35,000 defects per million to fewer than 4 defects is huge. It will require us to reduce defects rates

10,000 fold - about 84% per year for five consecutive years - an enormous task, one that stretches even the concept of stretch behavior.

Motorola has defined a rigorous and proven process for improving each of the tens of millions of processes that produce the goods and services a company provides. The methodology is called the Six Sigma process and involves four simple but rigorous steps: first, measuring every process and transaction, then analyzing each of them, then painstakingly improving them, and finally rigorously controlling them for consistency once they have been improved.

Following Motorola's experience closely, we have selected, trained and put in place the key people to lead this Six Sigma effort. We've selected our 'Champions' - senior managers who define the projects. We've trained 200 'Master Black Belts' - full-time teachers with heavy quantitative skills as well as teaching and leadership ability. We've selected and trained 800 'Black Belts' - full-time quality executives who lead teams and focus on key processes, reporting the results back to the Champions. We are beginning to train each of our 20,000 engineers so that all of our new products and services will be designed for Six Sigma production. And we have, at our Leadership Development Institute at Crotonville and at our businesses, an unmatched educational capability to train all 222,000 GE people in Six Sigma methodology.

We have a work-out culture in place at GE that is ideal for highly collaborative action-based team efforts, which will enhance our Six Sigma programs. To emphasize the importance of this initiative, we have weighted 40% of the bonus compensation for our managers on the intensity of their efforts and their progress toward Six Sigma quality in their operations. To date, we have committed $200 million to this effort, and we have the balance sheet that will permit us to spend whatever is required to get to our goal. The return on this investment will be enormous. Very little of this requires invention. We have taken a proven methodology, adapted it to a boundaryless culture, and are providing our teams every resource they will need to win.

Six Sigma - GE Quality 2000 - will be the biggest, the most personally rewarding and, in the end, the most profitable undertaking in our history. GE today is the world's most valuable company. The numbers tell us that. We are the most exciting global company to work for. Our associates tell us that. By 2000, we want to be an even better company, a company not just better in quality than its competitors - we are that today - but a company 10,000 times better than its competitors. That recognition will come not from us but from our customers.

Six Sigma - GE Quality 2000 - is a dream, but a dream with a plan behind it. It is a dream that is increasingly inspiring and exciting everyone in this company. We have the resources, the will, and above all, the greatest people in world business who will make it come true. Thanks for listening."

References

Akhtar, M. and Prescott, P. (1986) Response surface designs robust to missing observations, *Communications in Statistics. Simulation and Computing*, 15, 345-363.

American Society for Quality Control (1983) *Glossary and Tables for Statistical Quality Control*, Milwaukee, WI, ASQC.

Antony, J. (2005) Six Sigma for service processes, *Business Process Management Journal*, Vol. 12, No. 2, pp. 234-248.

Antony, J. (2003) *Design of Experiments for Engineers and Scientists*, Elsevier, Oxford, UK.

Antony, J. (2002) Robust design in new product development process: a neglected methodology in UK manufacturing organizations, *Work Study*, Vol. 51, No. 2, pp. 81-84.

Antony, J. and Banuelas, R. (2002) Critical success factors for the successful implementation of Six Sigma projects, *The TQM Journal*, Vol. 14, No. 2, pp. 92-99.

Antony, J., Banuelas, R. and Kumar A. (2006) *World Class Applications of Six Sigma: Real World Examples of Success*, Butterworth-Heinemann Ltd, Oxford.

Antony, J. and Taner, T. (2003) A conceptual framework for the effective implementation of statistical process control, *Business Process Management Journal*, Vol. 9, No. 4, pp. 473-489.

Arvidsson, M. and Gremyr, I. (2006) *Principles of Robust Design Methodology*, Quality and Reliability Engineering International, 2007.

Atkinson, A. C. and Donev, A. N. (1992) *Optimum Experimental Designs*, Clarendon Press, Oxford.

Atwood, J. (2005) Benchmarking: Using Design for Six Sigma, *iSixSigma Magazine*, July/August 2005, Vol. 1, No. 4.

Biles, W. E. (1975) A response surface method for experimental optimization of multi-response processes, *Industrial and Engineering Chemistry Process Design and Development*, 14, 152-158.

Bisgaard, S. and Freiesleben, J. (2004) Six Sigma and the bottom-line, *Quality Progress*, Vol. 37, No. 9, pp. 57-62.

Box, G. E. P. (1957) Evolutionary operation: a method for increasing industrial productivity, *Applied Statistics*, 6, 81-101.

Box, G. E. P. and Behnken, E. W. (1960) Simplex sum designs: a class of second order rotatable designs derivable from those of first order, *Annals of Mathematical Statistics*, 31, 838-864.

Box, G. E. P. and Draper, N. R. (1969) *Evolutionary Operation: A Statistical Method for Process Improvement*, Wiley, New York.

Box, G. E. P. and Draper, N. R. (1975) Robust designs, *Biometrika*, 62, 347-352.

Box, G. E. P. and Draper, N. R. (1987) *Empirical Model Building and Response Surfaces*, Wiley, New York.

Box, G. E. P. and Hunter, J. S. (1957) Multifactor experimental design for exploring response surfaces, *Annals of Mathematical Statistics*, 28,195-241.

Box, G. E. P. and Hunter, W. G. (1963) Sequential design of experiments for non-linear models, *Proceedings of the IBM Scientific Computing Symposium on Statistics, October* 21-23, 1963, 113-137.

Box, G. E. P., Hunter, W. G. and Hunter, J. S. (1978) *Statistics for Experimenters: An Introduction to Design, Data Analysis, and Model Building*, John Wiley, New York.

Box, G. E. P. and Jones, S. P. (1992) Designing products that are robust to the environment, *Total Quality Management*, 3, 265-282.

Box, G. E. P. and Wilson, K. B. (1951) On the experimental attainment of optimum conditions. *Journal of the Royal Statistical Society*, B, 13, 1-45.

Cambridge Advanced Learner Dictionary (2005) *Cambridge University Press, Cambridge*, UK.

Chen, X., Pickrell, G. and Antony, J. (2004) Experimental design in fiber optic sensor development, *International Journal of Productivity and Performance Management*, Vol. 53, No. 8, pp. 713-725.

Crosby, P. B. (1979) *Quality Is Free*, McGraw-Hill, New York.

Davies, A. J. and Kochhar, A. K. (2000) A framework for the selection of best practices, *International Journal of Operations and Production Management,* Vol. 20, No. 10, pp. 1203-1217.

Defeo, P. and Myers, R. H. (1992) A new look at experimental design robustness. *Biometrika*, 79, 375-380.

Deming, W. E. (1986) *Out of the Crisis*. Massachusetts Institute of Technology, Center for Advanced Engineering Study, Cambridge, MA.

Derringer, G. and Suich, R. (1980) Simultaneous optimization of several response variables, *Journal of Quality Technology*, 12, 214-219.

Draper, N. R. and Smith, H. (1981) *Applied Regression Analysis*, 2nd edn., John Wiley, New York.

Federov, V. V. (1972) *Theory of Optimal Designs* (translated and edited by W. J. Studden and E. M. Klimko), Academic Press, New York.

Feigenbaum, A. V. (1983) *Total Quality Control*, 3rd edn., McGraw-Hill, New York.

Finney, D. J. (1945) Fractional replication of factorial arrangement, *Annals of Eugenics*, 12, 291-301.

Fisher, R. A. (1935) *The Design of Experiments*, Oliver and Boyd, London and Edinburgh.

Gillies, A. C. (1997) *Software Quality: Theory and Management*, International Thomson Computer Press, UK.

Hader, R. J. and Park, S. H. (1978) Slope-rotatable central composite designs, *Technometrics*, 20, 413-417.

Hamada, M. and Wu, C. F. J. (1990) A critical look at accumulation analysis and related methods (with discussion), *Technometrics*, 32, 119-130.

Harry, M. J. and Schroeder, R. (2000) *Six Sigma, The Breakthrough Management Strategy Revolutionizing the World's Top Corporations*, Doubleday, New York.

International Organization for Standardization (1986) ISO 8402; Quality-Vocabulary.

Ishikawa, K. (1983) *Guide to Quality Control*, Asian Productivity Organization, Tokyo.

Jackenthal, E. (2004) Six Sigma Project Selection: Don't Follow — "Ready, Fire, Aim!". Available from http://www.staffing.org/articles/edjackenthal_acco0007ej.asp.

Janieszewski, S. and George, E. (2003) Six Sigma and Software Process Improvement, www.softwaresixsigma.com (accessed June 2007).

John, P. W. M. (1971) *Statistical Design and Analysis of Experiments*, Macmillan, New York.

Juran, J. J. (1964) *Managerial Breakthrough*, McGraw-Hill, New York.

Juran, J. J. (1981) *Juran on quality improvement: workbook*, Juran Enterprises, New York.

Kackar, R.N. (1985) Off-line quality control, parameter design and the Taguchi method. *Journal of Quality Technology*, 17(4), 176-188.

Kaye, M. and Anderson, R. (1999) Continuous improvement: the ten essential criteria, *International Journal of Quality and Reliability Management*, Vol. 16, No. 5, pp. 485-506.

Kempthrone, O. (1952) *The Design and Analysis of Experiments*, Wiley, New York.

Khuri, A. I. and Conlon, M. (1981) Simultaneous optimization of multiple responses represented by polynomial regression functions, *Technometrics*, 23, 363-375.

Khuri, A. I. and Cornell, J. A. (1987) *Response Surfaces: Designs and Analysis*, Marcel Dekker, New York.

Kiefer, J. (1984) *Collected papers, Vol. III, Design of experiments*, Springer Verlag, New York.

Kiefer, J. and Wolfowitz, J. (1960) The equivalence of two extremum problems, *Canadian Journal of Mathematics*, 12, 363-366.

Kiemele, M., Schmidt, S. and Berdine, R. (1997) *Basic Statistics: Tools for Continuous Improvement*, Air Academy Press, USA.

Korean Industrial Standards (1971) *Glossary and Tables for Quality Control*, KSA3001. Seoul, Korean Standard Association.

Khuri, A. and Cornell, J. (1996) *Response Surfaces: Design and Analysis*, Dekker, USA.

Lausby, R. G. (1994) Taguchi or classical experiments?, *Quality*, April, Vol. 33, pp. 55-57.

Leon, R. V., Shoemaker, A. C. and Kackar, R. N. (1987) Performance measures independent of adjustment, *Technometrics*, 29, 253-265.

Logothetis, N. and Wynn, H. P. (1989) *Quality through Design: Experimental Design, Off-line Quality Control and Taguchi's Contributions*, Clarendon Press, Oxford.

Magnusson, J., Kroslid, D. and Bergman, B. (2003) *Six Sigma: The Pragmatic Approach*, Studentlitteratur, Sweden.

Mann, D. (2002) *Hands on Systematic Innovation*, CREAX Press, Leper, Belgium.

McKee, B. and Kshirsagar, A. M. (1982) Effect of missing plots in some response surface designs, *Communications in Statistics. Theory and Methods*, II, 1525-1549.

Montgomery, D. C. (1997) *Design and Analysis of Experiments*, Wiley, USA.

Montgomery, D. C. (1991) *Design and Analysis of Experiments*, 3rd edn., Wiley, New York.

Montgomery, D. C. and Peck, E. A. (1981) *Introduction to Linear Regression Analysis*, Wiley, New York.

Montgomery, D. C. and Runger, G. (2003) *Applied Statistics and Probability for Engineers*, Wiley, USA.

Myers, R. H. (1971) *Response Surface Methodology*, Boston, Allyn and Bacon.

Myers, R. H. and Carter, W. H. (1973) Response surface techniques for dual response systems, *Technometrics*, 15, 301-317.

Myers, R. H. Khuri, A. I. and Vining G. (1992) Response surface alternatives to the Taguchi robust parameter design approach, *The American Statistician*, 46(2), 131-139.

Nair, V. N. (1992) Taguchi's parameter design: a panel discussion. *Technometrics*, 34(2), 127-161.

Oakland, J. S. (2003) *Total Quality Management Text with Cases*, 3rd edn., Elsevier Butterworth-Heinemann, Oxford, UK.

Pande, P., Neuman, R. and Cavanagh, R. (2000) *The Six Sigma Way: How GE, Motorola and Other Top Companies are Honing their Performance,* McGraw-Hill Professional, NY, USA.

Park, S. H. (1976) On robustness of response surface designs, *Journal of the Korean Statistical Society*, 5(2), 101-107.

Park, S. H. (1982) *Modern Design of Experiments* (in Korean), Min-young Sa, Seoul.

Park, S. H. (1987) A class of multifactor designs for estimating the slope of response surfaces, *Technometrics*, 29, 449-453.

Park, S. H. (1990) *Applied Design of Experiments* (in Korean), Young-ji Moonwha Sa, Seoul.

Park, S. H. (1993) *Quality Engineering* (in Korean), Min-young Sa, Seoul.

Park, S. H. (1996) *Robust Design and Analysis for Quality Engineering*, Chapman & Hall, London.

Park, S. H. (2003) *Six Sigma for Quality and Productivity Promotion,* Asian Productivity Organization, Tokyo.

Phadke, M. S. (1989) *Quality Engineering using Robust Design*, Prentice Hall, Englewood Cliffs, NJ.

Plackett, R. L. and Burman, J. P. (1946) The design of optimum multifactorial experiments, *Biometrika*, 33, 305-325.

Pukelsheim, F. (1993) *Optimal Design of Experiments*, Wiley, New York.

Pyzdek, T. (2003) *The Six Sigma Handbook*, The McGraw-Hill Companies, NY, USA.

Rao, C. R. (1947) Factorial experiments derivable from combinatorial arrangements of arrays, *Journal of the Royal Statistical Society* (Supplement), 9, 128–139.

Sen, A. and Srivastava, M. (1990) *Regression Analysis: Theory, Methods and Applications*, Springer-Verlag, New York.

Senge, P., Kleiner, A., Roberts, C., Ross, R., Roth, G. and Smith, B. (1999) *The Dance of Change: The Challenges of Sustaining Momentum in Learning Organizations*, Doubleday/Currency, NY.

Silvey, S. D. (1980) *Optimal Designs*, Chapman & Hall, London.

Snee, R. D. and Hoerl, R. (2003) *Leading Six Sigma: A step by step guide based on experience with GE and other Six Sigma Companies*, Prentice Hall, NJ, USA.

Struebing, L. and Klaus, L. A. (1997) Small businesses thinking big, *Quality Progress*, February, pp. 23–27.

Studt, T. (1993) DOE simplifies analysis, *R&D Mangazine*, March, pp. 73.

Taguchi, G. (1986) *Introduction to Quality Engineering*, Asian Productivity Organization, Tokyo.

Taguchi, G. (1987) *System of Experimental Design*, Vols 1 and 2, Quality Resources, White Plains, NY, and American Supplier Institute, Dearborn, MI.

Taguchi, G. and Wu, Yu-In (1979) *Introduction to Off-line Quality Control*, Central Japan Quality Control Association, Magaya, Japan (Available from American Supplier Institute, Inc., Dearborn, MI.)

Tague, N. R. (1995) *The Quality Toolbox*, ASQC Quality Press, Milwaukee, WI.

Vining, G. G. and Myers, R. H. (1990) Combining Taguchi and response surface philosophies: a dual response approach. *Journal of Quality Technology*, 22, 38–45.

Watson, G. H. (2005) *Design for Six Sigma: Innovation for Enhanced Competitiveness*, GOAL/QPC Publishing, Salem, NH.

Welch, W. J., Yu, T. K., Kang, S. M. and Sacks, J. (1990) Computer experiments for quality control by parameter design, *Journal of Quality Technology*, 22, 15–22.

Wu, Yu-In and Moore W. H. (1985) *Quality Engineering Products and Process Design Optimization*, American Supplier Institute, Dearborn, MI.

Wu, C. F. and Hamad, M. (2000) *Experiments: Planning, Analysis, and Parameter Design Optimization*, Wiley Inter-science, New York.

Xiao, H. (2000) Self-calibrated interferometric/intensity based fiber optic pressure sensors, Dissertation for Doctor of Philosophy, ECE dept., Virginia Polytechnic.

Yang, K. and El-Haik, B. S. (2003) *Design for Six Sigma: A Roadmap for Product Development*, McGraw-Hill, N.Y. USA.

Yoo, J. B. and Park, S. H. (1991) Robust designs to outliers for response surface experiments, *Journal of the Korean Statistical Society*, 20, 147-155.

Index